NATURALISM AND NORMATIVITY

COLUMBIA THEMES IN PHILOSOPHY

COLUMBIA THEMES IN PHILOSOPHY

Series Editor: Akeel Bilgrami,
    Johnsonian Professor of Philosophy, Columbia University

Columbia Themes in Philosophy is a new series with a broad and accommodating thematic reach as well as an ecumenical approach to the outdated disjunction between analytical and European philosophy. It is committed to an examination of key themes in new and startling ways and to the exploration of new topics in philosophy.

Edward Said, *Humanism and Democratic Criticism*

Michael Dummett, *Truth and the Past*

John Searle, *Freedom and Neurobiology: Reflections on Free Will, Language, and Political Power*

Daniel Herwitz and Michael Kelly, eds., *Action, Art, History: Engagements with Arthur C. Danto*

Michael Dummett, *The Nature and Future of Philosophy*

Jean Bricmont and Julie Franck, eds., *Chomsky Notebook*

NATURALISM

AND NORMATIVITY

EDITED BY

Mario De Caro and David Macarthur

COLUMBIA

UNIVERSITY

PRESS

NEW YORK

Columbia University Press
Publishers Since 1893
New York   Chichester, West Sussex
Copyright © 2010 Columbia University Press

Library of Congress Cataloging-in-Publication Data
Naturalism and normativity / edited by Mario De Caro and David Macarthur.
    p. cm. — (Columbia themes in philosophy)
   Includes bibliographical references and index.
   ISBN 978-0-231-13466-8 (cloth) — ISBN 978-0-231-13467-5 (pbk.) — ISBN
978-0-231-50887-2 (e-book)
  1. Naturalism. 2. Normativity (Ethics) I. De Caro, Mario.
II. Macarthur, David. III. Title. IV. Series.

   B828.2.N374   2010
   146—dc22

                        2009052544

We gratefully acknowledge the publishers for their permission to reprint the
following articles:
   Huw Price, "Truth as Convenient Friction," *Journal of Philosophy* 100 (2003):
167–90.
   Richard Rorty, "Naturalism and Quietism," *Philosophical Papers*, vol. 4:
Philosophy as Cultural Politics (Cambridge: Cambridge University Press, 2007),
147–59.
   Thomas Scanlon, "Metaphysics and Morals," *Proceedings and Addresses of the
American Philosophical Association* (presidential address) 77 (2003): 7–22.

CONTENTS

NATURALISM AND NORMATIVITY

INTRODUCTION

SCIENCE, NATURALISM, AND THE PROBLEM OF NORMATIVITY

Mario De Caro and David Macarthur

> *Normativity is ubiquitous*
> —Hilary Putnam

NORMATIVITY AND THE SCIENTIFIC NATURE

Normativity concerns what we should or ought to do and our evaluations of things or states of affairs. We normally say, for example, that one ought to keep one's promises, that if one accepts *p* and "If p, then q," one ought to accept *q*, or that Mozart was a better musician than Salieri. Plausibly, the sciences describe how things are, particularly the causal powers or causal regularities that exist in the world, lawlike or otherwise. Consequently, if one follows modern Scientific Naturalism in supposing that natural science, and only natural science, tells us what there is in the world, then there seems to be no room for the existence of normative facts—or at least this will be so insofar as they cannot be reduced to the kinds of objective, causal facts with which natural science deals. Such considerations set the stage for one of the fundamental issues confronting philosophers today: Are there any indispensable, irreducible normative facts involving, say, reasons, meanings, and values that are not, or cannot, be accommodated within the scientific image of the world? John McDowell provides a trenchant expression of the sense that this is indeed the case:

> Modern science understands its subject matter in a way that threatens, at least, to leave it disenchanted, as Weber put the point in an image that has become a

commonplace. The image marks a contrast between two kinds of intelligibility: the kind that is sought by (as we call it) natural science ["the kind we find in a phenomenon when we see it as governed by natural law"] and the kind we find in something when we place it in relation to other occupants of "the logical space of reasons" ["the kind of intelligibility that is proper to meaning"].[1]

The issue has been in the air since at least the late nineteenth century when there arose in the German-speaking world debates about the aims, subject matter, and methods of the social sciences: the so-called *Methodenstreit*.[2] Are human beings, their societies, and their cultural institutions open to the same kind of explanations as the natural objects studied by the natural sciences? If so, can they be *fully* explained in such terms? Does the fact that human beings engage with and respond to normative items such as meanings, values, and reasons require methods of investigation and interpretation (e.g., empathetic understanding, intentional explanation, introspection) that have no counterparts in the natural or, more broadly, social sciences?

In contemporary philosophy these issues often find expression in terms of the problem of "placing" or "locating" normative phenomena in the scientific image of the world. According to the most common form of naturalism, the image of the world provided by the natural sciences is all the world there is. Since this image seems, prima facie, not to include normative phenomena, the following question arises: What "place" can we find for the normative in the natural world? The question becomes urgent if, as seems highly plausible, we suppose that central normative phenomena are not going to be explained away or eliminated. Call this, for present purposes, the "placement problem." Similar placement problems arise with respect to intentional attitudes, consciousness, abstract entities, and so forth.

Although the placement problem is, in the first instance, posed in ontological terms, it has a semantic analogue or counterpart that is at least equally pressing. The aim of a great many programs of "naturalization" is to explain the role or function of normative concepts, given the supposed scientific discovery that the world lacks any autonomous normative facts. Theoretical options at this point include *error theories*, according to which normative concepts, although representational in intent, are true of nothing in the actual world; *reductionist theories*, according to which normative concepts are ultimately reducible to respectable scientific concepts, which themselves straightforwardly represent features of the world; and *nonfactualist theories* (e.g., classical expressivism), according to which normative concepts are not in the business of representing the world at all. Nonfactualist theories typically go on to offer alternative accounts of the roles normative concepts play in our lives, for example, being expressive of some noncognitive state such as desire.[3]

THE DOCTRINES OF SCIENTIFIC NATURALISM

In order to understand better why naturalism is thought to give rise to problems in accounting for normativity, we must enter into the difficult business of trying to provide at least a preliminary definition of the term "naturalism." It is a commonplace that this term has a long history and a variety of uses. The idea of philosophy having some important relation to, or conception of, "nature" has been a recurrent theme since at least the time of Aristotle. Our main concerns in this volume, however, will be with two varieties of naturalism that are of particular relevance to contemporary philosophy: Scientific Naturalism, which on a certain narrow understanding represents the current orthodoxy among the large majority of Anglo-American philosophers who fly the banner of "naturalism"; and Liberal Naturalism, a position or set of positions that, according to the authors of this volume, hold the promise of overcoming the impasses associated with scientific naturalist orthodoxy. These include the apparent intractability of the placement problem and the alleged unattractiveness of the alternatives open to a naturalist treatment of normative and evaluative concepts. Moreover, since orthodox Scientific Naturalism seems at odds with the idea of a distinctive philosophical method, Liberal Naturalism promises to allow for a more realistic and fruitful conception of the relation between philosophy and science.

What makes Scientific Naturalism and Liberal Naturalism both versions of *naturalism* is that neither countenances the supernatural, whether in the form of entities (such as God, spirits, entelechies, or Cartesian minds), events (such as miracles or magic), or epistemic faculties (such as mystical insight or spiritual intuition). The importance of this for the philosophical approach to normativity is that any form of naturalism will be opposed to Platonism about norms, where this is understood as the view that normative facts hold wholly independently of human practices (say, of reason giving) and are, as it were, simply there anyway waiting to be discovered. For similar reasons it will be opposed to a Moorean non-naturalism that holds that our access to normative facts is by way of a sui generis epistemic faculty of intuition directed at just this kind of fact. And of course it will be opposed to any theistic foundation for normative facts or our access to them.

However the very category of the supernatural is controversial since, for one thing, not everything that we take to be nonsupernatural, such as numbers or propositions, can be understood in unproblematically causal terms. One way of thinking about the difference between Scientific and Liberal Naturalism is as a dispute about how the complementary categories of the "natural" and the "supernatural" should best be understood. Scientific Naturalism interprets the natural strictly in terms of the scientific image of the world, narrowly or broadly conceived, whereas Liberal Naturalism, or some versions of it, offers a broader, more expansive conception of nature that makes room for a class of nonscientific, but nonetheless nonsupernatural, entities.

Let us now turn to consider the general category of Scientific Naturalism in some more detail. Scientific Naturalism can be seen as a cluster of positions that arise from an *ontological doctrine* and/or a *methodological doctrine*.[4] The ontological doctrine of scientific naturalism is famously, if vaguely, expressed by Wilfrid Sellars's remark that "science is the measure of all things, of what is that it is, and of what is not that it is not."[5] Somewhat more precisely, the view is the following:

> *Ontological doctrine of Scientific Naturalism.* The world consists of nothing but the entities to which successful scientific explanations commit us.

Scientific Naturalism is also definable in terms of a methodological commitment, often held in tandem with the ontological commitment above. Arthur Danto, some time ago, wrote that naturalism is "a species of philosophical monism according to which whatever exists or happens is natural in the sense of being susceptible to explanation through methods . . . paradigmatically exemplified by the natural sciences."[6] From a survey of the literature it seems that according to the majority of scientific naturalists, these explanations and methods are broadly empirical in character.[7] Consequently, another tenet of Scientific Naturalism is the following:

> *Methodological doctrine of Scientific Naturalism.* Scientific inquiry is, in principle, our only genuine source of knowledge or understanding. All other alleged forms of knowledge (e.g., a priori knowledge) or understanding are either illegitimate or are reducible in principle to scientific knowledge or understanding.[8]

Some immediate comments about these two doctrines are in order. First, the ontological doctrine is a metaphysical thesis not a scientific one and it plays a primary role in the way scientific naturalists give content to the otherwise rather vague antisupernaturalist claim shared by all contemporary forms of naturalism, scientific or nonscientific. Second, that scientific naturalists tend to have an ontology based on causal facts is evident from the way in which the ontological status of mathematical and abstract entities is somewhat of an embarrassment, and so generally avoided or quietly passed over. Arguably, these entities have to be accepted because of their indispensability in the explanations of the natural sciences, but their nature is not easily explained in scientific terms—especially if we think of science as primarily in the business of providing causal explanations.[9] Third, the interpretation of the ontological doctrine is importantly sensitive to how broadly the scope of the term "science" is conceived and, consequently, sensitive to how broadly one interprets the scientific image of the world and our methods for investigating it.

A further set of issues is raised by the fact that the philosophical method is traditionally supposed to be a priori, the discovery of necessary truths about the world by the use of reason alone. The scientific naturalists' methodological doctrine thus has strongly antitraditional implications for our conception of the method and role of philosophy in relation to the sciences. In particular, following Quine, many scientific naturalists claim that philosophy, far from being the final tribunal or foundation for scientific claims, must be methodologically *continuous* with science. David Papineau, for example, writes that "philosophy and science are engaged in essentially the same enterprise, pursuing similar ends and using similar methods." In Papineau's opinion, the "relatively superficial" differences between the two disciplines include philosophy's greater generality, differences in the ways the two disciplines gather their data, and the fact that the philosophical issues tend to generate "some kind of theoretical tangle."[10]

In "Naturalism and Quietism," which was originally written for this volume,[11] Richard Rorty follows Brian Leiter in dividing the Anglo-American philosophical world into "Naturalists" and "Wittgensteinian Quietists," a division that Rorty describes as "the deepest and most intractable difference of opinion within contemporary Anglophone philosophy." Rorty sees this split as, among other things, a battle between a resurgent ("scientific") metaphysical approach to philosophy associated with Saul Kripke, David Lewis, Kit Fine, Peter Van Inwagen, and David Armstrong and the alternative, typically nonmetaphysical, approaches proposed by Ludwig Wittgenstein, Wilfrid Sellars, Donald Davidson, John McDowell, and Robert Brandom. In further articulating this division, Rorty discusses three debates: between Huw Price and Frank Jackson over representationalism; between John McDowell and Timothy Williamson over whether there can be "elusive objects"; and between Robert Brandom and Jerry Fodor over inferentialist semantics.

Hilary Putnam in his chapter in this collection takes up the question of the relation between science and philosophy and casts a critical eye over the naturalistic attempt to absorb the latter into the former. In Putnam's reconstruction the growing secularization of philosophy in the past century or so has coincided with the rise of Scientific Naturalism, which involves a new conception of the relation between philosophy and science. According to the scientific naturalist, philosophy not only loses its traditionally preeminent role as a final court of appeal regarding matters of existence and knowledge; indeed, philosophy survives in name only, for it has no autonomous role to play. In Putnam's alternative image, philosophy is neither absolutely autonomous nor totally lacking in autonomy. In an age of science, Putnam urges us not to forget that philosophy has, and needs, both a theoretical face and a moral face. With regard to the former Putnam argues that philosophy retains a role as a source of (revisable) interpretations of the meanings of our theories. For example, in the case of fundamental physical theories there is, on Putnam's view, a fruitful interpenetration of science and metaphysics. Part of the problem in seeing this matter clearly, however, is that there are myths of what science is in both popular culture and philosophy, for example, the myth of "value-free facts" produced according to "canons" of induction.

In the background of the debate about whether philosophy has any legitimate and autonomous role to play distinct from that of the sciences is a wider ideologically driven debate about how we should think about the natural world that we inhabit. In "The Wider Significance of Naturalism: A Genealogical Essay," Akeel Bilgrami attempts to uncover the genealogical basis of "the deep division among philosophers today over naturalism," which he traces, ultimately, to a deep disagreement over whether irreducible values are part of the world or not. Bilgrami first gives an argument and motivation for the view that values are properties in the world, appealing to considerations of agency and the first-person point of view and exploiting some remarks in Gareth Evans about the relation between self-knowledge and intentionality, to show that seeing values as part of the world is essential to the fact of our possessing agency. He then traces genealogically why such a picture of value and the world we inhabit as agents came to be undermined by a metaphysics that grew around the new science in the late seventeenth century, making it seem inevitable to us that such a worldview was incompatible with that of science. Focusing on England, Bilgrami argues that this metaphysics of what is often too summarily described as "disenchantment" seems inevitable only because we write out of intellectual history; the vocal and lively dissenting scientific voices of the late seventeenth century, the deracination of value from nature and the world around us were neither compulsory nor the inevitable consequences of the new science and its laws. These voices were silenced not because of any scientific or philosophical superiority of the orthodox metaphysics they opposed, which was forming among the ideologues of the Royal Society around the new science. Rather, the dissenters lost out because of powerful political alliances that these ideologues forged with established Anglican interests, as well as with the newly emerging commercial interests that had a worldly interest in a vision of nature as brute and stripped of value. And such a vision of nature, according to Bilgrami, was realized in, and supported by, a transformation of the very conception of nature from the concept of an environment in which we "merely live" into the concept of *natural resources* to be systematically controlled and conquered for extractive economies. Bilgrami goes to extend his argument to trace and expose how later such a "disenchanted" worldview was modified to make it seem not merely a consequence of a commitment to the methods and metaphysics of the natural sciences, but also intrinsic to a dominant picture of the social sciences that prevails to this day.

In "Taking the Human Sciences Seriously," David Macarthur argues that three families of Scientific Naturalism can be usefully distinguished, depending on how they define what the legitimate ontology, methods, and scope of science are. The first, Extreme Scientific Naturalism, identifies science with physics alone. The second, Narrow Scientific Naturalism, allows that some of the other sciences of nature may be irreducible to physics but at the same time claims that science proper is to be limited to the natural sciences as opposed to the human sciences. Grouped together, these two views represent the naturalist orthodoxy in contemporary philosophy. However, it is important to see that Scientific Naturalism may also come in a third form, Broad Scientific Naturalism, which identifies science with *both* the natural sciences and the social and human sciences where at least some of the human sciences are understood as methodologically and ontologically irreducible to the natural sciences.

In light of these distinctions, Macarthur argues that the idea that there is a dualism of norm and nature is an overstatement—even from the limited perspective of a scientific naturalist. From the point of view of Broad Scientific Naturalism—a position that is not clearly visible in contemporary philosophy—it is plausible that certain indispensable and irreducible normative items that appear in, or are presupposed by, social scientific explanations can be admitted into the scientific image of the world.[12] Reasons and cognitive values are examples. This more liberal version of Scientific Naturalism promises not a solution to but a dissolution of the placement problem. That alone provides a strong motive for orthodox scientific naturalists to move toward a more liberal understanding of their own commitment to naturalism. However, Macarthur goes on to argue that Broad Scientific Naturalism faces the apparently intractable problem that it cannot scientifically explain rational normativity. This problem is particularly embarrassing if, as Macarthur supposes, the doctrines of Scientific Naturalism are themselves claims about what is rationally normative.

Macarthur's chapter raises a more general question. Is a Broad Scientific Naturalism that admits the irreducibility of the social sciences, with their peculiar ontologies and epistemologies, liberal enough? Many of the authors of this volume want to argue that it is not. It might be claimed, for example, that there are normative phenomena to which we have access *only* from the perspective of fully engaged agents in the world. And this is a perspective, it seems, that is incompatible with the perspective of a scientific observer, however empathetic. To admit this would be to accept the existence of non-scientific nonsupernatural phenomena. Or, it might be claimed that there are nonscientific modes of understanding or knowing (e.g., aesthetic or moral reasoning) that are incapable of being properly acknowledged or represented by scientific inquiries and discoveries. Both of these directions of thought fall within the category of what we have called Liberal Naturalism (more on this later).

In "Why Scientific Realism May Invite Relativism," Carol Rovane discusses another problem facing scientific naturalists. She argues that scientific realism (which is her label for the ontological component of Scientific Naturalism) is undermined by unnoticed *liaisons dangereuses* with relativism. In Rovane's view, in order to rule out relativism, scientific realists should establish a "view from everywhere," that is, the possibility of a super-knower who could know all the truths potentially known by every other possible knower. However, once one has assumed the realist conception of the facts as mind-independent (which strongly suggests that there may be knowers very different from us), it is doubtful that all the truths knowable in principle could be conjoined and embraced together by a single possible knower. In this light, Rovane claims that, far from possessing the conceptual resources that would be necessary to rule out relativism, scientific realism paradoxically encourages it.

Peter Menzies's chapter, "Reasons and Causes Revisited," is a reconsideration of the relation between reasons and causes in ordinary intentional explanation. Scientific Naturalism holds that intentional psychology is akin to a theory for explaining human behavior in terms of intentional states, but, according to Menzies, it has trouble acknowledging the normativity of intentional psychology. Kantian rationalism accepts this kind of normativity but mysteriously treats reasons as radically distinct from causes. For Menzies the problem is that both parties to this debate are wedded to the deductive-nomological account of scientific theorizing according to which causal explanations are one thing and rational explanations another. In Menzies's model-based conception of intentional psychology, we appeal to intentional states to predict and explain behavior because they both rationalize and cause behavior. Reasons are causes but, in contrast to Davidson's version of this position, only *one* set of generalizations is needed to provide both rationalizing and causal explanations.

THE DOCTRINE OF LIBERAL NATURALISM

Apart from raising questions about whether orthodox versions of Scientific Naturalism are viable, the papers collected here raise the question whether an alternative form (or forms) of naturalism—which fall within the field of what we have called Liberal Naturalism—are available and potentially more attractive than any version of Scientific Naturalism. One of the primary motivations for the present volume is the thought that the debate over which form of naturalism is best will depend to a considerable extent on which provides the best account of core normative phenomena such as reasons and values. As Putnam says, normativity is ubiquitous in our thought and talk. So it has some claim to being the central area for which philosophy must provide an account if we are to achieve the sort of self-understanding that Socrates, and many other philosophers since, have promised us.[13]

Liberal Naturalism, as we understand it, is not a precisely defined credo. It is better seen as a range of attempts to articulate a new form of naturalism that wants to do justice to the range and diversity of the sciences, including the social and human sciences (freed of positivist misconceptions), and to the plurality of forms of understanding, including the possibility of nonscientific, nonsupernatural forms of understanding (whether or not these also count as forms of knowledge). In addition, some of our authors want to allow for the possibility of nonscientific, nonsupernatural entities. Most of the papers in the current volume, therefore, can be thought of as investigations into the question of how best to characterize an alternative conception of naturalism along these lines. Rather than try to adjudicate this debate by prematurely deciding for a single alternative, we prefer to present these papers as contributions to a fruitful controversy that we want to recommend to the reader's attention.

If Liberal Naturalism cannot at this stage be characterized definitively, the question arises whether there is any useful way of characterizing what the authors of this volume have in common? Liberal Naturalism, as we have seen, is best thought of as occupying the typically overlooked conceptual space between Scientific Naturalism and Supernaturalism. A necessary condition for a view's being a version of Liberal Naturalism is that it rejects Scientific Naturalism, hence that it rejects the ontological doctrine or the methodological doctrine, or both. There are different ways of liberalizing naturalism beyond its orthodox understanding and also different ways of conceiving of, hence rejecting, the supernatural.

Let us start, then, with those who are liberal naturalists solely in virtue of rejecting the methodological doctrine. Consider Huw Price who rejects standard Scientific Naturalism (which he calls "object naturalism") on the grounds that it makes substantial representationalist assumptions about language *prior* to empirical study of the functions language actually serves. His idea of a *subject* naturalism starts by rejecting representationalism across the board and then showing how one can explain the functions of linguistic concepts or expressions in other terms. On one reading, Price holds to the orthodox naturalist ontology of the natural sciences in thinking that a commitment to the reality of values, say, does not require us to suppose that they are a special kind of entity. The vindication of our commonsense realist talk of values only requires the availability, from an anthropological point of view, of a nonrepresentational function for such talk. But even if that is so, Price departs from any kind of Scientific Naturalism in arguing that a science of linguistic functions can itself show us that there are nonscientific modes of knowing and understanding.

In "Truth as Convenient Friction," reproduced in the present volume,[14] Price considers the functional role of the normative concept of truth. He primarily takes issue with Rorty's pragmatist conviction that there is no significant practical distinction between justification and truth. Rorty claims that "obedience to a . . . commandment to seek the truth . . . will produce no behaviour not produced by the need to offer justification." In Price's view, this claim is false on the grounds that there is a widespread behavioral pattern that results from the fact that speakers take themselves to be subject to a commitment to seek the truth. Moreover, it is a behavioral pattern that Rorty, of all people, cannot afford to dismiss as a by-product of bad philosophy: it is conversation itself—something he champions. In order to account for ordinary conversational practice, in Price's view, we need to recognize that speakers take themselves to be governed by a norm stronger than that of justification—a norm that speakers acknowledge they may fail to meet, even if their claims are well justified. This norm provides the automatic and quite unconscious sense of engagement in common purpose that distinguishes conversation from a roll call of individual opinion. Truth, on Price's view, is the grit that makes our individual opinions engage with one another.

A debate between Rorty and Price over these matters is also included in this anthology. This debate is interesting for a number of reasons, not the least of which is that they are two of the leading neopragmatists in the world. Both Price and Rorty, for example, claim that truth is not a substantial property about which there could be an interesting philosophical theory. Their disagreement centers rather around the question of whether there is any need for a norm of truth over and above a norm of intersubjective agreement and the form that epistemic norms might take.

In "Two Directions for Analytic Kantianism: Naturalism and Idealism," Paul Redding criticizes Price's version of Liberal Naturalism—at least on the narrow reading sketched above—for a residual scientism, which threatens nihilism, a philosophical stance that overlooks or undermines our commitment to norms and values. On the positive side, Redding adopts Bernard Williams's idea that a view that avoids both scientism and Supernaturalism can be achieved only if philosophy is conceived as a "humanistic discipline"—one whose main goal is to contribute to our self-understanding through a critical reflection on the contingent, *historical* genesis of our ideas. From this perspective, Redding endorses a conception rooted in the lessons of Kant's German idealist successors. He suggests that we can only do justice to our normative practices if we follow Brandom's neo-Hegelian attempt to explain norms in terms of reciprocal recognition.

It is arguable, however, that Price does not restrict himself to a narrowly scientific ontology because he rejects the whole enterprise of ontology as a distinct philosophical explanatory project. By talking in certain ways we commit ourselves to the reality of certain kinds of objects—which ones being a matter of interpretation—but this is not a philosophical discovery about science; it is a truth about assertoric discourse in general. Indeed, from a subject naturalist perspective science itself can discover "that science is just one thing among many that we do with the linguistic tools of ontological commitment."[15] We cannot *start* (prior to empirical enquiry) with the metaphysically weighty view that only science can tell us anything about the range, function, and possible plurality of our ontological commitments. So perhaps Price is best read as rejecting *both* the ontological and methodological doctrines of Scientific Naturalism.

Another liberal naturalist position is articulated by John Dupré who, in addition to accepting nonscientific forms of knowing and understanding, holds to a scientific ontology of *both* the natural and human sciences.[16] He shares Price's view that in order to understand the human world, an irreducible plurality of insights, including more traditionally humanistic ones, is indispensable. In "How to be Naturalistic Without Being Simplistic in the Study of Human Nature," Dupré's rigorous empiricism assumes that if something exists *and* is made of something, then, in contrast to abstract entities, it is made of physical stuff. Nonabstract entities or properties *must* be accountable by a recognizable extension of the empirical methods of investigation employed by the sciences. But, on Dupré's view, the *empirical* evidence also shows us that some things have irreducible nonphysical properties and causal powers that do not figure in the explanations offered by physics. The moral is a pluralistic conception of science both in epistemic and ontological terms. According to Dupré, scientific naturalists are committed to a questionable metaphysical monism that we have no good *empirical* reason to accept. This questionable monism includes the reducibility of all natural sciences to physics, the *"principle of the causal closure of the physical world,"* and the idea that nonphysical properties are reducible to, or at least supervene (globally if not locally) on, physical properties.

In "Is Liberal Naturalism Possible?" Mario De Caro and Alberto Voltolini approach the issue of Liberal Naturalism by considering a dilemma posed by Ram Neta.[17] As Neta sees it, either Liberal Naturalism implies that controversial items (say, normative items) are in principle reducible to the ontology of the natural sciences or they are irreducible to it. If the former, then this view is simply orthodox Scientific Naturalism; if the latter, then the view becomes ipso facto a form of Supernaturalism. Thus, it seems, there is no logical space for Liberal Naturalism.

In order to contest this argument, De Caro and Voltolini offer the following characterization of Liberal Naturalism. First of all, like the vast majority of contemporary naturalisms, this view is committed to the claim (the "constitutive claim of naturalism") that no entity or explanation may be accepted whose existence or truth would contradict the laws of nature insofar as we know them. More specifically, Liberal Naturalism is characterized by two provisos—one epistemological and one ontological—that complement this claim. As to the first, even if some of the controversial entities were actually reduced or shown to be ontologically dependent on scientific entities, in order to account fully for the features of these entities, one might still have to turn to forms of understanding (such as conceptual analysis, imaginative speculation, or introspection) that are neither reducible to scientific understanding nor supernatural.[18] As to the ontological proviso of Liberal Naturalism, there may be entities (e.g., numbers) that do not and cannot causally affect the world investigated by the sciences and are both irreducible to and ontologically independent of entities accountable by science but are not supernatural either, since they cannot violate any laws of nature.[19]

By analyzing these two provisos, De Caro and Voltolini argue that Liberal Naturalism does not collapse into Scientific Naturalism, as the first horn of Neta's dilemma claims, because of its broader epistemological and ontological attitudes. The second horn of Neta's dilemma, the threat that Liberal Naturalism is Supernaturalism in disguise, is not serious, either. Supernaturalism is ruled out since no object, property, or event can be causally efficacious in the natural world and yet fail to be an object of scientific investigation (in principle, at least). In this light, liberal naturalists have no problem in ruling out, on scientific grounds, supernatural entities such as immaterial gods, infinite and perfect divine attributes, irreducibly miraculous events, or Cartesian minds—that is, causally efficacious immaterial particulars that cannot in principle be investigated scientifically. Moreover, supernatural entities (both causally efficacious and noncausally efficacious) would require special modes of understanding that would be irreconcilable with scientific explanation— and would thereby violate the fundamental claim of naturalism.

In "Metaphysics and Morals," reproduced here,[20] Thomas Scanlon articulates a liberal naturalist position that sees no clash between the world posited by the sciences and the normative phenomena that are registered in our accepting the truth of various moral and practical judgments. The latter are simply not in the business of purporting to refer to causal matters instantiated in space, according to Scanlon. Consequently, the obtaining of moral truths (or facts) is not to be thought of as part of or in competition with scientific accounts of the causal regularities of the world. One should not take the fact that, say, moral judgments are often based on causal claims concerning physical and psychological events as evidence that the claims made by moral judgments are themselves causal. Nor should one treat moral knowledge (supposing there is such) as analogous to perceptual knowledge, since the latter, in contrast to the former, involves the right kind of causal connections to items located in space. On Scanlon's view, moral judgments can be true without being about things investigated by the sciences and without needing any metaphysical support.

In "The Naturalist Gap in Ethics," Erin I. Kelly and Lionel K. McPherson are concerned with how to understand the authority or bindingness of moral reasons. Their chapter can be read as steering a middle course between the extreme positions of moral naturalism, on the one hand, and moral cognitivism (including that defended by Scanlon), on the other. According to Kelly and McPherson, the cognitivist may be right that moral judgments are reason-sensitive but is mistaken in thinking these judgments must have authority for all rational agents. At the other extreme, the naturalist may be right to challenge the rational authority of moral judgments as binding for all rational agents but wrong to suppose that the reason sensitivity in question can be reductively explained by appeal to psychology or biology. At the heart of their account, Kelly and McPherson draw a distinction between *acknowledging* and *accepting* a reason. Structural accounts of the normativity of reasons that appeal to presuppositions of reason, communication, or agency can at best show that we must acknowledge moral reasons in the sense of appreciating their normative significance relative to certain specific pursuits, goals, and attachments. They do not show that we must accept moral reasons in the sense of having to be guided or motivated by them ourselves on pain of irrationality. On this view, there is a gap between being aware of the normative significance of moral imperatives and accepting these same reasons in one's own deliberations. Frankfurt and Blackburn are right to think that a person's cares and commitments are needed to account for the rational force or bindingness of moral reasons. But their mistake is a failure to see that this concession is compatible with a cognitivist approach to the content of moral reasons. As Kelly and McPherson see it, moral judgments are not simply expressions of a person's sentiments or attitudes; rather, having certain sentiments or attitudes is a condition for having moral concern at all.

Another problem with moral naturalism concerns its account of moral psychology. In "Phenomenology and the Normativity of Practical Reason," Stephen L. White challenges the widely influential Humean moral psychology that supposes that practical reason only plays an instrumental role in finding the means to antecedently established ends (given by one's desires) and that what explains why the agent did what she did is simply some belief-desire pair. Although a sophisticated version of this doctrine can overcome various problems associated with traditional Humean moral psychology (extreme imprudence, pathological indifference, normativity), it still faces the problem of explaining what makes an action an *action* as opposed to something that merely happens to one. White argues that the basic problem plaguing the Humean account is that desires, given as objects, simply cannot explain action. For example, the Humean cannot adequately characterize the difference between what White calls "the passive subject"—one who agrees about all the objective facts stated in nonagential vocabulary but who finds the language of agency unintelligible—and ordinary agents like ourselves.

The Humean, on White's view, has no account of how a desire is given "subjectively" (at the personal level) in the normal way. White concludes that the representation of things as *desirable* (that is, where the objects of one's desires are given as desirable) is more basic and more primitive than the representation of things as desired by me. This requires a phenomenology that is both inflationary and deflationary relative to sense-data theories: inflationary, since seeing things as desirable (hence valuable) is the normal way desires are given to us; and deflationary, since we are often given external objects and their desirability directly without there being something else at the personal level (e.g., sense data) by virtue of which we have this access. White goes on to show that this perspective allows for a solution to "the moral problem" of Michael Smith and for a vindication of Aristotle's claim that the virtuous person is superior to the merely continent person.

Another version of Liberal Naturalism is that of John McDowell, whose "naturalism of second nature" involves a rejection of both the ontological and methodological doctrines of Scientific Naturalism. In this volume, both Peter Godfrey Smith and Marie McGinn want to retain something of the spirit or motivation of McDowell's version of Liberal Naturalism but take issue with its formulation, in particular, the lack of any substantial conception of the relation between first nature, the picture of the human provided by the natural sciences (particularly biology), and second nature, the picture of the human provided by our language and culture, including philosophy itself. McDowell seems to be content to say simply that our first nature (say, our physical, chemical, and biological nature) constrains the development and shape of our second nature: "the innate endowment of human beings must put limits on the shapings of second nature that are possible for them."[21]

In "Dewey, Continuity, and McDowell," Peter Godfrey-Smith sees Dewey, like McDowell, as attempting to overcome dualistic conceptions of the relation between normative thought and the natural world. In Dewey's account this dualistic inheritance of traditional philosophy can be traced all the way back to the Greeks and their prioritizing, in the realm of Being, the permanent and self-possessed and, in the realm of Knowing, the process of copying. If we do justice to the *changing* and *relational* from the perspective of a problem-solving epistemology, Dewey, on Godfrey-Smith's reading, shows us how we can accept the reality of normative phenomena left out of the natural scientific description of the world. Moreover, Dewey makes room for a philosophical project of relating normative phenomena to the world as described by the sciences, a "sideways-on" project that McDowell decisively eschews. From Dewey's perspective this is not a matter of reducing and then subsuming such phenomena under natural laws. As Godfrey-Smith puts it, "What the philosopher wants to do is ask general questions about how the 'habits of thought and action' involved in our use of normative concepts [especially the concept of a reason] relate to other facts about us, and how these habits function as human cognitive tools."

In "Wittgenstein and Naturalism," Marie McGinn raises related concerns by way of a discussion of two ways of reading the bearing of Wittgenstein's later philosophy on rule following: Wright's reading, according to which our responses to rules are grounded in primitive dispositions described in the language of disenchanted nature; and McDowell's "naturalized Platonist" reading, according to which when reasons come to an end or when we act blindly we are still able to conceive ourselves as within "the space of reasons."

Although McGinn comes down on the side of McDowell in this debate, she wants to press further the question of how we are to understand the relation of our natural history to our capacity to follow rules. McGinn is at pains to provide a conception of Wittgenstein's naturalism that avoids seeing him as claiming that the truth of our judgments depends on human agreements while at the same time being alive to the idea that the sense of our judgments does depend on certain general facts of nature, including certain constancies in human reactions, responses, and actions—including, in contrast to McDowell's naturalism, those that are preconceptual and untrained. In McGinn's view, what McDowell misses in assimilating his naturalism to Wittgenstein's is the way, in Wittgenstein's hands, a naturalism that involves attending to the details of things—that is, to the specific and variegated ways things act and work—can be fundamental to his method of philosophizing.

◎ ◎ ◎

The essays collected here open up the territory between, on the one hand, an exclusively scientific ontology and scientific modes of knowing and, on the other, all forms of Supernaturalism (including, of course, a Platonism that treats norms as intangible objects).[22] This is the contested terrain of what we are calling Liberal Naturalism, and it is an open question what the best way of conceiving this new form of naturalism is. Should we suppose that science, broadly conceived, settles all questions of ontology so that Liberal Naturalism simply takes the form of an epistemological commitment to nonscientific modes of knowing or understanding rationality, other minds, oneself, ethics, aesthetics, etc.? Or is there, in addition, a nonscientific, nonsupernatural ontology (of, say, actions, agents, values) that must be acknowledged if we are to make best sense of our practical and social lives? Furthermore, what kinds or categories of irreducible and categorical norms are there, and how should we go about conceiving and studying them if we are to avoid both a reductive scientism and an otherworldly Platonism? Is there a social phenomenology of normative phenomena such that the norms and values are simply those that we can mutually acknowledge as such?

Another important matter is the relation between what McDowell calls first and second nature, the puzzling relationship between the normatively structured "space of reasons," on the one hand—or perhaps we should say "spaces of reasons" if we suppose that reasons come in importantly different kinds—and the domain of naturalistic lawlike or causal explanations characteristic of modern science, on the other. The hope is that a Liberal Naturalism could bring the relations between these two senses of nature into better focus, thereby either endorsing McDowell's thesis of the autonomy of the space of reasons and the manifest image or revising it on the basis of more nuanced conceptions of the scientific image and the manifest image, thus, of the relation between the two images.

ACKNOWLEDGMENTS

The outlook of Liberal Naturalism was originally inspired by the writings of John McDowell and Hilary Putnam and we want to acknowledge a debt of gratitude to them for having blazed the trail. In the development of the outlook and in helping to gain a sense of its internal variety we have benefited from reviews of a previous collection of ours, *Naturalism in Question* (Harvard University Press; 2004/2008) and discussions over several years with Akeel Bilgrami, John Dupré, Huw Price, Hilary Putnam, and Stephen White. We would also like to gratefully acknowledge the encouragement and invaluable advise of Akeel and Stephen in the detailed conception and preparation of this work. Stephen is to be thanked, moreover, for providing very helpful comments on previous drafts of the introduction.

This book is dedicated with love to Giulia and Talia.

NOTES

1. John McDowell, *Mind and World* (Cambridge, Mass.: Harvard University Press, 1994), 70. The two inserted quotes are from p. 71.

2. Max Weber, *The Methodology of the Social Sciences*, trans. E. A. Shils and H. A. Finch (New York: Free Press, 1949). See also Richard Swedberg, *Max Weber and the Idea of Economic Sociology* (Princeton, N.J.: Princeton University Press, 1998).

3. For a discussion of the difficulties associated with these alternatives, particularly the second and third, see Michael Smith, *The Moral Problem* (Oxford: Blackwell, 1994).

4. Arguably, most scientific naturalists are committed to both the Ontological Doctrine and the Methodological Doctrines. However, one might argue for the Ontological Doctrine on purely a priori grounds which, depending on one's view of the a priori, might involve denying the Methodological Doctrines. Whether one could hold the Methodological Doctrine whilst denying the Ontological Doctrine will depend on one's understanding of the relation between scientific explanations and ontological commitment.

5. Wilfrid Sellars, "Empiricism and the Philosophy of Mind," in *The Foundations of Science and the Concepts of Psychoanalysis*, ed. H. Feigl and M. Scriven, *Minnesota Studies in the Philosophy of Science 1* (Minneapolis: University of Minnesota Press, 1956), 173.

6. Arthur Danto, "Naturalism," *Encyclopedia of Philosophy*, ed. Paul Edwards (New York: Macmillan, 1967), 5:448.

7. It should be noticed that, in a very different spirit, some liberal naturalists also endorse radical empiricist views of knowledge. John Dupré, for example, endorses empiricism as the "answerability of our belief to what we actually experience in the world" (*Human Nature and the Limits of Science* [Oxford: Oxford University Press, 2001], 163–64).

8. Some scientific naturalists, such as Frank Jackson, defend the idea of *revisable* a priori knowledge. See his *From Metaphysics to Ethics: A Defence of Conceptual Analysis* (Oxford: Clarendon Press, 1998).

9. An exception to this attitude is Hartry Field's mathematical fictionalism. See his *Science Without Numbers* (Oxford: Blackwell, 1980).

10. David Papineau, "Naturalism," *Stanford Encyclopedia of Philosophy*, http://plato. stanford.edu/entries/naturalism/.

11. Even though this paper was written especially for inclusion in the present volume, because of Rorty's untimely death in 2007 it first appeared in his *Philosophical Papers*, vol. 4: *Philosophy as Cultural Politics* (Cambridge: Cambridge University Press, 2007), 147–59.

12. There may be various reasons the human and social sciences are thought to be irreducible to the natural sciences. They include that there are no plausible reductions in the offing and that these sciences involve normative notions such as that of a reason or notions of subjectivity that are explanatorily autonomous.

13. It is relevant to note that the problem of how to accommodate the natural and the normative was seen to be at the heart of the German philosophical tradition in the late nineteenth century in the works of Rickert and Windelband.

14. This paper first appeared in *Journal of Philosophy* 100 (2003): 167–90.

15. Huw Price, "Quining Naturalism," *Journal of Philosophy* 104 (2007): 405.

16. It should be noticed that Dupré also believes that there is no clear-cut boundary between science and nonscience.

17. Ram Neta, "Review of M. De Caro and D. Macarthur, eds., *Naturalism in Question*," *Philosophical Review* 116 (2007): 662.

18. It is important to notice that this definition does not imply that Liberal Naturalism a priori denies the very possibility of reductions concerning, say, mathematical or phenomenological entities. What it rejects is the ideological assumption, characteristic of Scientific Naturalism, that such reductions *must* be possible.

19. An alternative liberal naturalist strategy has been proposed by Hilary Putnam in *Ethics Without Ontology* with regard to ethical and mathematical discourses. According to Putnam, these discourses are objective and irreducible to scientific discourses, but no distinctive set of objects is required for them. For present purposes, however, it can be argued that Putnam's nonontological view does not substantially differ from the pluralistic ontological view described in the text, since what is crucial for both views is that they conjugate objectivism with antireductionism and non-Supernaturalism.

20. This chapter first appeared in *Proceedings and Addresses of the American Philosophical Association* (presidential address) 77 (2003): 7–22.

21. John McDowell, *Mind, Value, and Reality* (Cambridge, Mass.: Harvard University Press, 1998), 190.

22. Hilary Putnam, "Was Wittgenstein Really an Anti-realist About Mathematics?" in *Wittgenstein in America*, ed. Timothy McCarthy and S. C. Stidd (Oxford: Oxford University Press, 2001), 184: "Our understanding of our concepts, and our employment of them in our richly conceptually structured lives, is not a mystery transaction with intangible objects, a transaction with something over and above the objects that make up our bodies and our environments."

PART I

CONCEPTUAL AND HISTORICAL BACKGROUND

1

THE WIDER SIGNIFICANCE
OF NATURALISM

A GENEALOGICAL ESSAY

Akeel Bilgrami

Some of the philosophical debates of our time are secular echoes, indeed secular descendants, of disputation some centuries ago that was no less intense and of measurably greater and more immediate public significance. If some of this sort of significance persists in our current debates, it is seldom on the surface. This is because of our tendency in analytic philosophy to view our metaphysical and epistemological concerns in relatively autonomous terms, unburdened by any political and cultural implication or fallout. Hence, such wider significance as might still exist can only be unearthed by paying some *genealogical* attention to the antecedent disputes in which the issues at stake loomed larger and more visibly in public and political life.

Though it is not by any means the only one that comes to mind, I will restrict my discussion to one example—the deep division among philosophers today over naturalism,[1] understood as the metaphysical claim that there is nothing in the world that is not countenanced by the methods of natural science.[2] Naturalism in this sense has evolved in recent years into a sophisticated doctrine, and with sophistication there has been a certain degree of acknowledgment that some concepts describing or expressing certain properties that are, on the face of it, nonnatural may not get a strict rendering into the conceptual vocabularies (physical, causal, functional) of the various natural sciences. Even so, naturalism posits various forms of *systematic dependency relations* in which these properties stand to the properties traversed by the explanatory methods of the natural sciences.[3] No properties are allowed that do not stand in these dependency relations. The primary focus of the debate has been over value properties—with intentional properties of mind plausibly thought to be, for reasons that I won't elaborate here, just a special case of value properties.[4] In a word, the debate is over whether values are or are not reducible to (do or do not stand in systematic dependency relations with) natural properties as defined above.

This debate has a well-studied history within the confines of philosophy, and in that history the chief protagonists have been Hume and Kant and their many successors down to this day. On the Humean side, there is a conception of values in which they are considered largely to be a refinement of our desires. They are mental states we possess that, though they may be more reared in and geared to social relations and social constraints than other passions (as, for instance, in Hume's elaboration of the notion of "sympathy" or in Adam Smith's account of them as "moral sentiments"), they are nevertheless *tendencies* of our mentality. On the other side, finding all this too psychologistic and tied to human inclination, Kant had relegated morals to a "noumenal" status within "pure practical reason" whose relation to the *perceptible* world was rendered at least prima facie problematic. I want to steer past this canonical dispute between Humeans and Kantians and in its stead make my subject John McDowell's conception of value because it helps to bring to the front much more specifically than either of those positions, a genealogy of the political and cultural significance of the vexed disagreement between "naturalists" (as I have defined the term) and their opponents.[5]

In the next section, I will motivate this conception of value and then, in the rest of the sections that follow, I will present the sort of genealogical analysis that displays the wider significance of the dispute about naturalism that this conception of value generates.

◉ ◉ ◉

Let me motivate this conception of value via a dialectic that begins with a familiar distinction.

It is a relatively familiar point, sometimes attributed to Spinoza, that one cannot both *intend* to do something and *predict* that one will do it at the same time.[6] When one predicts that one will do something, one steps outside of oneself and looks at oneself as the object of behavioral and causal and moti-vational tendencies. One looks at oneself as another might look at one, and so this is often called the "third-person point of view" on oneself. But when one intends to do something, one is asking "What should I do?" or "What ought I to do?"; one is being an agent not an observer of oneself, a subject rather than an object, and that is why this is sometimes known as a "first-person" perspective on oneself. Even when intentions to do something are formed without being deliberatively decisional answers to explicit questions of that form, they are distinctively within the first-person point of view by contrast with predictions of what one will do.

(A terminological aside: This vocabulary may be misleading since "first person" and "third person" can give the impression of being merely gram-matical categories involving the first- and third-person pronoun, while the perspectival categories that the distinction between intention and prediction invoke are *philosophical* categories that *do not coincide* with the *grammatical*. Proof of this failure of coincidence can be found in examples such as when someone says, "I predict that I will . . . ," where the first occurrence of the first-person pronoun, "I," is an agentive use and the second occurrence refers to oneself as an object of detached study or observation—raising hard ques-tions, incidentally, about breezy assumptions we make about unproblematic anaphora in such cases. For this reason it may be sensible to replace the terms the "first-person" point of view and the "third-person" point of view with "the agent's" or "the engaged" point of view and "the observer's" or "the detached" point of view, respectively.)

With whatever terminology we describe it, the crucial point is that though one can and does have both these points of view on oneself, one cannot have both these points of view on oneself *at the same time*.

The distinction, as I have presented it so far, is a distinction regarding two perspectives or points of view on *oneself*. But there ought also to be a similar distinction that holds for perspectives we have on the *world*. We can have a detached perspective on it, a perspective of study as is paradigmatically found in natural science (though that is just one highly systematic form that that perspective takes), and we can have a perspective of agency on the world, one of responding to it with practical engagement rather than with detached observation and explanatory purpose.

(Here again there is scope for being misled. The point is not that we are not agents when we are observing and explaining the world in scientific terms but that we, as agents, are taking a perspective of detached observation or study on it rather than one of practical engagement. A scientist in her scientific observation and study does engage with the world and is an agent when she does so, but she does so with a perspective on the world that is detached. This point was already visible in the example I gave above when I was speaking of a third-person perspective one can take on oneself. When I say, "I predict that I will . . . ," the first use of the personal pronoun is an agentive one, but the fact is that, qua prediction, the angle I have taken on myself is that of detached observation rather than of agency, by contrast with when I say "I intend to . . ." Exactly the same point holds of one's third-person perspective on the world. One does not cease to be an agent when one has a detached perspective on the world; one just treats the world as an object of detached study rather than as something that prompts our practical engagement.)

So, these contrasting points of view one has can apply to oneself *as well as* to *the world*. I want now to consider the *latter* and ask a crucial question: What must the world be like, what must the world contain, such that it moves us to such practical engagement, over and above detached observation and study? If the world prompts such engagement, it must contain elements over and above those we observe and study from a detached point of view. The obvious answer to the question is that over and above containing the facts that natural science studies it contains a special kind of fact, evaluative facts and properties, or, more simply, it contains *values*, and when we perceive them, they put normative demands to us and activate our practical engagement. Values, being the sort of thing they are, are not primarily the objects of detached observation; they engage with our first- rather than our third-person point of view on the world.

Thus if we *extend* in this way *onto the world* a presupposition of the fundamental distinction between intention and prediction (the presupposition of two contrasting perspectives that one can have *on oneself*), we get a conception of values that is neither Humean nor Kantian. We get a conception by which values are not merely something *we* generate with our mental tendencies and "*project*" onto the world (a favored metaphor among Humeans); instead, values are properties that are found in the world, a world of nature, of others who inhabit nature with us, and of a history and tradition that accumulates in the relations among these and within which value is understood as being "in the world." So conceived, values are not dismissible either as mere inclinations, as Kant did of Hume's psychologistic conception of values, nor (since they are perceptible properties in the world, precisely what Kant denied) are they dismissible as populating some gratuitous noumenal ontology of the pure and unencumbered will of "Practical Reason." It is not as if sympathy and moral sentiments, much stressed by Humeans, are left out of this picture, but sympathy and moral sentiments, in this picture, are our responses to the normative demands that we apprehend in our perceptions of the evaluative properties of the world.

I have tried to motivate a view of value that places it in the world as flowing from our commonsensical commitments to agency. The motivation was presented in two stages: first, I said that if the distinction between intention and prediction presupposes a distinction between a first-person or agent's point of view and a third-person or disengaged point of view that we can take on ourselves, then there ought to be a similar distinction of points of view that we can take upon the world; and second, if there is to be a first-person or agent's point of view we can take on the world of the sort that we can take on ourselves, then the *world* must contain values that prompt such a point of view of agency to be activated in our agentive responses to them.

The notion of agency and its presuppositions, derived from the initial Spinozist distinction between intention and prediction, play a crucial role in the motivation for such a view of value. But a question might be raised: Why can't agency consist in nothing more than the fact that we try and fulfill our desires, intentions, and so on. True, there is a first-person point of view that is activated and exercised in agency, but why can't it simply be exercised in our efforts to satisfy our desires and fulfill our intentions? Why do I insist that agency comes into play *only when our desires (and moral sentiments) are responding to the callings of something external, the evaluative properties in the world*? To put it in terms of my two-stage dialectic for the motivation, the question is: It is true that the distinction between intention and prediction points to a distinction between the first- and third-person points of view *on ourselves*, but why am I insisting that there actually be a replication or version of this distinction in points of view we have *on the world*.

These are good questions and fruitful ones. Rather than the conception of agency just presented in the second stage, the idea of the world's containing values that prompt such a point of view, they urge upon us much the more standard and much the more minimal and simple philosophical conception of practical agency, our capacity to act so as to fulfill our *desires* (on the basis of our *beliefs* about what will be a suitable available way to fulfill them). And by stressing this standard view of agency they resist the consequent of the conditional presented in the first stage, while granting the antecedent.

The questions, then, throw down the following challenge: The motivation I have presented for a conception of value that places values in the world depends on an unmotivated conception of agency as requiring an exercise of the first-person point of view conceived of as responses to normative demands from the world. According to this conception of agency, as I put it earlier, desires (including those desires that are loftier and amount to moral sentiments) are *not self-standing* but rather are responses to things in the world that have whatever it takes (evaluative properties) that prompt their activation. Why does this seem compulsory? Why can't desires be thought of as self-standing? How can we motivate the denial of their self-standingness, philosophically?

To answer this, we need to look a little harder at the relationship between desires and agency. Gareth Evans once said illuminatingly that questions put to one about whether one believes something, say, whether it is raining outside, do not prompt us to scan our mental interiority, they prompt us to look outside and see whether it is raining.[7] That is to say, one not only looks outside when one is asked, "Is it raining?" but also when one is asked, "Do you believe it is raining?"

Now, let's ask: Is this true of questions put to one about whether one *desires* something? When someone asks one, "Do you desire x?" are we prompted to ponder our own minds or are we prompted to consider whether x is desirable? There may be special sorts of substitutions for x where we might ponder our own minds but for most substitutions, I think, we would consider x's desirability. This suggests that our desires are presented to us as having desirabilities in the world as their objects.

If one thought this extension of Evans's point wrong, if one thought that a question of that sort prompted one to step back and consider by scanning our minds what we desired (rather than to consider what was desirable), that would suggest instead that our desires were presented to us in a way such that *what they were desires for* was available to us only as something that we could have access to when we stepped back and pondered our own minds—in the *third* person. But now, if the presupposition of Spinoza's point is right and if agency is present in the possession and exercise of the *first*-person rather than the third-person point of view, that makes it a question as to how this conception of our desires can square with the fact of our agency. To see our desires as reaching down all the way to desirabilities in the world places our desires squarely within the domain of our agency since now *what* we desire is presented to us in the experiencing of the desiring itself, rather than presented to us when we stepped back to observe our desires—thereby abdicating our agency.

This gives a decisive reason for resisting a self-standing view of desires— such a view cannot accommodate the fact of our agency—and in doing so it establishes two things: first, it establishes the deep and essential links between value and agency, and, second, it motivates the conception of value that resists naturalism about value by resisting (unlike Kantian forms of resistance to it) a purely "naturalistic" conception of *nature* and, more generally, the perceptible (phenomenal) *world*.

Putting it just this way as I have, following McDowell, in order to contrast it with the Kantian resistance to naturalism, might invite a confusion that needs to be preempted at the very outset. It would be a confusion to dismiss such an antinaturalist conception of value as taking an "unscientific" view of nature and the world. To say values are properties in the world (including nature) is to make the world (including nature) not comprehensively survey-able by the methods of natural science. That is the antinaturalism. How could this be an unscientific thing to say? Something is unscientific, one must assume, if it falls afoul of a claim of one of the natural sciences. What else could "unscientific" mean? But if this is what it means, then antinaturalism is not unscientific since no natural science contains the proposition that natural science has full coverage of the world (including nature). That is something that only a philosopher says (or a scientist, playing at being a philosopher). And one can find it to be bad philosophy, as McDowell no doubt does, without being accused of doing bad science in return. Claims such as those made by "creationism" or "intelligent design" are unscientific since they say something by way of answer to a question that is scientific (about the origins of the universe) and contradict what the best current science has to say on that question. But the antinaturalism we are considering, the claim that values are properties in the world, makes no attempt to answer any question of that kind. It is not a question within natural science. It is just a confusion to dismiss this antinaturalism as unscientific.

There is more to be said on this subject because there are sometimes concessions made by naturalism to antinaturalism that do little to redeem the deeper prejudices—accumulated over the centuries, as I will try and show in the next section in spheres of broader cultural and political significance than those that surface in current debates—that naturalism is prone to. Indeed, sometimes the concession only extends these prejudices.

Let me look a little more closely at one such concession that merely carries these prejudices over from the natural sciences to the social. Suppose one were to concede that the natural sciences do not have full coverage of the world (including nature). And suppose one does so specifically by conceding the importance of what I have placed on center stage, the notion of agency and the contrast of the point of view of agency with the point of view of detached observation and study. This can be done by allowing that the world contains such things as "*opportunities*." Thus, for instance, here in front of me in a glass there is a substance with the chemical composition H_2O, but right there, in the very same place, there is also something properly describable as "an opportunity" to satisfy a desire of mine, the desire to quench my thirst. The first is something that I study from a detached point of view, but the latter is necessarily something I respond to with practical engagement. The world, now, will not be comprehensively surveyed by the natural sciences since no natural science studies opportunities.

This is a concession that a naturalist, as I have defined him, might make. Has he conceded enough to the antinaturalist as I have defined him? Perhaps the answer has to be "yes" if all we care about is the letter and not the spirit of the antinaturalist's objections to the disenchantment that naturalism has wrought. And if so, we will need another label than naturalism for what the antinaturalist most deeply opposes (and a correspondingly different label for his own position). In note 2 I suggested that that label might be "scientism." Let me explain why that term is apt.

This concession by naturalism to antinaturalism can be made with the following theoretical and methodological aim in mind: Let it be that the world contains such things as opportunities that fall outside of the purview of *natural* sciences. They fall within the *social and behavioral* sciences, which now can be described as having the following as one of their goals (of course one among many other goals, though it may, in some sense that I won't try and elaborate here, be a very central and frameworking goal for these sciences within which more specific goals get their specific point): the goal of studying and explaining individual (possibly even eventually social) behavior as a kind of desire-satisfaction in human subjects in the light of their (probabilistic) apprehension of the desire-satisfying properties in the world, that is, *opportunities* that the world provides to satisfy our wants and preferences.

In this picture values themselves continue to be seen in entirely Humean terms, as generated by placing some internal constraints upon desires viewed as dispositions and tendencies in the subject. The subject, however, also has beliefs about what in the world is most likely to fulfill those desires. These may be described, as in the concession being considered, as opportunities in the world for desire-satisfaction. Though the world is now said to contain something (opportunities) that surpass the subject matter of the natural sciences, these contain nothing that is itself intrinsically normative. So what are allowed as properties in the world (the opportunities that prompt our first-person point of view of agency) are mere instruments for satisfying desires, but it is only these desires that (as Humeans insist) exclusively generate values when we put the right internal constraints upon them, and none of these constraints is in any way a normative constraint coming from the world. If there is any impression that values are in the world in this view, it is a phenomenological illusion brought about by our own "projections" onto the world. The perceptible world contains only *means* that human subjects perceive as having a measurable likelihood of satisfying their desires.

The social and behavioral sciences can see in this picture of the world the scope to extend the notion of scientific rationality. True, their angle on the world is less detached than the natural sciences. After all, one looks at the world with more practical engagement when one sees something not merely as H_2O but as an opportunity. Despite this concession to antinaturalism, the normative element in this picture of the practical domain is constructed entirely out of a normative void. It emerges only from within human causal and motivational tendencies and *dispositions*—our desires plus our beliefs about opportunities in the world that will be likely to gratify them. There are two aspects to the normative element, according to this view. The aspect of *value*, which is restricted to some sophisticated and constrained understanding of the desires, and the aspect of rationality, which is exhausted by the idea of acting so as to satisfy those desires on the basis of these beliefs regarding the likelihood of the opportunities in the world contributing to the satisfaction of the desires. No more intrinsic normative element is acknowledged, and that is the reason to think of this picture as rightly describable by the term "scientistic."

Sometimes a further concession toward antinaturalism is made by philosophers (such as Donald Davidson) whereby the normative element is seen as irreducible to human dispositions and causal tendencies, but it is not clear how, on this view, that concession can be ultimately grounded if those tendencies are not responsive to normative demands made by evaluative properties *in the world*. In this further concession, the irreducible normativity is supposed to enchant the human subject, but it remains mysterious how this is supposed to happen when the world the subject inhabits remains disenchanted. The human subject is supposed to be enchanted wholly from within. We may try to remove some of the mystery in this idea by saying that unlike nonhuman animals, human beings can ask of any one or more of their desires and inclinations and tendencies, "Is it good to have it?" a possibility that comes with language and a level of sophistication of thought that only linguistic creatures can possess. This question, "Is it good to have this desire, this disposition or tendency?" is a clear and intelligible one, and to the extent that it is intelligible, this use of "good" in a question of that form is proof that value is not simply reducible to desire and inclination and causal tendency.[8] It must be something over and above these, else that question cannot quite make sense.

This is all salutary and convincing, but the question remains as to what are the normative sources a subject can turn to in order to answer the question, "Is it or is it not good for me to have a certain desire, a certain disposition or tendency?" Davidson himself does not locate the source in anything other than desires themselves. There is no hint in his writing that the dispositions we have that are distinctly relevant to values are dispositions to respond to normative demands coming from the evaluative properties of the perceptible world that we inhabit with others, along the lines I tried to convey at the beginning of this section. Davidson is impervious to such a normative source because he is impervious to the considerations that I had raised earlier via Evans's insight, considerations that displayed the deep links between value and agency. Because it has no place for evaluative properties, for desirabilities that the world contains to which our desires are responses, the Davidsonian position must find another answer than the one I gave to the question about desires that parallels the question Evans raises about beliefs. And the only other answer there is, as I pointed out, forces an abdication of agency.

Thus, though Davidson was among the first to make an important concession to antinaturalism when he claimed an irreducibility for intentionality on the grounds that it is essentially caught up with normative considerations, his antinaturalism remains quite incomplete without the further claim that he fails to make—that normative considerations are grounded in the world, to whose demands our intentional states are responsive.[9]

There is something that needs to be qualified in my constant use of expressions such as "values in the world make normative *demands* on us; they *move* us to or *prompt* our engagement with the world." For someone like me, keen on making the evaluative enchantment of the world so much of a piece with our own capacities for agency, indeed, grounding the possibility of agency in such enchantment, this vocabulary might seem to betray a curious lapse, an undermining of the voluntaristic and decisional aspects of agency by the coercive force (betrayed in this rhetoric) of such an external calling from the world.

An external source of value that moves or prompts or makes demands of our agency is not coercive of the subject because it is only from within the first-person, agentive point of view that these external callings can so much as be recognized by the subject. That was the point of the appeal and the extension to desires of the insightful point by Gareth Evans about belief. If these callings' demands are recognized only from within the first-person point of view, there is no question of their being *coercive* forces. Rather, the subject, in such a recognition of the callings, acknowledges something on its own agentive terms, acknowledges by his agency the authority of those values to make those demands and calls. Agency, then, requires *two things at once*: a source of value from the outside and not merely from within our own causal tendencies and dispositions, as well as the human subject itself acknowledging this authoritative source of value from the first-person point of view and, therefore, allowing that authority to make its demands on the subject. Both points are implicit in the use I made of Evans's insight.

This essential role for the human subject in the understanding of values as properties of the world makes values a *distinctive* kind of property or fact in the world. One cannot aspire to apprehend such facts wholly without context and without interest weighing in. The subject is, in some sense, central to the properties that are in perceptible view to him. McDowell himself represents this in an analogy with secondary qualities. But that analogy is imperfect in some respects and may even mislead, if taken in the wrong direction. If the point of the analogy is to merely say that the human subject is not a cancellable element in the attempt to provide a complete characterization of the evaluative properties in the world, that is true and it is illuminating to have it pointed out. Value is more like "red" than "square." If one feels that a congenitally blind subject misses more of the redness of the tablecloth (in knowing merely the wavelength and other such specifications) than the squareness of the table (in knowing merely the geometrical properties of a square), then there is an important sense in which the human subject and its specific kind of visual sensibility is more relevant to the property of being red than it is to the property of being a square. Value, too, makes such an essential reference to the subject, though, obviously, the relevant sensibility here is a moral not a visual one. But in the case of value, the reference to the human subject goes deeper still since the perception of value properties may not be something that we *can* wholly aspire to have speak to us, independent of the context of social and other background factors that shape our perceptions, in a way that we might aspire to with secondary qualities such as red. Of course, *all* properties—including even primary qualities—are to some extent determined in our perceptions by our background conditions of thought. The familiar point of "the theory-ladenness of observation" was intended to acknowledge just that. But the further point about values is that to say that they are in the world (including nature) is to say something richer in assumption. The relations between the human subject and the world (including nature) he inhabits have a history and tradition within which his perceptions of the value properties in the world and nature at any given time speak to him and make normative demands. They will necessarily speak therefore in terms that are *contaminated* (I use a strong term such as that here to mark how much the point is supposed to exceed the acknowledgment of the mere theory-*ladenness* of observation in general) by a much richer set of background assumptions, and they may well therefore speak differently to subjects in different social and cultural contexts.

None of this should suggest cultural relativism—anymore than the theory-ladenness of the observation of nonevaluative natural facts suggests a conceptual relativism. Though no relativism is implied by it, I mention the possibility of differential responses to evaluative facts in the world partly at least to make clear that the *motivation* for insisting on a philosophical conception of values that views them as facts external to human subjects is *not* to provide some sort of argument against a relativism about values. If there are such arguments against relativism, they will not be found by any simple appeal to a conception of value as being in the world. The motivation and argument for such a conception of value rather is entirely as I stated it at the outset and then later consolidated with the Evansian considerations a little later: to get right the relations among human subjectivity (the first-person point of view), human agency, and value.

And if we have got these relations right, that is, if we have, via these considerations that first originated in a roughly Spinozist distinction and deepened in the Evansian argument I gave, given some genuine theoretical motivation for this conception of values as being in the world, then, without distraction from the debate between Humeans and Kantians, the dispute over naturalism can be recast as a dispute as to whether the world really does contain values, as McDowell claims, and therefore is not comprehensively surveyed by the methods of natural science. Naturalists (at least as I have narrowly defined them) deny that the world contains values. They don't deny that we may meaningfully talk of values, but they do deny that the talk's meaningfulness has to be understood in realist terms, whereby evaluative concepts describe real properties in the world that fall outside of the purview of natural science.

◎ ◎ ◎

I have sought to arrive at the realist picture of values via a dialectic that began with what seemed like a commonsense distinction between intention and prediction partly in order to convey how commonsensical it should seem to say that values are in the world. And a first pass at the genealogical issues I want to raise in this essay can be made by asking: Why has this natural way of thinking about values found so little place in the history of thought (and not just strictly philosophical thought in the narrow sense) about value in the last two or three hundred years? To answer this question would require one to get a sense of the interesting genealogy of our current debates about naturalism, and thereby a sense of the wider significance of that debate, which today is, at best, only highly implicit and, at worst, altogether missing in the idiom and the arguments in which the debate is conducted.

The answer to the question is to be found in one central strand in the intellectual and cultural history of the West in a phenomenon that can be traced, using a term that Weber put into currency and that McDowell, too, often uses to describe it: "disenchantment." For many centuries, this natural way of thinking about values as being in the world that I have presented within the secular terms of the more or less atheistic intellectual orientation I share with most philosophers today (and presumably with John McDowell, who has most explicitly revived this way of thinking) had its source in the presence of a *divinity* that was, in many a view, itself immanent in the world. And it is this source that was undermined in the modern period that Weber described "disenchantment."

This sort of point has, for sometime now—ever since Nietzsche's slogan—been made by summoning the image of the "dead father." And it continues to be made in this way in the current revival of tired Victorian debates about the irrationality of belief in a God and his creation of the universe in six days a few thousand years ago. It is common in the rhetoric wielded by those who speak and write today with scorn of such irrational beliefs to describe them in terms of one's continuing immaturity; one's persistence in an infantile reliance on a "father" whose demise was registered by philosophers (Nietzsche, but Hegel before him) much more than a century ago; and one's abdication of responsibility and free agency in the humbling of oneself to an authority that is not intelligible to human concepts and scientific explanatory methods, concepts and methods hard won in a struggle toward progress and enlightenment, after centuries of obscurantism.

All this may be true enough, but there is something concealing about making the point in just this way since it impoverishes the notion of "disenchantment" to one merely about loss of faith in God and his creation and his authority. What goes missing in this picture is the intellectual as well as cultural and political *prehistory* of the demise of such an authority figure. Well before his demise—brought about I suppose by the scientific outlook that we all now admire and that is rightly recommended by the authors of the string of recent, somewhat tedious books that have inveighed against such irrational belief—it was the metaphysics forming around the new science itself and *nothing less than science* that, far from registering his demise, proposed instead in the late seventeenth century a quite different kind of fate for "the father," a form of *migration*, an *exile* into inaccessibility from the visions of ordinary people to a place outside the universe, from where in the more familiar image of the clock winder, he first set and then kept an inert universe in motion. And much more than his "*death*," it is this exile and *deracination* of God from the world of matter and nature (and therefore from human community and perception) that reveals what is meant by "disenchantment."

There is no Latin expression such as "*deus deracinus*" to express the thought that needs expounding here. The expression for the God exiled by the ideologues of the Royal Society in England in the wake of the developments in science around Newton in the late seventeenth century is "*deus absconditus*," which may convey to the English speaker a fugitive fleeing rather than what I want to stress—the idea that it is from the roots of nature and ordinary perceptible life that God was removed. "*Racine*," or roots, is the right description of his immanence in a conception of a sacralized universe, from which he was torn away by the exile to which the metaphysical outlook of early modern science (aligned with thoroughly mundane interests) ushered him. Even so, "*conditus*" which literally means "put away for *safeguarding*," (with the "*abs-*" reinforcing the "awayness" of where God is safely placed) conveys something about the question I want to raise. What I want to ask is: Why should the authority figure need safeguarding in an inaccessibility? What dangers lay in his immanence, in his availability to the visionary temperaments of all those who inhabit his world? And why should the *scientific* establishment of *early* modernity seek this safekeeping in exile for a "father" whom its successor in *late*, more mature modernity would properly describe as "dead"?

These *genealogical* questions are crucial to the analysis I want to present about the wider significance of the debates around naturalism, first, because an answer to them would show that the "scientific rationality" that is so insistently extolled by these attacks on religious belief today did not emerge whole all at once but also because the answer reveals that—even if we allow it to be a gradual outcome of a triumphantly progressive intellectual history—to focus merely on the end point of that history as an ideal of rationality toward which we have sequentially and cumulatively progressed and converged in a long struggle against obscurantism is to give oneself an air of spurious innocence.

Narratives of progress have been much under attack for some time for their self-congratulatory triumphalism, but I think it is arguable that things are methodologically much worse than that. They are wrong—at any rate, deeply limited—on, and by, *their own* terms. In general, a sequence, especially when it is consecutively narrativized and dialectically and cumulatively conceived, as progressive ideals are bound to conceive it, cannot have started from the beginning of thought and culture itself. If a sequence is to aspire to conceptual and cultural significance (as the idea of progress suggests) it cannot have *its* beginnings at the very beginning of conceptual and cultural life. That would trivialize things—evacuate the notion of sequence of any of the substance and significance that progressivist narrative aspires to. It cannot be that we have been converging on the significant end point from the random inceptions of our intellectual and cultural existence. One assumes, rather, that there were many strands at the outset, endless false leads, but then at some point (what I am calling the *beginning* of the progressivist trajectory) we got *set* on a path, which we think of as the *right* path, from which point on the idea of *cumulative* steps toward a broadly specifiable end began to make sense, a path of *convergence* toward that end. Accumulation and convergence, then, don't start at the beginning of thought; rather, they start at some juncture that we think of as the *start* onto a *right* path.

This has many implications for intellectual historiography, some of them highly critical. Just to give you one example, I think it implies a real difficulty for philosophers such as Hilary Putnam when they say that scientific realism is true because it is the only explanation of the fact that there is a *convergence* in scientific theories—that is to say, the posits of science must be real because it is only their reality that would explain the *cumulative* nature of the claims of scientific theories over time.[10] What is the difficulty with this that I have in mind? It is this. Here, too, the fact is that these converging and cumulative trends have not existed since the beginning of theorizing about nature. In fact, Putnam would be the first to say that it is only sometime in the seventeenth century that we were set on the right path in science and from then on there has been a convergence that is best explained by the corresponding reality of what the converging scientific theories posit. But now a question arises: *What* makes it the case that *that* is when we were set on the *right* path? What is the notion of *rightness*, here? If we have an answer to this last question (about what makes the path the *right* path at *that* starting point), then that notion of rightness would *already* have established scientific realism and we don't need to wheel in scientific realism to explain the subsequent convergence.

Well, my subject is not scientific realism, so I give this example only to display the more general point that accounts of our rationality that stress our sequential development and progress towards a hard-won end cannot then just focus on the end point and avoid the importance of the *beginning* of the sequence, which may have greater power to illuminate than the end or even the sequence itself. If you wanted a slogan for what I have been saying, it is:

No teleology without genealogy! And, as I have said earlier, my own reason for stressing the early modern *origins* of our late modernity's proud embrace of scientific rationality is to make us less complacent about the ideal that we have embraced by uncovering in its genealogy the thick *accretions* to it that have had large implications for politics and culture.

Let me turn to these now and say more specifically why the scientific establishment of early modernity would have found it convenient to put the "father" in safekeeping, away from the visionary access of ordinary people.

◎ ◎ ◎

There are three things to observe at the very outset about this exile of the "father" for some two hundred years until Nietzsche announced his demise.

First, intellectual history of the early modern period records that there was a remarkable amount of dissent and explicit dissent against the notions that produced the exile, dissent by a remarkable group of intellectuals, who were most vocal first in England and the Netherlands and then elsewhere in Europe. For the sake of focus, I will restrict myself to England. Second, there was absolutely nothing unscientific about these freethinkers or their dissent. They were themselves scientists, then of course called "natural philosophers," fully on board with the new science and the Newtonian laws and all its basic notions, such as gravity. They were only objecting to the *metaphysical* outlook generated by official ideologues, who began to dominate the Royal Society, in which the much more complicated Newton of his private study was given a more orthodox public face by people such as Boyle and Samuel Clarke, a public move in which Newton himself acquiesced. And third, the metaphysical outlook of the dissenters was suppressed, the Royal Society ideologues won out, and their metaphysics became the orthodoxy not because of any superiority, either metaphysical or scientific, but because of carefully cultivated *social and political factors*, that is to say, alliances that the "Newtonians" formed with different social groups, such as the Anglicans and the commercial and mercantile interests of the time.[11]

To put a complex range of interweaving themes in the crudest summary, the dispute was at first sight about the very nature of nature and matter and, relatedly therefore, about the role of the deity and the broad cultural and political implications of the different views on these metaphysical and religious concerns. The metaphysical picture that was promoted by the exile of the "father" to a place outside the universe was that the world itself was, therefore, "brute" and "inert" and needed an external divine source for its motion. In the dissenting picture, by contrast, matter was *not* brute and inert but rather was shot through with an *inner* source of dynamism responsible for motion, which was itself divine. For the dissenters, God and nature were not separable as in the official metaphysical picture that was growing around the new science, and John Toland, for instance, to take just one example among the active dissenting voices, openly wrote in terms he proclaimed to be "pantheistic."[12]

This metaphysical disagreement, however, was caught up in a range of wider implications. One was this: some of the dissenters argued that it is only because one takes matter to be "brute" and "stupid,"[13] to use Newton's own term, that one would find it appropriate to conquer it with nothing but profit and material wealth as ends, and thereby destroy it both as a natural and a human environment for one's habitation. In today's terms, one might think that this point was a seventeenth-century predecessor to our ecological concerns, but though there certainly was an early instinct of that kind, it was embedded in a much more general point about how nature in an ancient and spiritually flourishing sense was being threatened and how this was in turn threatening our moral psychology of engagement with it, *including the relations and engagement among ourselves as its inhabitants.* This last point is vital to the significance of the issues at stake, which were not about nature in a purely self-standing sense. That is why the qualms expressed by the term "disenchantment of nature" were not by any means merely ecological qualms. The ideal of enchantment was (and is) an ideal of an unalienated life (to use Marx's later term), whether from nature or from one another as its inhabitants. Nature itself, therefore, was conceived in terms of its relations with its inhabitants and a history of those relations and a tradition that these engender in different societies, within which subjects engage with nature (broadly conceived in this way). All this went into the understanding of "nature" in what I have called the "ancient and spiritually flourishing sense" of that term.

Today, the most thoroughly and self-consciously secular sensibilities may recoil from the term "spiritually," as I have just deployed it, though I must confess that I feel no such self-consciousness despite being a secularist, indeed an atheist. The real point has not much to do with the rhetoric. If one had no use for the word, if one insisted on having the point made with words that we today can summon with confidence and accept without qualm, it would do no great violence to the core of their thinking to say this: the dissenters thought of the *world* not as brute but as *suffused with value.* That they happened to think the source of such value was divine may not be the deepest point of interest for us today. And, in fact, many of the dissenters were attacked by an inner circle of the Royal Society (consisting of Richard Bentley and Samuel Clarke, among others, and all approved by Newton himself) that had formed around the Boyle Lectures,[14] for having too tenuous a commitment to the divine on the grounds that the line between pantheism and atheism (as well as materialism) was much too thin. They argued that what was needed for the Protestant faith to flourish in a stable and abiding form was not merely an opposition to the Catholic sympathizers among the High Tories but an opposition to these "freethinkers" on the republican Left among the Whigs, who opposed their metaphysics and denied a providential role to God by making him co-eternal with matter itself and thereby too easily subtractable from it to yield materialism and atheism. So, though many of the freethinkers had an explicitly pantheistic commitment (Toland is said to have actually coined the term "pantheism"), they provided by this perceived approximation to materialism and atheism an excuse for those who opposed them to cast them as beyond the pale.[15]

I will return later to the wider political reasons for the Anglican establishment's insisting on the importance of a providential God keeping a universe in order from *without*. But for now, the point I am stressing is that to see God *within* was to see nature as sacralized, with the strict implication that it was thereby laden with *value*, making *normative* (ethical and social) demands on one, normative demands, therefore, that did not come merely from our own desires and subjective utilities. It is this sense of forming commitments by taking in, *in our perceptions*, an evaluatively "enchanted" world that therefore moved us to normatively constrained *engagement* with it, which many dissenters contrasted with the outlook that was being offered by the ideologues of the new science. A brute and disenchanted world could not move us to any such engagement since any perception of it would necessarily be a *detached* form of observation, and if one ever came out of this detachment, if there was ever any engagement with a world so distantly conceived, so external to our own sensibility, it could only take the form of mastery and control of something alien, with a view to satisfying the only source of value allowed by this outlook—our own desires and utilities and gain.

We are much used to the lament that we have long been living in a world governed by overwhelmingly commercial motives. What I have been trying to do is to trace this to its deepest *conceptual* sources, and that is why the seventeenth century is so central to a proper understanding of this world. Familiarly drawn connections and slogans, like "religion and the rise of capitalism," are only the beginning of such a tracing.

In his probing book *A Grammar of Motives*, Kenneth Burke says that "the experience of an impersonal outlook was empirically intensified in proportion as the rationale of the monetary motive gained greater authority."[16] This gives us a glimpse of the sources. As he says, one had to have an impersonal angle on the world to see it as the source of profit and gain, and vice versa. But I have claimed that the sources go deeper. It is only when we see the world as Boyle and Newton did, against the freethinkers and dissenters, that we understand further why there seemed no option but to stress this impersonality in our angle on the world. A desacralized world, to put it in the dissenting terms of that period, left us no other angle from which to view it but an impersonal one. There could be no normative constraint coming upon us from a world that was brute. It could not move us to engagement with it on *its* terms. All the term-making came from us. We could bring whatever terms we wished to such a world, and since we could only regard it impersonally, it being brute, the terms we brought in our actions upon it were just the terms that Burke describes as accompanying such impersonality, the terms of "the monetary" motives for our actions. Thus it is that the metaphysical issues regarding the world and nature, as they were debated around the new science, provide the deepest *conceptual* sources.

But why, one might ask, should the fact of the "father's" exile to an external place as a clock winder have led to an understanding of the universe as wholly brute and altogether devoid of value? Why was it not possible to retain a world suffused with values that were intelligible to all who lived in it, despite the inaccessibility of the figure of the father? Why must value require a sacralized site for its station, without which it must be relegated to proxy, but hardly proximate, notions of desire or utility and gain? It might seem that these questions are anachronistic, suited only to our own time when we might conceivably (though perhaps not with much optimism) seek *secular* forms of *reenchanting* the world. One cannot put them, at least not without strain and artificiality, to a period in which value was so pervasively considered to have a sacred source. But even if we cannot put these questions to a world view that, by our modern lights, was constricted by impoverished conceptual options, we can ask a diagnostic question about what forces prevented the *development* of the idea that the world is enchanted with evaluative properties whose normative demands on us, even if now purely secular, move our first-person point of view to a responsive moral agency? The diagnosis has many elements and needs more patient elaboration than I can possibly give it here, but one or two of the more straightforward points can be put down briefly.

◎ ◎ ◎

The core of the diagnosis is that an (alternative and more secular) ideal of enchantment never took hold because there were too many powerful social forces that were complicit in keeping it out.

The conceptual sources of disenchantment that I have traced are various, but they were *not* miscellaneous. The diverse conceptual elements of religion, capital, nature, metaphysics, rationality, and science were *tied together* in a highly *deliberate* integration, that is to say, in deliberately accruing worldly *alliances*. Newton and Boyle's metaphysical view of the new science won out over the freethinkers' and became official only because it was sold to the Anglican establishment and, in an alliance with that establishment, to the powerful mercantile and incipient industrial interests of the period in thoroughly predatory terms that stressed that nature may now be transformed in our conception of it into the *kind* of thing that is indefinitely available for our economic gain by processes of extraction, processes such as mining, deforestation, and plantation agriculture that were intended essentially as what we today would call "agribusiness." None of these processes could have taken on the *unthinking* and yet *systematic* prevalence that they first began to get in this period unless one had ruthlessly revised existing ideas of a world animated by a divine presence. From an *anima mundi*, one could not simply proceed to take at whim and will. Not that one could not or did not, till then, take at all. But in the past, in a wide range of social worlds, such taking had to be accompanied by ritual offerings of reciprocation that were intended to show respect toward nature as well to restore the balance with it; these offerings were made both before and after cycles of planting, and even hunting. The point is that, in general, the revision of such an age-old conception of nature was achieved in tandem with a range of seemingly miscellaneous elements that were brought together in terms that stressed a future of endlessly profitable consequences that would accrue if one embraced this particular metaphysics of the new science and build around it, in the name of a notion of "rationality," the institutions of an increasingly centralized political oligarchy (an incipient state) and an established religious orthodoxy of Anglicanism, which had penetrated the universities as well, to promote these specific interests. These were the very terms that the freethinkers found alarming for politics and culture, alarming for the local and egalitarian ways of life, which some decades earlier the radical elements in the English Revolution, such as the Levellers, Diggers, Quakers, Ranters, and other groups, had articulated and fought for.

These scientific dissenters themselves often openly avowed that they had inherited the political attitudes of these radical sectaries in England of about fifty years earlier and appealed to their instinctive, hermetic, neo-Platonist, and sacralized views of nature, defending them against the conceptual assaults of the official Newton/Boyle view of matter. In fact, the natural philosophies of Anthony Collins and John Toland (and of their counterparts in the Netherlands, who drew inspiration from Spinoza's pantheism, spread to France and elsewhere in Europe, and then, when strongly opposed, went into secretive Masonic Lodges and other underground movements) were in many details anticipated by the key figures of that most dynamic period of English history, the 1640s; these the radical groups had enjoyed hitherto unparalleled freedom of publication for a decade or more to air their subversive and egalitarian views based on a quite different conception of nature. Gerard Winstanley, one of the most well known among them, declared that "God is in all motion" and "the truth is hid in every *body*."[17] This way of thinking about the corporeal realm had for Winstanley, as he puts it, a great "leveling purpose." It allowed one to lay the ground, first of all, for a democratization of religion. If God was everywhere, then anyone may perceive the divine or find the divine within himself or herself, and therefore may be just as able to preach as a university-trained divine.[18] But the opposition to the monopoly of so-called experts was intended to aim at more than just the religious sphere. Through their myriad polemical and instructional pamphlets, figures such as Winstanley, John Lilburne, Richard Overton, and others reached out and created a radical rank-and-file population that began to demand a variety of other things, including the elimination of tithes, a leveling of the legal sphere by decentralizing the courts and eliminating feed lawyers, as well as the democratization of medicine by drastically reducing, if not eliminating, its costs and disallowing canonical and monopoly status to the College of Physicians. The later scientific dissenters were clear, too, that these were the very monopolies and undemocratic practices and institutions that would get entrenched if science, conceived in terms of the "Newtonianism" of the Royal Society, had its ideological victory.

Equally, that is to say, conversely, the Newtonian ideologues of the Royal Society around the Boyle lectures saw themselves—without remorse—in just the conservative terms that the dissenters portrayed them in.[19] They explicitly called Toland and a range of other dissenters, "enthusiasts" (a term of opprobrium ever since it had been deployed against the theology and politics of the radical elements of the revolutionary period) and feared that their alternative picture of matter was an intellectual ground for the social unrest of the pre-Restoration period, when the radical sectaries had had such great, if brief and aborted, popular reach. They were effective in creating with the Anglican establishment a general conviction that the entire polity should take the form of orderly rule (over a populace that had been unruly and restive for two decades) by a state apparatus centered around a monarch who served the propertied classes and that this was just a mundane reflection, indeed a mundane *version*, of an *externally* imposed divine authority that kept the universe (of brute matter) in orderly motion. This idea was opposed to the concept of a God *immanently* present in all matter and in all people, inspiring them with the enthusiasms to turn the "world upside down," in Christopher Hill's memorable phrase. To see God in every body and piece of matter, they anxiously argued, was to lay oneself open to a polity and a set of civic and religious institutions that were beholden to popular rather than scriptural and learned judgment and opinion. The Newtonian faction was just as effective in forging, with the commercial interests over the next century, the idea that a respect for a sacralized universe would be an obstacle to taking with impunity what one could from nature's bounty. By their lights, the only obstacles that now needed to be acknowledged and addressed had to do with the internal difficulties of advancing an economy geared to profit—the difficulties of transporting goods to markets, mobilizing labor, and so on. No other factors of a more metaphysical and ideological kind should be allowed to interfere with these pursuits once *nature* had been transformed in our consciousness to a set of impersonally perceived "*natural resources.*"

The alliances brought together by these anxieties ensured that the exile of the "father" from his immanent presence would leave in the world, thus desacralized, no residual evaluative properties that might provide an alternative, more secular source of enchantment. To repeat, it did so, *first*, with the argument that the exile would have the effect of creating a religious and metaphysical sensibility that could view nature as desacralized and ready for a predatory form of capitalist extraction, initially via a rapidly expanding system of "enclosures" and then over time in the next century via the industrial technologies that the new science had made possible for European economies. And, *second*, a distinct but supplementary argument arose that in this safeguarding of the father in inaccessibility, it was only a priestcraft emerging in a class of scripturally trained and learned divines from the universities that could fully comprehend a deity unavailable to the perception and comprehension of ordinary people and that this was to be integrated—by the very same economic, religious, and scientific alliances—with the elite possession of the *cognitive and informational* sources of power *quite generally,* whether in matters of law or medicine or the offices of government and administration. In a word, an oligarchic statecraft needed to ensure the profitable extractive economies that were being generated effectively by and for the propertied classes. From the point of view of this ideology emerging around the new science, the idea that values to live by are available to the ordinary perceptions of the world would demote the privileged knowledge possessed by the elites to something more arcane by making the sources of political morality much more democratic.

It was precisely the threat of the *democratization* of value that was arrested in the early modern developments I have briefly—much too briefly—sketched. And this democratization was replaced instead by ideals of civility generated by the courts of a monarch and the propertied classes,[20] a phenomenon well studied by scholars such as Norbert Elias,[21] though I would add one functional gloss to his illuminating survey of its historical importance. These courtly civilities did not merely contrast with the rude social turmoil of a brute populace; they formed themselves into a screen that hid from the early modern European courts and elites the cruelties of their own perpetration, recognizing cruelties only in the behavior and lifestyles of the brute populace against whom they defined themselves.[22] This went on to lay the ground for the abstract morphing of these civilities into the codifications of rights and constitutions of later modernity in orthodox liberal frameworks that, despite all the enormous good they have done and are deservedly admired for, similarly hide from us the cruelties of our own perpetration on distant lands, allowing one to recognize cruelties only in societies where they are unaccompanied by the concealing formalities of such liberal codifications.[23]

◎ ◎ ◎

The considerations unearthed in the last three sections give a sense of the wider significance that historically grounded the dispute over whether value may be seen as being in the world, and they show how genealogically loaded the term "disenchanted" is, despite McDowell's rather bland use of it in the contemporary version of the dispute.

It is important to record that the diverse elements in these considerations—metaphysics, theology, politics, political economy, and culture—were integrated by these alliances I mentioned in a recurring rhetoric of "*rationality*" and "*science*." This thickly laden ideal of scientific rationality is entirely missing in the story that is told by contemporary writers such as Dawkins and Dennett when they present their much "thinner" ideal of rationality as the outcome of a struggling modernity against a long history of reactionary obscurantism.[24] In our own current philosophical idiom, "rationality" is a rather "thin" and circumscribed ideal, referring mainly to the codifications of inductive, deductive, and decision-theoretic reasoning, with perhaps some more or less elaborated notion of "coherence" thrown in. This is just how it should be. But the term in the earlier, much "thicker" sense that I have been outlining was meant to identify much more than the principles that relate observational evidence to theoretical conclusions (principles, that is, that would show the hypothesis of creationism, for instance, to be spectacularly false), or the principles of logical deduction or of practical reason as we now think of it; it was meant to mark an entire way of thinking of nature and it's relation to our economic and political interests. This was most evident when these mercantile, political, scientific, and religious alliances extended over time the ideas that I have outlined and justified the colonial conquest of distant lands, which, too, were to be viewed as brute nature available for conquest and control—but only *so long as one was able to portray the inhabitants of the colonized lands in infantilized terms,* as a people who were as yet unprepared, precisely because of a *mental lack* of such a notion of *scientific rationality,* to have the right attitudes toward nature and commerce and the statecraft that allows nature to be pursued for commercial gain. It is this integral linking of the new science through its metaphysics with these attitudes that was conveyed by the earlier, thicker understanding of "scientific rationality."

It is not as if one cannot find in the writing of philosophers and scientists of an earlier time the "thinner" and more circumscribed ideal of rationality, and scientific rationality in particular. But part of the point of my tracing the work of a range of worldly alliances in this genealogy of the notion of "disenchantment" is to show how many of them were also, in the name of "science" and the metaphysics growing around the new science, "thickening" what would otherwise have been an innocuous and "thin" notion of scientific rationality.

Once that point is brought to the center, a whole tide of confusingly ambiguous disputation regarding the Enlightenment subsides. The fact is that there was more than one strand in the Enlightenment, and some of these strands were so different from one another that it is perhaps best to say that there was more than one Enlightenment. This has come to be recognized in the recent writing on the idea of a "radical Enlightenment," a label that suits well the late-seventeenth-century and eighteenth-century freethinkers as well as some of the later figures in the British and German romantic tradition.[25] What distinguishes this radical tradition from the more orthodox and canonical strand of the Enlightenment has partly to do with differing attitudes toward what I have called the "thick" notion of scientific rationality. In fact, once we disambiguate the notion of scientific rationality in its thick and thin meanings, a standard strategy of the orthodox Enlightenment against fundamental criticisms raised against it is exposed as defensive posturing. It would be quite wrong and anachronistic to dismiss this initial and early intellectual—and, as I said, perfectly *scientific*—source of critique in the seventeenth and eighteenth centuries that I have expounded in the last three sections, from which later critiques of the orthodox Enlightenment derived, as being irrational (as was sometimes done), once one disambiguates the term "rational" as I just have. Far from being irrationalist, these early opponents of the incipient values of the Enlightenment, whose ideas have clear affinities with recurring heterodox traditions in the West since their time, constitute what can rightly be thought of as the early phase of "the Radical Enlightenment".

To dismiss the dissenters pantheistic tendencies as being unscientific and in violation of the norms of rationality, as was done by their orthodox contemporaries, would be to run together in a blatant slippage the general and "thin" use of terms like "scientific" and "rational" with just this "thick" notion of scientific rationality that I have identified and that played a justificatory role in the development of both a predatory form of capitalism and later colonial conquest, and it is only this "thick" notion that the dissenters were so jittery about. They had nothing against any more attenuated notion of rationality whatever and were themselves, as I said, quite on board with the details of the scientific laws. Dismissals of later critiques of the orthodox Enlightenment exploit the same slippage—and the entire appeal to "scientific rationality" as a defining feature of our modernity trades constantly on just such a slippage, subliminally appealing to the hurrah element of the general and "thin" terms "rational" and "scientific," which we all applaud, to tarnish serious criticism of the orthodox Enlightenment, while ignoring in their critique that the opposition is to the thicker notion of scientific rationality, which was defined in terms of specific scientific, religious, and commercial alliances.

Were we to apply the *thin* conception of "scientific" and "rational" (the one that is widely accepted among philosophers today), the plain fact is that *nobody* in that period was, in any case, getting prizes for leaving God out of the worldview of science. That one should think of God as voluntaristically affecting nature from the outside (as the Newtonians did) rather than sacralizing it from within (as the freethinkers insisted), was not in any way to improve on the *science* involved. Both views were therefore just as "unscientific," just as much in violation of scientific rationality, in the "thin" sense of that term that we would now take for granted. What was in dispute had nothing to do with science or rationality in that sense at all. What the early dissenting tradition was opposed to is the *metaphysical* orthodoxy that grew around Newtonian science and its implications for broader issues of culture and politics. This orthodoxy with all of its wider implications is what successive critiques of the mainstream of the Enlightenment have opposed, and the sleight of hand in the frequent dismissals of all opposition to the Enlightenment as being irrationalist lies precisely in the hope that accusations of irrationality, because of the *general* stigma that the term imparts in its "*thin*" usage, will disguise the very specific and "thick" sense of rationality and irrationality that are actually being deployed by the opposition. Such (thick) *irra-tionalism* is precisely what the dissenters yearned for and hindsight shows what an admirable yearning it was.

Part of my motivation for giving this genealogy of the debates around naturalism is to help bring out how that genealogy provides for a disambiguation of the term "scientific rationality" and exposes this sleight of hand. If Dawkins and Dennett and others, rightly inveighing against our current irrationalism of clinging to the figure of a father whose demise followed upon our embrace in late modernity of a "thin" notion of rationality, had also acknowledged the wide range of issues that centered on the exile of the father in early modernity and the "thick" notion of rationality that it engendered, their books might have a much greater interest for those who are trying to come to a deep understanding of the widespread religiosity in our own time.

◎ ◎ ◎

I have sketched why, for considerations that have a significance well outside of philosophy, indeed strictly outside of science as well, a *sacralized version* of a certain conception of value was aggressively opposed in the late seventeenth century. Certain alliances won, and their philosophical position, which Weber described with the term "disenchantment," was consolidated over the next two or three hundred years. Such a philosophical position may accurately be described in hindsight and in our own vocabulary as an early form of "naturalism" that would not countenance *in* the world anything that was not susceptible to study by the methods of natural science. It is not that in that earlier period nothing was countenanced to exist at all that was not susceptible to scientific inquiry as we conceive of it now. But since the "father" had been exiled to a place outside the universe, no such countenanced thing or property was *in the world* (except under the exceptional and occasional category of "miracles").[26] The subsequent "death" of the "father" then transformed the philosophical doctrine of naturalism to its more current forms that will not brook even the *secular version* of enchantment that lead to a position like the one McDowell urges. That lands us with our current disputation over naturalism, with which I began.

I pointed out that some key concessions that were made to antinaturalism by naturalists in the history of this dispute between these two philosophies of value merely extended the reach and scope of naturalism from the natural to the social sciences. Far from conceding anything deep to antinaturalism, this form of concession to our agency and the first-person point of view (which allows the world to contain "opportunities," that is, properties that fall out-side of the reach of the *natural* sciences because they prompt our agential, first-person responses to the world) promoted a certain scientistic concep-tion of the *social* sciences. *And this too had its wider significance.* It extended the alienating elements of a disenchanted world by explicitly making it a site of instrumentality. With this concession, the ideal of scientific rationality in its *"thin"* version can now be presented as being more than the codifications of confirmation theory and inductive logic. It can be presented as the frame-working ideal of the disciplinary regimes of the social sciences as follows: rational human behavior consists in acting so as to satisfy our desires on the basis of these perceptions of opportunities in the world. So the world, in con-taining opportunities, contains things that go beyond what the natural sci-ences study, but it contains nothing that this specific understanding of the social and behavioral sciences cannot bring within *their* purview. And now the surrounding metaphysical or philosophical picture in which the world is viewed as a site of opportunities can generate its own more specifically *"thick"* version of rationality a century and more after the early modern period I was discussing. The thickening this time is also, as always, via worldly alliances, though with far less input from the Protestant religious establishment than in the earlier period and with more developed industrial technologies to tap the "opportunities" the world contains. The effects on us and the world wrought by this thick ideal deserve—and get—a more detailed specification than is given by Weber's earlier and general term "disenchantment" with the later rhetoric of terms such as "commodification," "alienation," and so on that one finds in Marx and those influenced by him.

But the very fact that Weber and Marx were able to mobilize terms such as "disenchantment" and "commodification" and "alienation" at all against these thick notions of rationality makes absolutely clear the deep connections that exist between value and agency and a certain conception of the perceptible world that we inhabit as agents. These are all terms that describe how our relations to the world were impoverished in ways that desolate us, when we severed these deep connections in our conceptual and material lives. This was the wider significance of the dispute about naturalism in the early mod-ern period that I have tried to excavate genealogically.

The extent to which that wider significance survives *in our own time* is a fascinating question, but it is a very hard question to answer in depth and detail with full attention to the range of different interests that it integrates. It would be no bad thing for analytic philosophers, who are engaged with issues of naturalism, to allow themselves to be mobilized by these broader terms that Weber and Marx deployed and to come out of their more cramped focus and idiom to do their bit in answering it.

NOTES

I am grateful, as always, to Carol Rovane and Stephen White for helpful discussions on the themes of this essay.

1. Though the debate has spread widely across the discipline of Anglophone philosophy in the last few years, naturalists continue to be the overwhelming majority in the discipline, with volumes like the present one and its predecessor, Mario De Caro and David Macarthur, eds., *Naturalism in Question* (Cambridge, Mass.: Harvard University Press, 2004), gathering the voices of a recently emerging opposition.

2. One caveat: I am using the term "naturalism" in a rather restricted way, limiting the term to a scientistic form of the philosophical position. So, the naturalism of Wittgenstein or John McDowell or even P. F. Strawson falls outside of this usage. In fact all three of these philosophers are explicitly opposed to naturalism in the sense that I am using the term. Perhaps "scientism" would be the better word for the philosophical position that is the center of the dispute I want to discuss. It accommodates a certain attitude exemplified in naturalism that is present in a certain way of understanding the nature of the *social and behavioral* sciences. I will say more about this later. David Macarthur, an editor of this volume, explicitly asked me if I had any objection to a term he favors and uses in his introduction, "*liberal* naturalism," to describe a position distinct from "naturalism" in the narrow scientistic sense that I mean, a position that accommodates the kind of naturalism that McDowell and Strawson and Wittgenstein embrace. I have no objection to that term or the position it describes, no more than I do to Marxists aspiring to "*people's* democracy."

3. The range of such dependency relations goes from various versions of what is called "non-reductive materialism" to versions of what is called "supervenience." The former doctrine and label surfaces more in naturalism about intentional states in particular rather than value, though it is perhaps extendable in its proponents' ambitions to the latter. My own commitment to antinaturalism that will emerge *in the present essay* is opposed to most, if not all, of these views. I have argued elsewhere—see *Self-Knowledge and Resentment*, chapter 5 (Cambridge, Mass.: Harvard University Press, 2008)—that even the weakest form of the supervenience thesis is not quite coherently assessable. But that is a strong claim (stronger than anything even in John McDowell, with whom I wrestle in that chapter on this very subject), and I doubt that anything I say in the present essay presupposes the truth of that claim. So its diagnosis of these issues should be acceptable to those like McDowell, who do embrace some version of supervenience of evaluative properties on natural properties, understood as properties countenanced by the methods of natural science.

4. Donald Davidson in a number of articles—see especially his *Essays on Actions and Events* (Oxford: Oxford University Press, 1980)—was perhaps the first to introduce value or normative considerations in the understanding of intentionality, though he was much less clear than he might have been about the extent to which intentional states are just a special instance of values (or commitments), that is, are themselves values or commitments of a special sort. Once one is clear about this, one can see the naturalistic irreducibility of intentional states as just a special case of the naturalistic irreducibility of value. There is, then, no need to think of there being *two* irreducibilities, one of intentionality to the properties of the central nervous system, and the other of value to the natural properties studied by the natural sciences. The latter irreducibility subsumes the former. See chapter 5 of my *Self-Knowledge and Resentment* for a detailed discussion of this issue.

5. McDowell's position is presented in a number of essays, some of which are gathered in his *Mind, Value, and Reality* (Cambridge, Mass.: Harvard University Press, 2001), but see especially "Values and Secondary Qualities." Among contemporary Humean positions on value, the most resolute is Simon Blackburn's, which is presented in some of the writings in his *Essays in Quasi-Realism* (Oxford: Oxford University Press, 1993). There are several contemporary Kantians too, of course, such as Nagel, Scanlon, Parfit, and Korsgaard. The classic statement is Nagel, *The Possibility of Altruism* (Princeton, N.J.: Princeton University Press, 1978). For Hume and Kant themselves, see, respectively, any of the various editions of *A Treatise of Human Nature* and *Groundwork of the Metaphysics of Morals*.

6. See especially Stuart Hampshire's book on Spinoza, first published in 1951 and republished as *Spinoza and Spinozism* (Oxford: Oxford University Press, 2005). See also Hampshire's *Freedom of the Individual* (Princeton, N.J.: Princeton University Press, 1975)

7. Gareth Evans, *Varieties of Reference* (Oxford: Oxford University Press, 1983), 225.

8. The alert reader will recognize this to be an extension of G. E. Moore's open question argument. For more on this subject, see chapter 5 of my *Self-Knowledge and Resentment*.

9. This failing in Davidson has a clear antecedent in Weber himself, who was one of the very first to have made this further concession I am discussing, saying that the human subject, individual and collective, cannot be studied by methods that fail to acknowledge that any domain of study in which human subjectivity is to be found is value-laden. But he nowhere linked this influential observation with his own influential lament about the disenchantment of the world. Such concessions to the irreducibility of the human subject and to the value-laden-ness of the human sciences that study it, though not false, remain incomplete and shallow without a full acknowledgment of the fact that the world itself (including nature) is naturalistically irreducible, that it is enchanted with evaluative properties that move us to practical engagement with the world.

10. See Hilary Putnam, *Meaning and the Moral Sciences* (London: Routledge and Kegan Paul, 1978).

11. I am deliberately focusing on a later period and not tracing some of these tendencies to the more general and somewhat earlier *mechanistic* turn that is so often emphasized in the literature of early modern intellectual history partly because the *worldly* motivations to support a certain incipiently "naturalist" metaphysics was most explicitly formulated by figures in the Royal Society *after* Newton and it is they who formed the alliances with the Anglican religious interests as well as the commercial and mercantile interests that I want to emphasize. I am grateful to Mario De Caro for insisting that I explain in a note why I focus on this later period.

12. John Toland's pantheistic views were presented in a series of works, starting with *Christianity Not Mysterious* in 1696; more explicitly in the discussion of Spinoza in *Letters to Serena* (1704); and then in the late work *Pantheisticon* (1724). These writings are extensively discussed in Margaret Jacob's excellent work, *The Radical Enlightenment* (London: George Allen and Unwin, 1981).

13. The point is not that there was contempt for nature after the exile of God. God's creation and ward could be something that one could respect with wondrous awe, even as miraculous, as scientists often did. The point, rather (see below for more on this), was that the exile, by removing from nature any ingredient *within it* that would prevent its being viewed primarily as natural resources, gave sanction via a set of worldly alliances to a certain form of political economy and political culture.

14. In his will, Boyle endowed these lectures, saying that they were to be given by a chosen London clergyman eight times a year and they were "for proving the Christian religion against notorious infidels, viz., Atheists, Theists, Pagans, Jews and Mahometans, not descending lower to any controversies among Christians themselves" (quoted in John J. Dahm, "Science and Apologetics in the Early Boyle Lectures," *Church History* 39, no. 2 [1970]: 172–86).

15. The history of this heterodox pantheism is disparate and not without its inner transformations. The leaders of radical groups such as the Levellers and Diggers during the English revolution (figures such as Overton and Winstanley) wrote and spoke of the presence of God in all things and all matter. For them, to resist the gap between God and His Creation was part of their resistance to the privileged place given to an elite clergy that could claim to mediate that gap. This entire resistance both in its metaphysical and its political aspects was anxiously dismissed by the Anglican establishment as "enthusiasm" preached on behalf of the aspiring lower classes that would destabilize the order that had finally come with the Restoration. Successive Boyle lecturers inveighed against the legacy of "enthusiasm" and the unrest of the revolutionary period that was generated in its name. I say more about this below. The dissenters against the Newtonians' "holy alliance" (as it was called) with the Anglican establishment also invoked pantheism some decades after the revolutionary period of the 1640s. Toland had studied Bruno's Italian works of a century earlier and was much influenced by its neo-Platonist, hermeticist ideas, but possibly because of the tremendous hostility to "enthusiasm" in the previous

few decades, he had begun to eschew the mystical elements in Bruno and seemed to subscribe to a more domesticated, more "rational" form of hermeticism, though he still declared himself a pantheist.

16. Kenneth Burke, *A Grammar of Motives* (Berkeley: University of California Press, 1969).

17. Cited in Christopher Hill, *The World Turned Upside Down* (London: Penguin, 1975), 293, from *The Works of Gerard Winstanley*, ed. G. H. Sabine (Ithaca, N.Y.: Cornell University Press, 1941); my italics.

18. The significance of this is not to be run together with the cliché about the Protestant Reformation's sustained opposition to the priestcraft enshrined in popery. The later scientific dissenters who appealed to Winstanley's metaphysics and politics found themselves opposing precisely the *Protestant* establishment, which, in explicit alliance with the dominant ideologues of the Royal Society, had exiled God to a place inaccessible to all but the learned scriptural judgment of its university-trained divines.

19. I say "conservative," but the label may be misleading. On the political landscape, these figures are best described as "moderates." They used arguments derived from the metaphysics around the new science to oppose both the Tory Jacobites, who were supporters of the "Catholic king" (now exiled), James II, and they also opposed the pro-Revolution republican section among the Whigs. I use the term "conservative" only to mark their vehement opposition to the latter. Over the next few decades, these "moderates" dominated the Whig party itself and were entrenched in the ruling oligarchy.

20. This is the wider political outcome regarding value that came out of the repudiation of the more democratic possibilities that Winstanley and others had hoped for from their quite different conception of value, which emerged from a quite different understanding of the relations between nature and the human community. It is what gets lost in the genealogical recesses of our more abstract philosophical understanding of the Humean outcome regarding value that comes from the repudiation of "enchantment" that McDowell speaks of. I will not try and make the links between the more abstract and the more political levels of discussion regarding value here—I will simply rest with having tried to reveal some of the wider significance *in the genealogy*.

21. Norbert Elias, *The Civilizing Process* (Oxford: Blackwell, 1982).

22. This attitude towards the mass of working people is manifest, for instance, in Boyle, who wrote of the radical sects and their "vulgar" followers with great anxiety and determined opposition. For a good discussion of Boyle's attitudes in this regard and his highly complicit role in the alliances I have mentioned above, see James R. Jacob, *Robert Boyle and the English Revolution*, Studies in the History of Science Series (New York: Burt Franklin, 1977).

23. The word "morph" might be misleading here, so let me warn against it. I am not suggesting by any means that these civilities are all that there is to the codifications

we find in modern constitutions. The arguments given for rights and constitutions are quite distinct from those that ground notions of civility, and so is their content and substance.

I am saying rather that the screening function of "civility" that I had mentioned, which blinds one to the cruelties that one perpetrates, is carried over to "rights" in modern, liberal ideology as it is practiced by advanced, industrial liberal-democratic nations of what we have taken to calling "the North." The idea of rights, which these countries are rightly proud of having achieved, nevertheless has the function of often blinding them to the cruelties they have perpetrated upon distant lands (the countries of "the South") because they recognize cruelty only in the form that occurs in those southern nations—Saddam's Iraq or Mugabe's Zimbabwe, as it might be. I want to stress, too, that when I make this point about this function that rights have in our time, I mean merely what I say and no more. The point is ripe for misunderstanding and should not be taken to express any hostility to rights, whose important and beneficial achievements for modern society is undeniable.

24. Richard Dawkins, *The God Delusion* (Boston: Houghton and Mifflin, 2006); Daniel Dennett, *Breaking the Spell: Religion as a Natural Phenomenon* (New York: Viking, 2006)

25. See Margaret Jacob, *The Radical Enlightenment: Pantheists, Freemasons, and Republicans* (London: George Allen and Unwin, 1981); and Jonathan Israel, *Radical Enlightenment* (Oxford: Oxford University Press, 2002). Israel emphasizes much more than Jacob does the influence of Spinoza. I should also add that what is meant by "radical" and who the exemplars are of "the radical" in the two works do not by any means coincide.

26. Why do I emphasize the category of the "exceptional"? Because though miracles were, of course, countenanced as occurring in the world over these centuries, these were essentially considered to be punctuations in an otherwise disenchanted world. They were not pervasively present in the form of enchantment as in the hylozoic or pantheistic picture of the dissenting deists.

2

NATURALISM AND QUIETISM

Richard Rorty

Philosophy is an almost invisible part of the contemporary intellectual life. Most people outside of philosophy departments have no clear idea of what philosophy professors are supposed to contribute to culture. Few think it worth the trouble to inquire.

The lack of attention that our discipline receives is sometimes attributed to the technicality of the issues currently being discussed. But that is not a good explanation. Debates between today's philosophers of language and mind are no more tiresomely technical than were those between interpreters and critics of Kant in the 1790's.

The problem is not the style in which philosophy is currently being done in the English-speaking world. It is rather that many of the issues discussed by Descartes, Hume and Kant had cultural resonance only as long as a significant portion of the educated classes still resisted the secularization of moral and political life.[1] The claim that human beings are alone in the universe, and that they should not look for help from supernatural agencies, went hand-in-hand with the admission that Democritus and Epicurus had been largely right about how the universe works. The canonically great modern philosophers performed a useful service by suggesting ways of dealing with the triumph of mechanistic materialism.

But as the so-called "warfare between science and theology" gradually tapered off, there was less and less useful work for philosophers to do. Just as medieval scholasticism became tedious once Christian doctrine had been synthesized with Greek philosophy, so a great deal of modern philosophy began to seem pointless after most intellectuals either abandoned their religious faith or found ways of rendering it compatible with modern natural science. Although rabble-rousers can still raise doubts about Darwin among the masses, the intellectuals—the only people on whom philosophy books have any impact—have no such doubts. They do not require either a sophisticated metaphysics or a fancy theory of reference to convince them that there are no spooks.

After the intellectuals had become convinced that empirical science, rather than metaphysics, told us how things work, philosophy had a choice between two alternatives. One was to follow Hegel's lead and to become a combination of intellectual history and cultural criticism—the sort of thing offered by Heidegger and Dewey, as well as by such people as Adorno, Strauss, Arendt, Berlin, Blumenberg, and Habermas. This way of doing philosophy flourishes mainly in the non-Anglophone philosophical world, but it is also found in the books of such American philosophers as Robert Pippin.

The other alternative was to imitate Kant by developing an armchair research program, thereby helping philosophy win a place in universities as an autonomous academic discipline. What was needed was a program that resembled Kant's in having no place for observation, experiment or historical knowledge. German neo-Kantians and British empiricists agreed that the core of philosophy was inquiry into something called "Experience" or "Consciousness." An alternative program was launched by Frege and Peirce, this one purporting to investigate something called "Language" or "the Sign."

Both programs assumed that, just as matter can be broken down into atoms, so can experience and language. The first sort of atoms include Lockean simple ideas, Kantian unsynthesized intuitions, sense-data, and the objects of Husserlian *Wesenschau*. The second include Fregean senses, Peircean signs, and Tractarian linguistic pictures. By insisting that questions concerning the relation of such immaterial atoms to physical particles were at the core of their discipline, philosophers in Anglophone countries shoved social philosophy, intellectual history, culture criticism, and Hegel out to the periphery.

Yet there have always been holists—philosophers who were dubious about the existence of either atoms of consciousness or atoms of significance.

Holists often become skeptics about the existence of shadowy surrogates for Reality such as "Experience," "Consciousness" and "Language." Wittgenstein, the most celebrated of these skeptics, came close to suggesting that the so-called "core" areas of philosophy serve no function save to keep an academic discipline in business.

Skepticism of this sort has come to be labeled "quietism." Brian Leiter, in his introduction to a recently-published collection titled *The Future for Philosophy*, divides the Anglophone philosophical world into "naturalists" and "Wittgensteinian quietists." The latter, he says, think of philosophy as "a kind of *therapy*, dissolving philosophical problems rather than solving them."[2] They are, Leiter is glad to report, in the minority, having the upper hand in only four major graduate departments (Harvard, Berkeley, Chicago and Pittsburgh). "Unlike the Wittgensteinians," Leiter writes, "the naturalists believe that the problems that have worried philosophers (about the nature of the mind, knowledge, action, reality, morality, and so on) are indeed real."[3]

I think Leiter's account of the stand-off between these two camps is largely accurate. He has identified the deepest and most intractable difference of opinion within contemporary Anglophone philosophy. But his account is misleading in one respect. Most people who think of themselves in the quietist camp, as I do, would hesitate to say that the problems studied by our activist colleagues are *unreal*. They do not divide philosophical problems into the real and the illusory, but rather into those that retain some relevance to cultural politics and those that do not. Quietists, at least those of my sect, think that such relevance needs to be demonstrated before a problem is taken seriously. This view is a corollary of the maxim that what does not make a difference to practice should not make a difference to philosophers.

From this point of view, questions about the place of values in a world of fact are no more unreal than questions about how the Eucharistic blood and wine can embody the divine substance, or about how many sacraments Christ instituted. Neither of the latter problems are problems for *everybody*, but their parochial character does not render them illusory. For what one finds problematic is a function of what one thinks important. One's sense of importance is in large part dependent on the vocabulary one employs. So cultural politics is often a struggle between those who urge that a familiar vocabulary be eschewed and those who defend the old ways of speaking.

Consider Leiter's assertion that "Neuroscientists tell us about the brain, and philosophers try to figure out how to square our best neuroscience with the ability of our minds to represent what the world is like."[4] The quietist response is to ask whether we really want to hold on to the notion of "representing what the world is like." Perhaps, they suggest, it is time to give up the notion of "the world," and of shadowy entities called "the mind" or "language" that contain representations of the world. Study of the history of culture helps us understand why these notions gained currency, just as it shows why certain theological notions became as important as they did. But such study also suggests that many of the central ideas of modern philosophy, like many topics in Christian theology, have become more trouble than they are worth.

Philip Pettit, in his contribution to *The Future for Philosophy*, gives an account of the naturalists' metaphilosophical outlook that is somewhat fuller than Leiter's. Philosophy, he says, is an attempt to reconcile "the manifest image of how things are," and the "ideas that come to us with our spontaneous everyday practices" with "fidelity to the intellectual image of how things are."[5] In our culture, Petit says, the intellectual image is the one provided by physical science. He sums up by saying that "a naturalistic, more or less mechanical image of the universe is imposed on us by cumulative developments in physics, biology and neuroscience, and this challenges us to look for where in that world there can be room for phenomena that remain as vivid as ever in the manifest image: consciousness, freedom, responsibility, goodness, virtue and the like."[6]

Despite my veneration for Wilfrid Sellars, who originated this talk of manifest and scientific images, I would like to jettison these visual metaphors. We should not be held captive by the world-picture picture. We do not need a synoptic view of something called "the world." At most, we need a synoptic narrative of how we came to talk as we do. We should stop trying for a unified picture, and for a master vocabulary. We should confine ourselves to making sure that we are not burdened with obsolete ways of speaking, and then insuring that those vocabularies that are still useful stay out of each other's way.

Narratives that recount how these various vocabularies came into existence help us see that terminologies we employ for some purposes need not link up in any clear way with those we employ for other purposes—that we can simply let two linguistic practices co-exist peaceably, side by side. This is what Hume suggested we do with the vocabulary of prediction and that of assignment of responsibility. The lesson the pragmatists drew from Hume was that philosophers should not scratch where it does not itch. When there is no longer an audience outside the discipline that displays interest in a philosophical problem, that problem should be viewed with suspicion.

Naturalists like Pettit and Leiter may respond that they are interested in philosophical truth rather than in catering to the taste of the day. This is the same rhetorical strategy that was used by seventeenth-century Aristotelians trying to fend off Hobbes and Descartes. Hobbes responded that those who were still sweating away in what he called "the hothouses of vain philosophy" were in the grip of a obsolete terminology, one that made the problems they discussed seem urgent. Contemporary quietists think the same about their activist opponents. They believe that the vocabulary of representationalism is as shopworn and as dubious as that of hylomorphism.

This anti-representationalist view can be found in several contributions to a recent collection of titled *Naturalism in Question*, edited by Mario De Caro and David Macarthur. It is most explicit in Huw Price's essay, "Naturalism without Representationalism." Price makes a very helpful distinction between object naturalism and subject naturalism. Object naturalism is "the view that in some important sense, all there is is the world studied by science."[7] Subject naturalism, on the other hand, simply says that "we humans are natural creatures, and if the claims and ambitions of philosophy conflict with this view, then philosophy needs to give way."

Whereas object naturalists worry about the place of non-particles in a world of particles, Price says, subject naturalists view these "placement problems" as "problems about human linguistic behavior."[8] Object naturalists worry about how non-particles are related to particles because, in Price's words, they take for granted that "substantial 'word-world' semantic relations are a part of the best scientific account of our use of the relevant terms."[9] Subject naturalists are semantic deflationists: they see no need for such relations—and, in particular, for that of "being made true by." They think once we have explained the uses of the relevant terms, there is no further problem about the relation of those uses to the world.

Bjorn Ramberg, in an article called "Naturalizing Idealizations," uses "pragmatic naturalism" to designate the same approach to philosophical problems that Price labels "subject naturalism." Ramberg writes as follows:

Reduction, says the pragmatist, is a meta-tool of science; a way of systematically extending the domain of some set of tools for handling the explanatory tasks that scientists confront. Naturalization, by contrast, is a goal of philosophy: it is the elimination of metaphysical gaps between the characteristic features by which we deal with agents and thinkers, on the one hand, and the characteristic features by reference to which we empirically generalize over the causal relations between objects and events, on the other. It is only in the context of a certain metaphysics that the scientific tool becomes a philosophical one, an instrument of legislative ontology.[10]

Pragmatic naturalism, Ramberg continues, "treats the gap itself, that which transforms reduction into a philosophical project, as a symptom of dysfunction in our philosophical vocabulary." The cure for this dysfunction, in Ramberg's words, is to provide "alternatives to what begins to look like conceptual hang-ups and fixed ideas . . . [and to explain] how our practice might change if we were to describe things . . . in altered vocabularies."[11]

Frank Jackson's book *From Metaphysics to Ethics* is a paradigm of object naturalism. Jackson says that "serious metaphysics . . . continually faces the location problem." The nature of this problem is explained in the following passage:

> Because the ingredients are limited, some putative features of the world are not going to appear *explicitly* in some more basic account. . . . There are inevitably a host of putative features of our world which we must either eliminate or locate.[12]

Subject naturalists, by contrast, have no use for the notion of "merely putative feature of the world," unless this is taken to mean something like "topic not worth talking about." Their question is not "What features does the world *really* have?" but "What topics are worth discussing?" Subject naturalists may think that the culture as a whole would be better off if a certain language-game were no longer played, but they do not argue that some of the words deployed in that practice signify unreal entities. Nor do they urge that some sentences be understood as about something quite different from what they are putatively about.

For Jackson, the method of what he calls "serious metaphysics" is conceptual analysis, for the following reason:

> Serious metaphysics requires us to address when matters described in one vocabulary *are made true by* matters described in another vocabulary. But how could we possibly address this question in the absence of a consideration of when it is right to describe matters in the terms of the various vocabularies? . . . And to do that . . . is to do conceptual analysis.[13]

But conceptual analysis does not tell the serious metaphysician which matters make which statements about other matters true. He already knows that. As Jackson goes on to say, "Conceptual analysis is not being given a role in determining the fundamental nature of the world; it is, rather, being given a central role in determining what to say in less fundamental terms given an account of the world stated in more fundamental terms."[14]

As I have already emphasized, subject naturalists have no use for Jackson's key notion—that of "being made true by." They are content, Price says, with "a use-explanatory account of semantic terms, while saying nothing of theoretical weight about whether these terms 'refer' or 'have truth-conditions.'"[15] The subject naturalist's basic task, he continues, is "to account for the uses of various terms—among them, the semantic terms themselves—in the lives of natural creatures in a natural environment."

If you think that there is such a relation as "being made true by" then you can still hope, as Jackson does, to correct the linguistic practices of your day on theoretical grounds, rather than merely cultural-political ones. For your a priori knowledge of what makes sentences true permits you to evaluate the relation between the culture of your day and the intrinsic nature of reality itself. But subject naturalists like Price can criticize culture only by arguing that a proposed alternative culture would better serve our larger purposes.

Price confronts Jackson with the following question: "[if we can explain] why natural creatures in a natural environment come to *talk* in these plural ways—of 'truth', 'value', 'meaning', 'causation', all the rest—what puzzle remains? What debt does philosophy now owe to science?"[16] That question can be expanded along the following lines: If you know not only how words are used, but what purposes are and are not served by so using them, what more could philosophy hope to tell you?

If you want to know about the relation between language and reality, the quietist continues, consider how the early hominids might have started using marks and noises to coordinate their actions. Then consult the anthropologists and the intellectual historians. These are the people who can tell you how our species progressed from organizing searches for food to building cities and writing books. Given narratives such as these, what purpose is served by tacking on an account of the relation of these achievements to the behavior of physical particles?

Both Jackson and Price pride themselves on being naturalists, but different things come to their minds when they speak of "nature." When Jackson uses that word he thinks of particles. A subject naturalist like Price thinks instead of organisms coping with, and improving, their environment. The object naturalist expresses his fear of spooks by insisting that everything be tied in, somehow, with the movements of the atoms through the void. The subject naturalist expresses his fear of spooks by insisting that our stories about how evolution led from the protozoa to the Renaissance should contain no sudden discontinuities—that it be a story of gradually increasingly complexity of physiological structure facilitating increasingly complex behavior.

For the subject naturalist, the import of Price's dictum that "we are natural creatures in a natural environment" is that we should be wary of drawing lines between kinds of organisms in non-behavioral and non-physiological terms.

This means that we should not use terms such as "intentionality," or "consciousness" or "representation" unless we can specify, at least roughly, what sort of behavior suffices to show the presence of the referents of these terms.

For example, if we want to say that squids have intentionality but paramecia do not, or that there is something it is like to be a bat but nothing it is like to be an earthworm, or that insects represent their environment whereas plants merely respond to it, we should be prepared to explain how we can tell—to specify what behavioral or physiological facts are relevant to this claim. If we cannot do that, we are kicking up dust and then complaining that we cannot see. We are inventing spooks in order to make work for ghost-busters.

This emphasis on behavioral criteria is reminiscent of the positivists' verificationism. But it differs in that it is not the product of a general theory about the nature of meaning, one that enables us to distinguish sense from nonsense. The subject naturalist can cheerfully admit that any expression will have a sense if you give it one. It is rather that traditional philosophical distinctions complicate narratives of biological evolution to no good purpose. In the same spirit, liberal theologians argue that questions about the number of the sacraments, though perfectly intelligible, are distractions from the Christian message.

Fundamentalist Catholics, of course, insist that such questions are still very important. Object naturalists are equally insistent that it is important to ask, for example, how collocations of physical particles manage to display moral virtue. Quietist Christians think that the questions insisted on by these Catholics are relics of a relatively primitive period in the reception of Christ's message. Quietist philosophers think that the questions still being posed by their activist colleagues were, in the seventeenth century, reasonable enough. They were a predictable product of the shock produced by the New Science. By now, however, they have become irrelevant to intellectual life. Christian faith without sacramentalism and what Price calls "naturalism without representationalism" are both cultural-political initiatives.

◎ ◎ ◎

So far I have been painting the object naturalist vs. subject naturalist opposition with a fairly broad brush. In the time that remains I shall try to show the relevance of this opposition to a couple of current philosophical controversies.

The first of these is a disagreement between Timothy Williamson and John McDowell. The anthology edited by Brian Leiter to which I have already referred includes a lively polemical essay by Williamson titled "Past the Linguistic Turn?" Williamson starts off by attacking a view that John McDowell takes over from Hegel, Wittgenstein and Sellars: viz., "Since the world is everything that is the case . . . there is no gap between thought, as such, and the world." Williamson paraphrases this as the claim that "the conceptual has no outer boundary beyond which lies unconceptualized reality" and again as the thesis that "any object can be thought of."[17]

Williamson says that "for all that McDowell has shown, there may be necessary limitations on all possible thinkers. We do not know whether there are elusive objects. It is unclear what would motivate the claim that there are none, if not some form of idealism. We should adopt no conception of philosophy that on methodological grounds excludes elusive objects."[18]

I think that McDowell, a self-professed quietist, might respond by saying that we should indeed adopt a conception of philosophy that excludes elusive objects. We should do so for reasons of cultural politics. We should say that cultures that worry about unanswerable questions like "Are there necessary limitations on all possible thinkers?" "Could God change the truths of arithmetic?" "Am I dreaming now?" and "Is my color spectrum the inverse of yours?" are less advanced than those that respect Peirce's pragmatic maxim. Superior cultures have no use for what Peirce called "make-believe doubt."

Williamson is wrong to suggest that only idealism could motivate McDowell's thesis. The difference between idealism and pragmatism is that between metaphysical or epistemological arguments for the claim that any object can be thought of and cultural-political arguments for it. Pragmatists think that the idea of necessary limitations on all possible thinkers is as weird as Augustine's thesis about the inevitability of sin—*non posse non peccare*. Neither can be refuted, but healthy-mindedness requires that both be dismissed out of hand.[19]

The clash of opinion between McDowell and Williamson epitomizes the opposition between two recent lines of thought within analytic philosophy. One runs from Wittgenstein through Sellars and Davidson to McDowell and Brandom. The other is associated with what Williamson calls "the revival of metaphysical theorizing, realist in spirit . . . associated with Saul Kripke, David Lewis, Kit Fine, Peter van Inwagen, David Armstrong and many others."[20] The goal of such attempts to get past the linguistic turn is, Williamson says, "to discover what fundamental kind of things there are, and what properties and relations they have, not how we represent them."[21] The contrast between these two lines of thought will become vivid to anyone who flips back and forth between the two collections of articles from which I have been quoting—Leiter's *The Future for Philosophy* and De Caro's and Macarthur's *Naturalism in Question*.

Quietists think that no kind of thing is more fundamental than any other kind of thing. The fact that, as Jackson puts it, you cannot change anything without changing the motions or positions of elementary physical particles, does nothing to show that there is a problem about how these particles leave room for non-particles. It is no more philosophically pregnant than the fact that you cannot mess with the particles without simultaneously messing with a great many other things. Such expressions as "the nature of reality" or "the world as it really is" have in the past, quietists admit, played a role in producing desirable cultural change. But so have many other ladders which we would be well advised to throw away.

Quietists who have no use for the notion of 'the world as it is apart from our ways of representing it' will balk at Williamson's thesis that "What there is determines what there is for us to mean." But they will also balk at the idealists' claim that what we mean determines what there is. They want to get beyond realism and idealism by ceasing to contrast a represented world with our ways of representing it. This means giving up on the notion of linguistic representations of the world except insofar as it can be reconstructed within an inferentialist semantics. Such a semantics abjures what Price calls "substantial word-world relations" in favor of descriptions of the interaction of language-using organisms with other such organisms and with their environment.

◎ ◎ ◎

The controversy about inferentialist semantics is the second of the two I want briefly to discuss. The best-known objection to Brandom's inferentialism is Fodor's. The clash between Fodor and Brandom epitomizes not only the difference between representationalist and inferentialist semantics but the larger atomist-holist conflict to which I referred earlier. Fodor thinks that philosophy can team up with cognitive science to find out how the mechanisms of mind and language work. Brandom is skeptical about the idea that there are any such mechanisms.

Brandom takes Davidsonian holism to the limit. As Davidson did in "A Nice Derangement of Epitaphs," he repudiates the idea that there is something called "a language"—something that splits up into bits called "meanings" or "linguistic representations" which can then be correlated with bits of the physical world. He tries to carry through on the Quine-Davidson hope for, as Kenneth Taylor has put it, "a theory of meaning in which meanings play no role."[22] So he abandons the notion of a sentence having a "cognitive content" that remains constant in all the assertions it is used to make.

Brandom cheerfully coasts down what Fodor derisively describes as "a well-greased and well-traveled slippery slope" at the bottom of which lies the view that "no two people ever mean the same thing by what they say."[23]

Brandom does this because he wants to dismiss the idea that I get what is in my head—a cognitive content, a candidate for accurate representation of reality—into your head by making noises that effectuate this transmission. He hopes to replace it with an account of what he calls "doxastic scorekeeping"—keeping track of our interlocutors' commitments to perform certain actions in certain conditions (including assent or dissent to certain assertions).

Such commitments are attributed by reference to social norms. These norms authorize us to gang up on people who, having said "I promise to pay you back," or "I will join the hunt," make no move to do so. The same goes for people who, having uttered "p" and "if p then q," obstinately refuse to assent to "q." We, unlike the brutes, can play what Brandom calls the "game of giving and asking for reasons." Our ability to play this game is what made it possible for us to assume lordship over the other animals. To say that we, unlike the brutes, have minds is just another way of saying that we, but not they, play that game. Fodor to the contrary, finding out how the brain works will not help us find out how the mind works.[24] For the mind is not a representational apparatus, but rather a set of norm-governed social practices.

Brandom does not call himself a "naturalist," perhaps because he thinks the term might as well be handed over to the fans of elementary particles. But the whole point of his attempt to replace representationalist with inferentialist semantics is to tell a story about cultural evolution—the evolution of social (and, in particular, linguistic) practices—that focuses on how these practices gave our ancestors an evolutionary edge. Unless one is convinced that particles somehow enjoy an ontological status superior to that of organisms, that will seem as naturalistic as a story can get.

Brandom cheerfully admits that "A word—'dog', 'stupid', 'Republican'—has a different significance in my mouth that it does in yours, because and insofar as what follows from its being applicable, its consequences of application, differ for me, in virtue of my different collateral beliefs."[25] But this difference is not a problem for anybody except philosophers who, like Fodor, take the notion of "cognitive content" seriously.

We are likely to look for substantive word-world relations as long as we ask Fregean questions about little atoms of linguistic significance such as "Does the assertion that the morning star is the evening star have the same cognitive content as the assertion that the thing we call the morning star is the same thing as the one we call the evening star?" If "same cognitive content" just means "will do as well for most purposes," then the answer is yes,

But Fregeans, invoking Church's Translation Test, brush aside the fact that either sentence can usually be used to get the job done. The real question, they say, is not about uses but about senses, meanings, intensions. Sense, these philosophers say, determines reference in the same way that the marks on the map determine which slice of reality the map maps. Meanings cannot be the same thing as uses, for there is a difference between semantics and pragmatics. It is semantics that determines sameness and difference of cognitive content.

But we shall have a use for the notion of "same cognitive content" only if we try to hold belief and meaning apart, as Frege thought we should and Quine told us we should not. If we continue on along the path that Quine and Davidson cleared, we shall come to agree with Brandom that "particular linguistic phenomena can no longer be distinguished as 'pragmatic' or 'semantic'."[26] Brandom has no more use for a distinction between these two disciplines than Davidson did for a distinction between knowing a language and knowing our way around the world generally.

◎ ◎ ◎

So much for the two controversies on which I wanted to comment. I hope that my discussion of the disagreements between McDowell and Williamson and between Brandom and Fodor has helped make clear why I think that Price's distinction between two forms of naturalism is so useful. Subject naturalists like Price, Ramberg and I urge our activist colleagues to stop talking about great big things like Experience or Language, the shadow entities that Locke, Kant, and Frege invented to replace Reality as the subject-matter of philosophy. Doing so might lead, eventually, to evacuating the so-called "core areas" of philosophy. Object naturalists like Jackson, Leiter, Petit, and Fodor fear that philosophy might not survive if it purged itself in this way. But subject naturalists suspect that the only thing our discipline would lose would be its insularity.

NOTES

1. The most important change produced by secularism was a shift from thinking of morality as a matter of unconditional prohibitions to seeing it as an attempt to work out compromises between competing human needs. This change is well described in a famous article by Elizabeth Anscombe called "Modern moral philosophy." She contrasts hard, unconditional, prohibitions of such things as adultery, sodomy and suicide with the soft, squishy consequentialism advocated by, as she

says, "every English academic moral philosopher after [Sidgwick]." That conse-quentialism is, Anscombe says, "quite incompatible with the Hebrew-Christian ethic": Elizabeth Anscombe, *Ethics, Religion and Politics* (University of Minnesota Press, 1981), 34.

In the United States we are currently experiencing a return to the latter ethic—a revolt of the masses against the consequentialism of the intellectuals. The current red-state vs. blue-state clash is a flare-up of the old struggle about the seculariza-tion of culture. But almost nobody now looks to philosophy for help with this struggle. In the seventeenth and eighteenth centuries, they did. Writers like Spi-noza and Hume did a great deal to advance the secularist cause. In the course of the nineteenth and twentieth centuries, however, the baton was passed to art and literature. Novels whose characters discussed moral dilemmas without reference to God or Scripture took the place of moral philosophy.

2. Brian Leiter, ed., *The Future for Philosophy* (New York: Oxford University Press, 2004), 2.
3. Ibid., 2–3.
4. Ibid., 3.
5. Philip Pettit, "Existentialism, Quietism and Philosophy," in Leiter, *The Future for Philosophy*, 306.
6. Ibid. Pettit adds that "philosophy today is probably more challenging, and more difficult, than it has ever been." This is probably true, but the same can be said of Christian theology.
7. Mario De Caro and David Macarthur, eds., *Naturalism in Question* (Cambridge, MA: Harvard University Press, 2004), 73.
8. Ibid., 76.
9. Ibid., 78.
10· Bjorn Ramberg, "Naturalizing Idealizations: Pragmatism and the Interpretive Strategy," *Contemporary Pragmatism* 1, no. 2 (2004): 43.
11. Ibid., 47.
12. Frank Jackson, *From Metaphysics to Ethics: A Defence of Conceptual Analysis* (Ox-ford: Oxford University Press, 1998), 5.
13. Ibid., 41–42; emphasis added.
14. Ibid., 42–43.
15. Huw Price, "Naturalism without Representationalism," in De Caro and Macar-thur, eds., *Naturalism in Question*, 79.
16. Ibid., 87.
17. Timothy Williamson, "Past the Linguistic Turn," in Leiter, ed., *The Future for Phi-losophy*, 109.
18. Ibid., 110.
19. Pragmatism takes its stand against all doctrines that hold, in the words of Leo Strauss, that "Even by proving that a certain view is indispensable to living well, one merely proves that the view in question is a salutary myth: one does not prove it to be true": *Natural Right and History* (Chicago: Chicago University Press, 1968),

6. Strauss goes on to say that "Utility and truth are two entirely different things." Pragmatists do not think they are the same thing, but they do think that you cannot have the latter without the former.

20. Williamson, "Past the Linguistic Turn," 111.

21. Ibid., 110–111.

22. Kenneth Taylor, *Truth and Meaning: An Introduction to the Philosophy of Language* (Oxford: Blackwell, 1998), 147. Taylor thinks of Davidson's distaste for meanings as a result of his preference for extensional languages. This may have played a role in Davidson's (early) thinking, but it plays none in Brandom's. Once one gets rid of the "making true" relation, there will be reason to think non-extensional languages fishy.

23. Jerry Fodor, "Why Meaning (Probably) Isn't Conceptual Role," in Stephen Stich and Ted Warfield, eds., *Mental Representations* (Oxford: Oxford University Press, 1994), 143.

24. I have argued to this effect in more detail in "The Brain as Hardware, Culture as Software," *Inquiry* 47, no. 3 (2004): 219–235.

25. Robert Brandom, *Making It Explicit* (Cambridge, MA: Harvard University Press, 1994), 587.

26 Ibid., 592.

3

IS LIBERAL NATURALISM POSSIBLE?

Mario De Caro and Alberto Voltolini

According to a commonly held view, Liberal Naturalism is not a genuine metaphilosophical option. This view is often supported by an argument in the form of a dilemma. If Liberal Naturalism grants that the most philosophically controversial items (things, properties, and events that prima facie appear to be beyond nature) are reducible to or are ontologically dependent on the entities accountable by science, then this view is not liberal enough to be distinguished from Scientific Naturalism. If, on the other hand, Liberal Naturalism denies that possibility for at least some of the aforementioned items, then, by being *too* liberal it loses its naturalistic credentials and cannot be accepted by the philosophers who are committed to taking the scientific view of the world seriously. The conclusion of this argument is that between Scientific Naturalism and Supernaturalism there is no logical space that could be occupied by Liberal Naturalism. This dilemma is potentially fatal for Liberal Naturalism, so it is one that its sympathizers cannot afford to ignore.

This argument, we believe, reflects an overly simple view of the ontological, epistemological, and methodological possibilities open to the naturalist. In this paper, we defend a form of Liberal Naturalism according to which there may be philosophically legitimate entities that, on the one hand, are ineliminable and, on the other hand, are not only irreducible to scientifically accountable entities but may also be ontologically independent from them. It follows that those entities cannot be exhaustively explained or explained away by the natural sciences, so they cannot be considered natural in the scientific-naturalist sense. We argue, however, that such entities need not be supernatural either and that they can be accepted without abandoning a scientific account of the world. Indeed, such entities can be accepted as long as, first, they cannot causally interfere with the course of natural events as studied by science and, second, the investigation of such entities does not require any special modes of understanding that would be irreconcilable with rational forms of understanding. Such entities can therefore be accommodated by a liberal form of philosophical naturalism, conceived as essentially different from both Scientific Naturalism and Supernaturalism.

THE DILEMMA OF LIBERAL NATURALISM

A clear, if concise, presentation of the anti–Liberal Naturalism dilemma has been offered by Ram Neta:

> What if digestion, or respiration, or reasoning are natural kinds, their nature consisting simply in the mechanisms that enable them to occur? Is the liberal naturalist committed to denying this possibility? If so, then I confess I can see no good reason to accept Liberal Naturalism. And if not, then I confess I do not understand just what Liberal Naturalism is.[1]

In this passage, the assumption that reasoning can be exhaustively described or explained away by science is only one of the many philosophically controversial cases of this kind. (Other examples would include mathematical entities, normativity, intentionality, modality, responsibility, freedom, and consciousness.) The dilemma sketched by Neta involves two subarguments. According to the first, if Liberal Naturalism is understood as implying that, at least in principle, the controversial items can be explained, or explained away, by science (that is, if they are "natural kinds" or are reducible to those), then Liberal Naturalism proves to be no different from Scientific Naturalism. According to the second argument, if Liberal Naturalism sees these items as irreducible to the entities that can be explained by science, it becomes ipso facto a form of Supernaturalism. Therefore the challenge posed by this dilemma is to show that, between the Scylla of Scientific Naturalism and the Charybdis of Supernaturalism, a logical space exists for Liberal Naturalism.[2]

Before discussing this issue, however, some preliminary steps are in order. First, a potential objection has to be addressed. Does one really have to care about how naturalism should be defined and how broad and inclusive the definition should be? Is this not merely an empty terminological question? David Papineau seems to suggest this when he writes that it would be fruitless to try to adjudicate some official way of understanding the term "naturalism."[3] Different contemporary philosophers interpret the term differently, and this disagreement about usage is, in Papineau's opinion, no accident. According to Papineau, for better or worse, "naturalism" is widely viewed as a positive term in philosophical circles—few philosophers are happy to be considered "nonnaturalists" nowadays. This inevitably leads to differences in how the term is understood. Philosophers with relatively weak naturalist commitments are inclined to understand "naturalism" in a less restrictive way, in order not to be disqualified as "naturalists," while those who uphold stronger naturalist doctrines are happy to set the bar for "naturalism" higher.

Of course, Papineau has a point here, since there may easily be pointless hairsplitting about the connotation of the word "naturalism." However, one should not think that *every* attempt to precisely qualify it must be fruitless. In particular, as we have seen, Liberal Naturalism has been accused of being too disrespectful toward science for it to be considered a legitimate form of "naturalism." This charge is a substantive one. The metaphilosophical constraint that philosophical views should not be at odds with science is both attractive and well established. We will argue, however, that there are philosophical views whose ontological and epistemological liberality does not make them unacceptable for philosophers who sincerely respect science, and thus they still deserve the honorific title of "naturalism."

It should also be noticed that "naturalism" is a vague philosophical term of art. For example, it means different things when applied to the views of the Ionian philosophers, Aristotle, Epicurus, Bruno, Spinoza, Hume, Goethe, Schelling, the positivists, Dewey, or to what G. E. Moore labeled "ethical naturalism." Most of those views claimed that nothing existed beyond nature, but they interpreted the term "nature" in dramatically different ways—so different that some of those views are now considered forms of straightforward Supernaturalism for both ontological and epistemological reasons, which we shall examine below.[4]

The most common forms of contemporary naturalism, however, involve more than the bare claim that nothing exists beyond nature. According to these views, no entity or explanation should be accepted whose existence or truth could contradict the laws of nature, insofar as we know them.[5] This thesis can be called the "constitutive claim of contemporary naturalism." In this case, however, questions of interpretation arise very soon. It is true that everybody would agree that this claim implies the denial of intelligent designers, prime movers unmoved, and entelechies. (This, though, need not be construed as a denial *in principle*, since we can imagine circumstances in which we could get evidence for the existence of some of these entities.) However, things become much more contentious as soon as one moves beyond these simple cases toward more problematical ones, such as values, abstract entities, modal properties, free agents, and conscious phenomena. In fact, as long as these items are viewed as resisting all attempts to explain them or explain them away in a scientific fashion (to treat them as "natural kinds," as Neta puts it), two questions immediately arise: first, whether their existence is incompatible with the laws of nature and, second, whether they should consequently be considered *entia non grata* and discarded from the "first-grade conceptual system" of philosophy.[6]

BETWEEN SCYLLA AND CHARYBDIS: SCIENTIFIC NATURALISM AND SUPERNATURALISM

The dilemma facing liberal naturalists, then, is whether they can maintain the constitutive claim of naturalism (according to which all items or explanations whose existence or truth would contradict the laws of nature are unacceptable) alongside the idea that some items may exist that science cannot fully explain or explain away, even in principle.

In order to assess this issue, one must has to consider what distinguishes Liberal Naturalism from Scientific Naturalism, on the one hand, and from Supernaturalism, on the other. The latter views must therefore be considered. It makes sense to begin with Scientific Naturalism and, more specifically, with two related claims, accepted by most scientific naturalists, that compel a narrower reading of the constitutive claim of naturalism mentioned above.[7]

The first of these claims is both ontological and methodological in character. It states that ontology should be shaped by the natural sciences alone and that, in principle, the natural sciences can account for reality in all its aspects. In this light, for example, Alan Lacey wrote that "everything is natural, i.e. . . . everything there is belongs to the world of nature, and so can be studied by the methods appropriate for studying that world, and the apparent exceptions can be somehow explained away."[8] This implies that in order to be philosophically acceptable, a potential entity should either be accounted for, at least in principle, by conceptual apparatus of the natural sciences or eliminated altogether from the repertoire of ontology. In this perspective, a legitimate account of an entity means either that all facts about that entity can be explained or explained away by the sciences or, more moderately, that all facts about that entity are shown to be ontologically dependent on scientific facts.[9] At any rate, a question that faces scientific naturalists is what they should do with regard to the entities that resist all attempts at reducing them to or showing that they are ontologically dependent on entities that are accountable by science, but that still appear ineliminable since they play an essential role in our epistemic practices. According to some philosophers who sympathize with Scientific Naturalism (such as Colin McGinn, who with regard to this issue was inspired by Noam Chomsky), the only acceptable strategy for those who want to conform to the requirements of Scientific Naturalism would be to acknowledge that those items represent unsolvable mysteries, in the sense that the human species is intellectually unequipped to account for them in scientifically acceptable terms.[10]

The second claim with which scientific naturalists often restrict the interpretation of the constitutive claim of naturalism is a metaphilosophical corollary of the first. According to it, since science is, at least in principle, the only legitimate source of knowledge, philosophy has to take a scientific form, since it is seen as "a part of one's system of the world, continuous with *the rest of science*," as Quine famously wrote.[11] In this perspective, philosophy should therefore be conceived, methodologically and ontologically, as part of the scientific view of the world.

On the opposite side of the metaphilosophical spectrum, Supernaturalism rejects the constitutive claim of naturalism and, a fortiori, the two other claims that characterize Scientific Naturalism proper. This is because this view is committed to the existence of entities that are "above nature"—or, more precisely, to the existence of entities that can violate the laws of nature. In most cases, this violation is evident since these entities are supposed to interfere preternaturally with the natural course of events. The prototypical supernatural entity with which many advocates of this view are concerned is a personal divinity. It is important for us to analyze the conceptual articulation of this view both because it is shared by other forms of nontheistic Supernaturalism (which are less known but more relevant here) and because, undeniably, part of the pathos of the second branch of the dilemma clearly depends on the idea that Liberal Naturalism may turn out to be no more supportable than theism.

Theistic supernaturalism has therefore been defined as the view according to which,

(1) Something else besides nature exists: namely, God; (2) nature depends for its existence upon God; (3) the regularity of nature can be, and sometimes is,

interrupted by God; and (4) such divine interruptions are in natural terms quite unpredictable and inexplicable.[12]

In this view, God has the faculty to intervene in the natural course of events in ways that are by definition preternatural, that is, in principle unpredictable and unexplainable by science. Moreover, the epistemology connected with Supernaturalism is by definition antirational ("faith, revelation, and the authority of Scripture take the place of reason")[13] and therefore irreconcilable with all rational forms of understanding.

Unsurprisingly, this view has relevant consequences for philosophy, as long as (following a long tradition) the purpose of addressing the issue of the divine is thought within its scope. From this point of view, the constitutive claim of naturalism that philosophy should not accept any entity that could violate the laws of science (in a way that could not be captured, in principle, by any higher-order scientific account) is rejected, and so are, a fortiori, the ontological/methodological claim and the metaphilosophical claim that characterize Scientific Naturalism.

In order to make this definition of Supernaturalism useful for the discussion, however, two remarks are necessary. First, a philosophical view can be supernaturalist even if it does *not* assume the existence of a personal divinity that preternaturally interferes with the natural course of events. Indeed, what really matters for judging the Supernaturalism of a philosophical view is whether the view is committed to the existence of *any* entity or force that is in principle unaccountable by science, ineliminable from our ontology, and contradictory to scientific knowledge. In this sense, one could, for example, imagine an ontologically dualistic view that—without appealing to God's existence—assumes that mental properties are ontologically ineliminable, in principle unexplainable by science, and able to alter causally the natural course of events (something like a godless Cartesianism, then). Certainly such a view should be considered a form of Supernaturalism.

Second, an advocate of Supernaturalism may conceive of supernatural entities or forces that are utterly detached from the natural world and therefore do not interfere in any way with natural causal processes (Parmenides's eternal and unchanging Being, the absolutely self-sufficient God characteristic of some forms of Neoplatonism, or even the Buddhist Nirvana, conceived as the complete extinction of the flame of the self may be examples of this view). Unsurprisingly, however, these views appeal to special cognitive powers—which typically include some extreme forms of mystic illumination—in order to account for the human capacity to grasp those noncausal and supernatural entities or forces. For the purpose of our discussion, it is crucial to notice that those cognitive powers are absolutely irreconcilable with anything we could intuitively regard as natural forms of understanding. This is enough to rank the views that appeal to such kinds of entities within Supernaturalism.

The challenge posed by the dilemma for Liberal Naturalism is whether it can differentiate itself from both Scientific Naturalism and Supernaturalism simultaneously. This challenge would be answered if it was shown that one can legitimately hold the constitutive claim of naturalism (according to which all items or explanations whose existence or truth could contradict the laws of nature are unacceptable) alongside the idea that some items may exist that science cannot fully explain or explain away, even in principle. In order to assess this question, let's therefore consider what the main features of Liberal Naturalism are.

THE VERY POSSIBILITY OF LIBERAL NATURALISM

In the first place, liberal naturalists are committed to accept the constitutive claim of naturalism, according to which no entity or explanation should be accepted whose existence or truth would contradict the laws of nature, insofar as we know them. Liberal naturalists, however, have a more open-minded attitude than scientific naturalists with regard to both the epistemology and the ontology of the controversial entities (such as moral, abstract, phenomenological, modal, or intentional entities) that the latter *assume* have to be scientifically explained or explained away.

It is not that liberal naturalists have unrevisable a priori reasons to deny that the controversial entities can be scientifically explained or explained away. The liberal naturalist views, however, are characterized by two provisos—one epistemological and one ontological—that complement the constitutive claim of naturalism. Let's suppose that some of the controversial entities were actually reduced or shown to be ontologically dependent on scientific entities; nevertheless, according to the epistemological proviso of Liberal Naturalism, in order to account fully for the features of these entities, one might still have to turn to forms of understanding (such as conceptual analysis, imaginative speculation, or introspection) that are neither reducible to scientific understanding nor supernatural.[14] Moreover, according to the ontological proviso of Liberal Naturalism, there may be entities that do not and cannot causally affect the world investigated by the sciences and that are both irreducible to and ontologically independent of entities accountable by science but are not supernatural either, since they do not and cannot violate any laws of nature (an example of such entities will be discussed in this chapter's last section).[15]

The latter point clearly shows that liberal naturalists tend to privilege an ontological criterion according to which we are committed to accepting not only the entities that pull their own weight in our best causal/explanatory theory of the world (as claimed by the scientific naturalists),[16] but also the entities that are implicit in other sound and successful ordinary transactions of ours. In general, it has been convincingly argued that the causal requirement cannot be applied to domains whose relation to human beings is essentially not spatiotemporal.[17]

In this light, liberal naturalists tend to disagree strongly with both the ontological/methodological and the metaphilosophical claims through which scientific naturalists qualify the constitutive claim of naturalism. Regarding the ontological claim, to begin with, they deny that entities that cannot be eliminated, reduced to, or shown to be ontologically dependent on scientifically accountable entities are ipso facto supernatural and should therefore be treated as fictions, illusions, or unsolvable mysteries. In this sense, it may be said that, compared with Scientific Naturalism, Liberal Naturalism widens the realm of the natural.

In order to understand this point, it is crucial to understand the difference between the claim that no naturalistic philosophy can accept anything that could contradict the laws of nature (which is the constitutive claim of naturalism) and the claim that ontology should be shaped by the natural sciences alone (which is the ontological/methodological claim of Scientific Naturalism). In fact, the former claim leaves a logical space open for something that the latter instead rules out, that is, things whose existence would not be incompatible with the nomological-causal structure of the universe, as investigated by science, but would not be explainable with reference to such a structure either. Such things thus would not be explained or "explained away" by science. Of course, there being a logical space for something does not imply that that something actually exists (specific arguments should be added in order to support this). However, the acceptance of this possibility is what characterizes the branch of Liberal Naturalism that we defend in this article.

A good example of the liberal naturalists' refusal of the ontological/methodological claim of Scientific Naturalism concerns the issue of whether real credit should be given to the concepts of the so-called "manifest image" or the "space of reasons" (to use Wilfrid Sellars's terminology) or the "agential perspective" (to use Stephen White's terminology), which are all conceived of in opposition to the view of the world as governed by scientific laws.[18] Are such concepts really in need of reduction or elimination? To put it differently, how seriously should those concepts be taken by philosophers? One of the champions of Liberal Naturalism, John McDowell, for example, explicitly recognizes that the concepts of knowing and thinking, as well as normative properties in general, cannot be subsumed under scientific laws: that is, they do not belong to the space of the "natural" as defined by the natural sciences. Still, according to McDowell, this does not mean that those concepts should be viewed as supernatural. What one should say, instead, is that as animals, human beings are part of nature, but by sharing a culture with other human beings (by participating in the space of reasons), they also acquire a "second nature." And second nature is still a form of nature and should not be considered supernatural.

It should be noted that in itself Liberal Naturalism is neutral with regard to whether the irreducible and scientifically unexplainable entities supervene globally on entities that can be scientifically explained or explained away. Rather, whether the various liberal naturalist views accept this thesis depends on their specific theoretical features. Precisely defined, global supervenience is simply a relation of *covariance*, according to which the subvenient entities vary when the supervenient entities vary. This, however, does not imply the stronger relation of *determination*, in which the supervenient entities are ontologically dependent, either for their existence or for their individuation, on their physical character.[19] Once supervenience is defined in this narrow sense, some liberal naturalists (such as John McDowell) are happy to defend the idea that, say, mental entities or values globally supervene on physical entities;[20] others (such as Akeel Bilgrami and Stephen White) claim instead that even in its weakest form the supervenience thesis is "unassessable" since the attempt to relate, say, evaluative facts to nonevaluative facts hinges upon something akin a categorical mistake;[21] finally, other liberal naturalists (such as John Dupré) argue that the supervenience thesis is empirically vacuous.[22]

Regarding the metaphilosophical claim of Scientific Naturalism, liberal naturalists reject the idea of the methodological and ontological continuity between science and philosophy even if, as said, they endorse the constitutive claim of naturalism, that is, the idea that no entity or explanation should be accepted whose existence or truth could contradict the laws of nature, insofar as we know them. This claim, in fact, does not imply that philosophy should adopt the methods of science or should not cover issues that are of no concern to science. In this spirit, for example, Hilary Putnam, one of the leading liberal naturalists, claims that philosophy should consist of "a modest non-metaphysical realism squarely in touch with the results of science"—where "non-metaphysical" also means "non-scientistic."[23]

Ontological tolerance plus methodological discontinuity explains how Liberal Naturalism may escape the first horn of the dilemma. In allowing for entities that cannot be studied by the methods of natural sciences, Liberal Naturalism shows how it differs from Scientific Naturalism. For liberal naturalists, reality is just more diverse than the way philosophers oriented strictly by science take it to be. The idea is that there are different aspects of reality—only some of which involve spatiotemporal relations—that are irreducible to one another. This is to say that only some aspects of reality are explicable by referring to the laws of science.

The question raised by the other horn of the dilemma, however, remains. Is it possible to accept ontological tolerance and methodological discontinuity and still remain naturalists? Is Liberal Naturalism *too* liberal to be called naturalism justifiably? Or, to put it differently, is this view just a form of Supernaturalism in disguise? The answer is negative both for an ontological and an epistemological reason.

From an ontological point of view, the entities that according to many liberal naturalists may exist over and above scientifically explainable entities are precisely the entities that make no difference in the causal order of the world—that is, entities that in principle do not violate any scientific laws. These are entities that cannot induce any downward causation: that is, they cannot produce any break in the causal closure of the world investigated by the sciences (as might well happen, on the contrary, whenever a supernatural property were instantiated). In Jaegwon Kim's terms, those properties "[float] freely, unconstrained by the physical domain."[24] Yet, contrary to what Kim claims, floating free is not necessarily indicative of what is supernatural—or, in his words, of what is not "minimally physical"—since it may also be a feature of *liberally* natural entities. Indeed, in a liberal naturalist perspective, the fact that a controversial kind of entity has no causal power, far from being a problem, is a necessary condition (albeit, of course, not a sufficient one) for accepting it as real. The existence of entities of that kind, however, would be perfectly compatible with the claims of the natural sciences, since their existence would not imply any violation of the causal closure of the natural world. So, by accepting the possibility of such entities, liberal naturalists are not *eo ipso* driven toward Supernaturalism.

From an epistemological point of view, the nonscientifically explainable properties that may exist according to liberal naturalists are only those that, in order to be grasped, do not require any special modes of understanding that would be irreconcilable with rational understanding. And this shows that Liberal Naturalism cannot be considered a form of crypto-supernaturalism not only from the ontological but also from the epistemological point of view.

Summing up, a feature of many liberal naturalist views—one that, by differentiating those views from both Scientific Naturalism and Supernaturalism, refutes the dilemma—is that there may be entities that cannot be scientifically explained or explained away and yet are not supernatural, since they cannot contravene the laws of the world investigated by the sciences and are not grasped in an antiscientific way. In the final section of this essay, we will briefly consider an example of properties that fit this description, that is, modal properties.

THE FEASIBILITY OF LIBERAL NATURALISM: THE CASE OF MODAL PROPERTIES

Modal properties are a good example of the kind of properties to which liberal naturalists can appeal in order to differentiate themselves from both Scientific Naturalism and Supernaturalism. First, there are excellent reasons to think that modal concepts are essential to human thought. As Anand Vaidya has written, "Humans have a natural tendency to modalize. A tendency to think about, assert, and evaluate statements of possibility and necessity. To modalize is to either entertain a modal thought or to make a modal judgment. Modal thoughts and judgements either explicitly or derivatively involve the concept of possibility, necessity, or essence."[25] Well-known examples of how relevant modal talk is for us are offered by the epistemology of mathematics, practical and theoretical reasoning, and philosophical methodology (in particular, with regard to the role played by thought experiments).[26]

In this respect, it is has been convincingly stressed that we have a natural tendency to give a realist interpretation of the modal properties our ordinary transactions seem to commit us to—that is, we spontaneously tend to give a *de re* interpretation of modal discourse.[27] More important, persuasive theoretical arguments have been offered in order to justify this tendency: so, for example, the distinction between accidental and essential features of physical reality, which is proper of modal essentialism, has been influentially revitalized by reflections on the semantics of natural language.[28] Another clear example of the crucial role played by *de re* modality in human practices is offered by the discussion on free will and moral responsibility. Most philosophers who study the free will issue take the so-called possibility of doing otherwise (clearly a modal property) as a necessary condition of free will, and many conceive it, in a Kantian spirit, as a requirement of rationality as well.

According to John McDowell, for instance, there is "a deep connection between reason and freedom; we cannot make sense of a creature's acquiring reason unless it has *genuinely* alternative possibilities of action, over which its thought can play".[29]

Unsurprisingly—given their respect for the manifest image and their predilection for the idea that in this ontology, one should accept the entities implicit in all sound and successful ordinary transactions (not only the causally efficacious ones)—liberal naturalists may easily be realists with regard to modal properties. Indeed, as we have seen, they may allow that modal properties supervene globally on physical properties. Yet, in their views, modal properties may well be independent of physical properties not only for their existence but also for their individuation—as, for example, the property of being necessarily odd, which obviously is neither *instantiated because of* nor *individuated by* any physical property.

Scientific naturalists, on the contrary, tend to conceive of modal properties as *logical fictions*—in the eliminativist sense of this phrase as originally put forward by Bertrand Russell.[30] According to this view, there is no reason these properties should belong to the overall inventory of what exists.[31] However, just speaking of logical fictions in the eliminativist sense does not solve the problem, since one still needs an argument that justifies this eliminativist attitude toward modal properties (as Papineau puts it, "fictionalists need to show how these claims can be rephrased so as to eliminate the fictional elements, and moreover they need to show that the rephrased versions constitute a cogent world view").[32]

In general, as we have said, scientific naturalists tend to claim that the only entities that exist are those that are part of the causal network. Yet this claim (besides encountering serious problems with the ontological status of abstract entities) begs the question liberal naturalists pose about the legitimacy of more inclusive ontological criterion. At any rate, other, more specific attempts to dispense with modal properties have hitherto proved to be rather unsatisfying as well. William Lycan, for instance, has convincingly suggested that if one tries to rule out modal properties by fictionalizing them (i.e., by first reducing modal properties to possible worlds and then by dealing with possible worlds ultimately in terms of a make-believe discourse grounded in the imaginative faculty of human beings), one risks falling into a vicious explanatory circle.[33] In this light, philosophers who want either to eliminate modal properties from ontology or to reduce them to nonmodal properties still owe liberal naturalists a convincing account. Thus, it is not unjustified for liberal naturalists to hold a realist attitude toward modal properties. And this shows that Liberal Naturalism does not collapse into

Scientific Naturalism for it may allow for properties that the latter would not accept.

This shows that the first horn of the dilemma as regards the special case of modal properties can be addressed. At this point, however, one may wonder whether, where modal properties are concerned, the second horn of the dilemma mentioned above could be problematic. That is, one might wonder whether, by accepting into its ontology entities for which science cannot provide an account, Liberal Naturalism proves itself to be *too* liberal—that is, nothing more than a form of crypto-Supernaturalism.

As seen above, the most common branch of Supernaturalism is characterized by the acceptance of entities that could break the causal closure of the world studied by natural sciences (for instance, Cartesian mental properties). With regard to modal properties, however, it is generally agreed that they cannot alter the causal order of the world (the fact that a man *possibly* stands in the doorway cannot make the doorway occupied; the fact that the *Mona Lisa* has *necessarily* been painted by Leonardo cannot make any difference in what the physical csonstitution of the *Mona Lisa* is; and the fact that the number two is *necessarily* even cannot cause any change in the physical world). So, undoubtedly, Liberal Naturalism is very different from this form of Supernaturalism.[34]

However, one could still wonder whether Liberal Naturalism does not collapse into the second form of Supernaturalism described earlier—the form that allows for an ontology of noncausal entities that require what we would intuitively regard as supernatural powers for their apprehension. The answer is negative again. It is true, as was said earlier, that according to Liberal Naturalism there is a methodological difference between philosophical reasoning and scientific reasoning—a difference that is evident with regard to how philosophers deal with noncausal properties such as the modal ones. As opposed to what happens with the supernatural noncausal entities, however, modal properties—as well as the other noncausal entities accepted by liberal naturalists—do not require any special, antirational epistemic power in order to be grasped (they do not require faith, revelation, the authority of Scripture, or mystical intuitions, for example). In order to grasp modal properties, one merely has to have the capacity for carrying out imaginative speculation—and, more specifically, the capacity to wonder whether, if the world were different from how it actually is, an object would still possess the corresponding nonmodal properties. Nothing makes this peculiar epistemic capacity incompatible with scientific forms of knowledge—even if it may be difficult to reconcile it with some prominent philosophical theories such as causal theories of justification and knowledge. And this crucial epistemological difference shows that liberal naturalism is different from the second form of Supernaturalism, as well as from the first.

◎ ◎ ◎

The version of Liberal Naturalism we have defended here is characterized by the claim that noncausal properties (such as modal properties) may exist that are ontologically ineliminable; irreducible to, and ontologically independent of, scientifically explainable properties; causally inefficacious, so that they do not, and cannot, violate any scientific laws; accountable in ways that, if different from scientific understanding, do not contradict it. These ontological and epistemological views show how Liberal Naturalism can be different from Scientific Naturalism without being a form of Supernaturalism in disguise.[35]

NOTES

1. Ram Neta, "Review of M. De Caro and D. Macarthur, eds., *Naturalism in Question,*" *Philosophical Review* 116 (2007): 662.

2. Of course, the two horns of the dilemma are conceptually independent, so they can be, and in fact sometimes are, presented independently. In *Naturalism* (Grand Rapids, Mich.: William B. Eerdmans , 2008), for example, Stewart Goetz and Charles Taliaferro argue that between Liberal Naturalism and Scientific Naturalism there is no relevant difference as to the important philosophical issues. Conversely, James Lenman mocks John McDowell's self-presentation as a naturalist of a liberal form, by writing that "many philosophers would find his use of the term 'naturalist' here somewhat Pickwickian"—which is a way of saying that McDowell's Liberal Naturalism is nothing more than Supernaturalism, in disguise! ("Moral Naturalism," in *Stanford Encyclopedia of Philosophy*, ed. E. Zalta, http://plato.stanford.edu/entries/naturalism-moral/notes.html#18).

3. David Papineau, "Naturalism," *Stanford Encyclopedia of Philosophy*, ed. E. Zalta, http://plato.stanford.edu/entries/naturalism.

4. "If 'nature' is understood in the older, broader sense in which everything that has a nature is natural, then God, angels, and departed souls may be classified as natural" (Goetz and Taliaferro, *Naturalism*, 95).

5. It is not, of course, that we know with certainty which laws of nature we are actually acquainted with. In a naturalist perspective, however, we are *justified* in believing that at least some of those laws are correctly expressed by the claims of contemporary science. These claims, therefore, constrain which entities and explanations one can accept if one wants to be a naturalist.

6. On the idea of first- and second-grade conceptual systems, see W. V. Quine, *Word and Object* (New York: Columbia University Press, 1960), and "Epistemology Naturalized," in *Ontological Relativity and Other Essays* (New York: Columbia University Press, 1969), 69–90.

7. A more complete analysis of Scientific Naturalism is offered in the introduction of this volume. Here we will concentrate on the doctrines connected with this view that are particularly relevant with regard to Neta's dilemma.

8. Alan Lacey, "Naturalism," in *The Oxford Companion to Philosophy*, ed.Ted Honderich (Oxford: Oxford University Press, 2005), 640. Some religiously oriented philosophers use the term "naturalism" in a different sense, to label the view according to which science and religion, including standard Hebraic-Christian theism, can be epistemically well integrated: see the discussions in Paul Draper, "God, Science, and Naturalism," in *The Oxford Handbook of Philosophy of Religion*, ed. William Wainwright (Oxford: Oxford University Press, 1999), 272–303; and Robert Audi, "Philosophical Naturalism at the Turn of the Century," *Journal of Philosophical Research* 25 (2000): 27–45. This way of using the term, however, is idiosyncratic. Here we will follow the more orthodox use.

9. For a defense of the more radical view, see, for example, Sandro Nannini and H. J. Sandkühler, eds., *Naturalism in Cognitive Sciences and the Philosophy of Mind* (Frankfurt: Peter Lang, 2000). The notion of ontological dependence is at this stage being used in a deliberately vague, though intuitive, way. The notion and its relation to supervenience will be clarified below.

10. Colin McGinn, *Problems in Philosophy: The Limits of Enquiry* (Oxford: Blackwell, 1993).

11. W. V. Quine, "Epistemology Naturalized," in *Ontological Relativity and Other Essays* (New York: Columbia University Press, 1969), 26. Another epistemological claim commonly made by scientific naturalists is that there is no a priori knowledge (a claim that deeply affects the conception of the philosophical method). On this, see the introduction to this volume.

12. S. T. Davis, "Is It Rational for Christians to Believe in Resurrection?" in *Contemporary Debates in Philosophy of Religion*, ed. M. L. Peterson and R. J. Vanarragon (Oxford: Blackwell, 2004), 165. It is interesting to notice that most of the best recent dictionaries of philosophy (including the ones edited by Robert Audi, Simon Blackburn, and Ted Honderich) do not have a specific entry for the term "Supernaturalism."

13. W. L. Reese, *Dictionary of Philosophy and Religion* (Atlantic Highlands, N.J.: Humanities Press, 1996), 744.

14. On this, cf. the introduction to this volume and David Macarthur's article in this anthology. John Dupré is a liberal naturalist who grants that the natural sciences can in principle account for all *the stuff* of which the world is made, but this does not imply that the natural sciences can explain all features of the things that exist in the world. It is interesting to notice that Dupré maintains allegiance to the

constitutive claim of naturalism, even if he advocates downward causation. This is because in his view the thesis of the causal closure of the physical (which is at odds with downward causation) should be rejected for *empirical* reasons. See Dupré's article in this anthology and *The Disorder of Things: Metaphysical Foundations of the Disunity of Science* (Cambridge, Mass.: Harvard University Press, 1993).

15. In *Ethics Without Ontology* (Cambridge, Mass.: Harvard University Press, 2004), Hilary Putnam, a leading liberal naturalist, claims that ethical and mathematical discourses are both objective and irreducible to scientific discourses, but no ontological commitment simply follows from them because ontology as such is a dead philosophical project. It could be argued, however, that the difference between Putnam's nonontological view and the pluralistic ontological view described in the text is less substantial than it may appear since what is crucial for both views is that they conjugate objectivism with antireductionism and nonSupernaturalism. For a similar view, see Ernest Sosa, "Experimental Philosophy and Philosophy Intuition," in *Experimental Philosophy*, ed. Joshua Knobe and Shaun Nichols (Oxford: Oxford University Press, 2008), 232.

16. "The driving motivation for ontological naturalism is the need to explain how different kinds of things can make a causal difference to the spatiotemporal world" (David Papineau, "Naturalism")

17. On this point, see Thomas Grundmann, "The Nature of Rational Intuitions and a Fresh Look at the Explanationist Objection," *Grazer Philosophische Studien* 74 (2007): 69–87; and Rick Tieszen, "Gödel and the Intuition of Concepts," *Synthese* 33 (2002): 363–91.

18. Cf. W. Sellars, "Philosophy and the Scientific Image of Man," in *Frontiers of Science and Philosophy*, ed. R. Colodny (Pittsburgh: University of Pittsburgh Press, 1962), 35–78; John McDowell, *Mind and World* (Cambridge, Mass.: Harvard University Press, 1994); Stephen White, "Subjectivity and the Agential Perspective," in *Naturalism in Question*, ed. Mario De Caro and David Macarthur (Cambridge, Mass.: Harvard University Press, 2004), 201–30; White, "Empirical Psychology, Transcendental Phenomenology, and the Self," in *Cartographies of the Mind*, Massimo Marraffa, Mario De Caro, and Francesco Ferretti (Dordrecht: Springer, 2007), 243–54. Of course, we are not claiming that there are no differences in how Sellars, McDowell, and White conceive of the human perspective on the world; however, these differences are not relevant here.

19. Cf. Jaegwon Kim, *Philosophy of Mind* (Boulder, Colo.: Westview Press, 1996), 11; E. J. Lowe, "Ontological Dependency," *Philosophical Papers* 23 (1994): 31–48; Lowe, *The Possibility of Metaphysics* (Oxford: Clarendon Press, 1998).

20. John McDowell, "Response to Akeel Bilgrami", in *McDowell and His Critics*, ed. Cynthia Macdonald and Graham Macdonald (Oxford: Blackwell, 2006), 66–72.

21. Cf. Akeel Bilgrami, "Some Philosophical Integrations," in *McDowell and His Critics*, ed. Cynthia Macdonald and Graham Macdonald (Oxford: Blackwell, 2006),

50–66; Stephen L. White, "The Transcendental Significance of Phenomenology," *Psyche* 13, no. 2 (April 2007), http://psyche.cs.monash.edu.au/symposia/siegel/White.pdf.

22. Cf. Dupré's article in the present collection.

23. Hilary Putnam, "The Content and Appeal of 'Naturalism,'" in *Naturalism in Question*, ed. Mario De Caro and David Macarthur (Cambridge, Mass.: Harvard University Press, 2004), 286n. 1.

24. Jaegwon Kim, *Mind in a Physical World* (Cambridge, Mass.: MIT Press, 1998), 15.

25. Anand Vaidya, "The Epistemology of Modality," in *Stanford Encyclopedia of Philosophy*, ed. E. Zalta, http://plato.stanford.edu/entries/modality-epistemology.

26. Ibid. According to Robert Audi, philosophical naturalism *as such* seems committed to the possibility of modal discourse: "Naturalism in any plausible form requires positing causal relations, and it is doubtful that these can be understood apart from subjunctives; those, in turn, seem at best difficult to comprehend without appeal to possible worlds or something comparably abstract" ("Philosophical Naturalism at the Turn of the Century," 39).

27. Cf., for example, David Wiggins, *Sameness and Substance* (Oxford: Blackwell, 1980), 106–7.

28. See Saul Kripke, *Naming and Necessity* (Oxford: Blackwell, 1980); Hilary Putnam, "The Meaning of 'Meaning,'" in *Philosophical Papers*, vol. 2: *Mind, Language, and Reality* (Cambridge: Cambridge University Press, 1975), 215–71. For a discussion of these topics and bibliographical references about it, see Vaidya, "The Epistemology of Modality." It has been convincingly argued, by the way, that the reevaluation of essentialism, more than on Kripke's semantics of rigid designators, has depended on the fact that Quines's qualms against modalities are actually ineffective once the distinction between *de re* and *de dicto* modality is accepted. Cf., e.g., Baruch A. Brody, *Identity and Essence* (Princeton, N.J.: Princeton University Press, 1980), 85–87; John L. Mackie, "*De* What Re Is *De Re* Modality?" *The Journal of Philosophy* 71 (1974): 553.

29. John McDowell, "Two Sorts of Naturalism" (1995), reprinted in *Mind, Language, and Reality* (Cambridge, Mass.: Harvard University Press, 1998), 170; our italics.

30. Cf. Bertrand Russell, "The Philosophy of Logical Atomism" (1919), reprinted in *Logic and Knowledge* (London: Routledge, 1956), 177–281.

31. Cf. (all from E. Zalta, ed.. *Stanford Encyclopedia of Philosophy*): Daniel Nolan, "Modal Fictionalism," http://plato.stanford.edu/archives/sum2002/entries/fictionalism-modal/2002; Matti Eklund "Fictionalism," http://plato.stanford.edu/entries/fictionalism/2007; William Ramsey, "Eliminative Materialism," http://plato.stanford.edu/entries/materialism-eliminative/.

32. Papineau, "Naturalism." For a realist interpretation of fictional entities in general, cf. Alberto Voltolini, *How Ficta Follow Fiction: A Syncretistic Account of Fictional Entities* (Dordrecht: Springer, 2006).

33. See William Lycan, *Modality and Meaning* (Dordrecht: Springer 1994), which crit-

icizes the fictionalism about modality defended by David Armstrong in *A Combinatorial Theory of Possibility* (Cambridge: Cambridge University Press, 1989).

34. On this issue, cf. also Daniel Andler, "Is Naturalism the Unsurpassable Philosophy for the Sciences of Man in the Twenty-first Century?" in *The Present Situation in the Philosophy of Science*, ed. F. Stadler et al. (Berlin: Dordrecht, forthcoming).

35. We thank Martine Nida-Rümelin, Ben Schupman, Jesús Vega Encabo, and Stephen White for their comments and suggestions on previous versions of this paper.

PART II

PHILOSOPHY AND THE
NATURAL SCIENCES

4

SCIENCE AND PHILOSOPHY

Hilary Putnam

WHY DO WE NEED PHILOSOPHY?

I want to begin by considering why and how the very need for philosophy became a question (in our times, at least). A reason often given for the contemporary debate concerning the need or role for philosophy is that philosophy for so long—from the Middle Ages until the end of the nineteenth century, in fact—was so heavily invested in two "ontotheological" (i.e., metaphysical-*cum*-theological) ideas, namely, the idea of God (although the "God of the philosophers" was always very different from the God of the celebrated "man—or woman—on the street"),[1] and the idea of the immateriality of the soul.[2] Although most nineteenth-century scientists were still churchgoers, the posture of twentieth-century (and now twenty-first-century) science has been decidedly secular, and analytic philosophy, for the most part, has the same posture. That is not to say there aren't scientists or analytic philosophers who are religious. But the idea of God as an entity we need to postulate in order to account for the existence of the natural world or to postulate as a foundation for morality is no longer widely accepted. Indeed, although I am a practicing Jew, I myself don't believe in either "ontotheology" or in the idea that ethics requires a religious "foundation." In fact, in my forthcoming book, *Jewish Philosophy as a Guide to Life: Rosenzweig, Buber, Levinas, Wittgenstein,*

I describe my current religious standpoint as "somewhere between John Dewey in *A Common Faith* and Martin Buber." And I understand Dewey to be saying that the kind of reality God has is the reality of an ideal, and that is the way I, too, conceive of God. But that is *not* the subject of this essay.

For the large number of people, including many religious people, who have lost the belief in the God of ontotheology, the metaphysical First Cause and *ens necessarium*, and now find themselves unable to believe in the God of the unreflective believer or in life after death, it *can* seem that the whole of philosophy between, say, Plato and Hegel, was vast mistake. That is what Heidegger's proclamation of the end of "ontotheology" seemed to amount to, and Wittgenstein's description of the new task of philosophy as "showing the fly the way out of the fly-bottle" has often been read as pronouncing a similar verdict on traditional philosophy. Moreover, not only has belief in God, as God was traditionally conceived, ceased to be something that educated people take for granted in the West, but the successes of evolutionary biology, genetics, computer science, and brain science have demolished the idea that the "seat" of our mental faculties must be an immaterial soul. In addition, many natural scientists are hostile to traditional philosophy and are glad to dismiss it as obsolete. (Nor are they particularly friendly to analytic philosophy, which they often regard as scientifically uninformed hairsplitting.)

In sum, even if the enormous prestige and the enormous success of science have not jeopardized the position of the humanities as important disciplines, they have called into question the *raison d'être* of philosophy. Is philosophy really just a relic from past ages that we need either to discard or to replace with something else?—even if we disguise the fact that the latter is what we are doing by retaining the *word* "philosophy"?

LOGICAL POSITIVISM AS A FAILED RESPONSE TO THE QUESTION OF PHILOSOPHY'S FUNCTION

Logical positivism was a product of the combined influence of the great physicist Ernst Mach's version of empiricism and of Bertrand Russell's belief that, in the new logic that he and Whitehead did so much to perfect, a tool had been discovered that would either solve or dissolve the traditional problems of philosophy once and for all. It was thus a decidedly twentieth-century movement. In fact, Carnap's first great book, *The Logical Construction of the World*, was not published until 1928. In that book, there is still an interest in epistemology, and Husserl is even cited as a forerunner.[3] But in 1934, Carnap wrote the following: "All statements belonging to Metaphysics, regulative Ethics, and (metaphysical) Epistemology have this defect, are in fact unverifiable and, therefore, unscientific. In the Viennese Circle, we are accustomed to describe such statements as nonsense."[4]

In these words we have the essence of logical positivism. In fact, they express in an extremely condensed way the two chief principles of the movement: First, the meaningful statements in our language are exactly the statements that can be tested and "verified" (established to be true or false by the methods of science).[5] All other statements are *nonsense*—totally devoid of discussable content. Second, every one of the fields into which traditional philosophy was divided—metaphysics, ethics, and even epistemology (referred to in this passage as "(metaphysical) epistemology")—must be abandoned because they consist entirely of such "nonsense."

So what is left for philosophers to do? How could the logical positivists continue to teach in philosophy departments, train graduate students, publish in philosophy journals, and even create new philosophy journals? The answer Carnap went on to give, especially in his 1935 *The Logical Syntax of Language*, was that they weren't really doing what had traditionally been *called* "philosophy" at all; they were engaged in studying the "logic of science."

This might lead one to expect that Carnap must then have proceeded to prove a number of results in a scientific field called by that name. But one would be wrong. In all the books and papers that Carnap wrote—books and papers that are rightly prized as brilliant contributions to important philosophical debates—there is, to be blunt, not one contribution to anything that anyone today recognizes as logic of science. There are attempts to "reduce" nonobservational terms in science (terms such as "atom," "gene," and "gravitational field tensor") to observation terms (an attempt that was a total failure, as Carnap came to recognize). There was an attempt, in *The Logical Syntax of Language* itself, to provide a syntactic characterization of "analytic" sentences (which category was supposed to include all the sentences of mathematics)—another failure. There were, as mentioned, brilliant contributions to philosophical debates—particularly debates in the philosophy of language and the philosophy of mathematics. And finally, there was an attempt to formalize inductive logic, with results universally regarded as disappointing. In sum, to the extent that Carnap achieved anything important and impressive it was in philosophy and not in the logic of science, and to the extent that he tried to obtain impressive and important results in the logic of science, he met with complete failure. Carnap and his followers did not succeed in folding philosophy into science, as they hoped to do.

POSTMODERNISM AS ANOTHER FAILED RESPONSE

If logical positivism sought to counter the perceived danger to philosophy from science, the danger that science had rendered philosophy *obsolete*, postmodernism sought to restore the prestige of philosophy by retorting that science itself, and, indeed, everything we think of as a description of "facts," is just a form of fiction—useful fiction, to be sure, but in the end just more of the many webs of ideology Western culture keeps spinning. It is not, I hasten to add, that the postmodernists thought there is some other kind of discourse that is free of deceit. Rather, at least in the hands of Jacques Derrida and his followers (but to a greater or lesser extent in the hands of the other gurus of the movement as well), discourse itself is seen as *inherently* deceptive, if only because it tempts us to believe that here is such a thing as a truthful representation of reality while actually there is not and cannot be any such thing. Or so postmodernists claim.

One might think that such a revolutionary claim would be backed by strong arguments. In fact, even Richard Rorty, the most intelligent analytically trained philosopher to be won over to the movement, described "a lot" of Derrida's arguments as "just awful."[6] When he defended postmodernist views, Rorty did so by adapting arguments from analytic philosophers (not one of whom subscribed to Rorty's use of their arguments, however!), namely Sellars, Quine, Davidson, and myself. But, rather than discuss this "can of worms" here, let me simply say that "representationalism"—the view that we can and often do succeed in *representing* parts and aspects of reality in language—is just what used to be called "realism" and realism does not become a terrible fallacy to be looked down upon with scorn just because a number of professors have given it the new name "representationalism" and *declared* it to be such.

THE IMPORTANCE AND VALUE OF PHILOSOPHY

If neither the positivist response (fold philosophy into science) or the "postmodernist" response (declare science to be fiction) is tenable, what remains? Looking back over fifty-six years of "doing philosophy," I find that two definitions of philosophy appeal to me the most and also that each definition requires supplementation by the other.

The first definition comes from Stanley Cavell's famous book, *The Claim of Reason*, and I should like to quote from the passage in which it occurs at some length:

But if the child, little or big, asks me: "Why do we eat animals? or Why are some people poor and others rich? or What is God? or Why do I have to go to school? or Do you love black people as much as white people? or Who owns the land? or Why is there anything at all? or How did God get here? I may feel my answers thin, I may feel run out of reasons without being willing to say, 'This is what I do'... and honor that.

Then I may feel that my foregone conclusions were never conclusions *I* had arrived at, but were merely imbibed by me, merely conventional. I may blunt that realization through hypocrisy or cynicism or bullying. But I may take the occasion to throw myself back on my culture, and ask why we do what we do, judge as we judge, how we have arrived at these crossroads. . . . In philosophizing I have to bring my own language and life into imagination. What I require is a convening of my culture's criteria, in order to confront them with my words and life as I pursue them and as I may imagine them; and at the same time to confront my words and life with the life my culture's words may imagine for me: to confront the culture with itself along the lines in which it meets in me.

This seems to me a task that warrants the name of philosophy. . . . In this light, philosophy becomes the education of grownups.[7]

The second definition comes from Wilfrid Sellars' essay "Philosophy and the Scientific Image of Man."[8] "The aim of philosophy," he wrote, "is to understand how things in the broadest possible sense of the term hang together in the broadest possible sense of the term."

Cavell's definition, "the education of grownups," and the examples he gives of the "child's" questions, which may show the "grownup" that he or she needs education, point to what I will call the moral face of philosophy, the face that interrogates our lives and our cultures as they have been up to now and that challenges us to reform both. Sellars's definition points to what I will call the theoretical face of philosophy, the face that asks us to clarify what we think we know and to work out how it all "hangs together". Indeed, Cavell's comprehends Sellars's, since a grownup who does not care if his view of things hangs together hardly counts as "educated."

Logical positivism, in a limited way, sought to preserve the theoretical face of philosophy (albeit dismissing many important theoretical issues as "metaphysical") while banishing the moral face entirely, while postmodernism wants to preserve the moral face of philosophy—albeit often reducing it to what Richard Rorty once described as "the hallucinatory effects of Marxism, and of the post-Marxist combination of De Man and Foucault currently being smoked by the American Cultural Left"[9]—at the expense of the theoretical face.

My own view is that philosophy at its best has always, in every period, included some philosophers who brilliantly represent the moral face of the subject and some philosophers who brilliantly represent the theoretical face, as well some geniuses whose insights span and unite both sides of the subject. To renounce either the moral ambitions of philosophy or its theoretical ambitions is not just to kill the subject of philosophy; it is to commit intellectual and spiritual suicide.[10]

DOES SCIENCE NEED PHILOSOPHY?

That philosophy is not to be identified with science is not to deny the intimate relation between science and philosophy. The positivist idea that all science does is predict the observable results of experiments is still popular with some scientists, but it always leads to the *evasion* of important foundational questions. For example, the recognition that there is a problem of *understanding* quantum mechanics, that is, a problem of figuring out just how physical reality must be in order for our most fundamental physical theory to work as successfully as it does, is becoming more widespread, but that recognition was *delayed* for decades by the claim that something called the "Copenhagen Interpretation" of Niels Bohr had solved all the problems. Yet the "Copenhagen interpretation," in Bohr's version, amounted only to the vague philosophical thesis that the human mind couldn't possibly understand how the quantum universe was in itself and should just confine itself to telling us how to use quantum mechanics to make predictions *stateable in the language of classical, that is to say, non-quantum-mechanical, physics!* (In my lifetime, I first realized that the "mood" had changed when I heard Murray Gell-Man say in a public lecture sometime around 1975 "There is no Copenhagen Interpretation of quantum mechanics. Bohr brainwashed a generation of physicists!")

Only after physicists stopped being content to regard quantum mechanics as a mere machine for making predictions and started taking seriously what this theory actually *means* could real progress be made. Today many new paths for research have opened as a result:[11] string theory, various theories of quantum gravity, and "spontaneous collapse" theory are only the beginning of quite a long list. And Bell's famous theorem, which has transformed our understanding of the "measurement problem," would never have been proved if Bell had not had a deep but at the time highly unpopular interest in the meaning of quantum mechanics.[12]

In cosmology, however, there has unfortunately been somewhat of a revival of the positivist contempt for the question of the meaning of general relativity in recent years, but, owing to the influence of Einstein, who always recognized that physical theories are not mere formal systems, the great majority of astrophysicists continue to try to *understand* the nature of cosmic spacetime and of the forces that shape the destinies of astronomical objects (including black holes), and not simply to say, as Steven Weinberg now appears to urge, that sometimes it is more convenient to use one theory, and sometimes it is more convenient to use another, and there is no reason to ask which is really true.[13] It is precisely at the level of fundamental physical science, in fact, that it becomes clear that the sharp separation that the positivists thought they saw, and that our culture often takes for granted, between metaphysics and physics is most untenable. Both physics and metaphysics flourish most when they interact and interpenetrate, that is, when they push Sellars's question, "how things in the broadest possible sense of the term hang together in the broadest possible sense of the term."[14]

SCIENCE AND NORMATIVITY

Part of the difficulty of seeing the relation between science and philosophy clearly is that false ideas about science abound in ordinary life and philosophy, including the influential idea that there is something like an "algorithm" for doing science. Yet this idea is not one that any scientist I know believes in. A few years ago, speaking to an audience which contained at least fifty Nobel Prize winners, I said the following:

I have argued that even when the judgments of reasonableness are left tacit, such judgments are presupposed by scientific inquiry. (Indeed, judgments of *coherence* are essential even at the observational level: we have to decide *which* observations to trust, which scientists to trust (sometimes even which of our *memories* to trust). I have argued that judgments of reasonableness can be objective. And I have argued that they have all of the typical properties of "value judgments." In short, I have argued that my pragmatist teachers were right: "knowledge of facts presupposes knowledge of values."[15] But the history of the philosophy of science in the last half century has largely been a history of attempts—some of which would be amusing, if the suspicion of the very idea of justifying a value judgment which underlies them were not so serious in its implications—to *evade* this issue. Apparently any fantasy—the fantasy of doing science using only deductive logic (Popper), the fantasy of vindicating induction deductively (Reichenbach), the fantasy of reducing science to a simple sampling algorithm (Carnap), the fantasy of selecting theories given a mysteriously available set of "true observation conditionals" or, alternatively, "settling for psychology" (both are Quine's)—is regarded as preferable to rethinking the whole dogma—the last dogma of empiricism?—that facts are objective and values are subjective and never the twain shall meet.

And not one of those scientists disagreed with me!

Nevertheless, if what we might call "the myth of inductive logic" is not one to which scientists themselves subscribe, it has had a powerful influence on the way laypersons and philosophers (including philosophers of science) think about science—an influence that, I would argue, it is one of the functions of philosophy, at its best, to combat. In fact, the idea that beliefs about "facts" must be fundamentally different from "attitudes" toward values was supported by the most influential defender of the idea that values are "noncognitive," the father of "emotivism" (as the doctrine that value judgments are mere modes of emotional "persuasion" was called), Charles Stevenson, precisely by employing the claim that value judgments can't be verified by "induction and deduction" (while "beliefs about facts" supposedly could be). Thus what began as a technical issue in epistemology, the issue about "induction," came to play a crucial role in debates over the possibility of the rational discussion of value issues—an issue of immense importance to our culture.

The case for the idea that facts and values are deeply entangled draws on some of the best philosophical work of the last hundred years:[16] on Quine's attack on the positivists' analytic/synthetic distinction, on the work done by some of Wittgenstein's best followers (notably Iris Murdoch, Phillippa Foot, and John McDowell) on the way "thick" ethical concepts such as "cruel" resist "factorization" into a "purely descriptive" component and an "expressive" or "emotive" component, and on my own observations on the way in which the epistemic values that inform science are, after all, *value* concepts, too. For a number of years, in fact, I have argued that in science, and particularly in the social sciences, we are unavoidably dealing with an entanglement of facts, theories, and values. It is like a three-legged stool—all three legs are needed, or it falls over. The all-too-popular idea that if something is a "value judgment" then it must be wholly "subjective" rests not just upon *shaky* foundations but on foundations that have completely collapsed. In my opinion, understanding this is vital if we are to regain faith in the possibility and the importance of *rational debate about values*.

THE SCIENTIFIC AND MANIFEST IMAGES

The need to "save the appearances"—that is, to do justice to the manifest image—in our theorizing about the world has been an important aspiration of philosophy from its inception. Aristotle's insistence that metaphysics must "save the appearances" can be viewed as a form of respect for ordinary language against Platonic speculation. If "ordinary language" is just something to be sneered at, then is the whole vocabulary we have for describing the world of human agents to be either despised or else replaced by the "Newspeak" of some social science?

To me it seems clear that the descriptions of human life we find in the novels of Tolstoy or George Elliot are not mere *entertainment*; they teach us to perceive what goes on in social and individual life. And such descriptions require the many subtle distinctions that ordinary language has made available to us. The question of the relevance or irrelevance of "how we speak" is not just a question for philosophers, although it is that, too. It is a question for philosophers because once ordinary language is laughed out of the room, philosophical theories are no longer held responsible at all to the ways we actually speak and actually live. But it is a question for more than just philosophers because, at bottom, contempt for ordinary language is contempt for all the humanities.[17]

ANOTHER LOOK AT THE HISTORY

I began this essay with a particular account of the "crisis" of philosophy, a view made popular in Europe by Heidegger's *Sein und Zeit*. In that view, a natural one for a former seminarian like Heidegger, traditional philosophy was seen as "ontotheology." (Heidegger's solution to the crisis was that we should all become Heideggerians.) Another account of the "crisis," one I might have begun with instead, can be found in the writings of Bertrand Russell and also in the logical positivists. On that account, it was the realization that philosophy led to debates that are never settled, coupled with the supposed fact that "the new logic" could resolve the old unsettleable problems, that required that traditional philosophy be entirely replaced.

"Progress" in philosophy need not consist of "settling" issues once and for all. Indeed, it does not consist of that in any serious area of human endeavor or inquiry that I know of. And those problems have quite old roots, which suggests that philosophy was never just a handmaiden to theology, even if in the Middle Ages it was often urged to be that. To give only one of many possible examples: the issue about whether talk of unobservables in physical science is really "representational" already appeared at the time of Berkeley and Hume, and is thus centuries old.[18] How to interpret quantum mechanics is simply a contemporary spin-off, occasioned by the puzzling character of that theory. Philosophy was never just ontotheology, and even when philosophers were concerned with ontotheology, they were concerned with much more than that. That is a first reason that the idea of a fundamental "crisis" in philosophy and the "end of philosophy" is deeply mistaken. And if the questions of philosophy are indeed "unsettleable," in the sense that they will always be with us, that is a wonderful thing, not something to be regretted.

NOTES

1. The God of the philosopher is very different from the God of the ordinary believer whether the philosopher be, for example, Aquinas or Spinoza or Kant or Hegel.

2. Spinoza was an exception here. For Kant, however, the immaterial soul existed only in the noumenal realm, as the "transcendental unity of apperception"; Hegel, characteristically wanted matter and soul to be in some sense identical and also nonidentical.

3. There are references to Husserl and Husserl's student Becker in Carnap's *Der Logische Aufbau der Welt*, as well as uses of the Husserlian term "epoché", in the original (1928) edition, most of which were removed by the translator in the English (1967) edition. See: http://philsci-archive.pitt.edu/archive/00002858/.

4. Rudolf Carnap, *The Unity of Science* (London: Kegan Paul, Trench, Hubner, 1934), 26–27.

5. Later this was softened to the requirement that meaningful statements must be "confirmable or disconfirmable," and still later by still more complicated requirements. For a brief history, see the first chapter of my *Collapse of the Fact/Value Dichotomy* (Cambridge, Mass.: Harvard University Press, 2002).

6. "Searle is, I think, right in saying that a lot of Derrida's arguments (not to mention some of Nietzsche's) are just awful" (Richard Rorty, *Philosophical Papers*, vol. 2: *Essays on Heidegger and Others* [Cambridge: Cambridge University Press, 1991], 93n).

7. Stanley Cavell, *The Claim of Reason* (Oxford: Oxford University Press, 1979), 125.

8. Wilfrid Sellars, "Philosophy and the Scientific Image of Man," in Robert Colodny, ed., *Frontiers of Science and Philosophy* (Pittsburgh, PA: University of Pittsburgh Press, 1962). Reprinted in *Science, Perception and Reality* (London: Routledge & Kegan Paul Ltd and New York: The Humanities Press, 1963); reissued in 1991 (Atascadero, CA, Ridgeview Publishing Co.), 37.

9. Richard Rorty, *Philosophical Papers*, 2:137.

10. Philosophy, when it conceives of itself as a purely theoretical subject, first, tends to become overambitious (e.g., it then tries to realize the dream of a grand metaphysical system); and, second, it is held up by the public to the standard of cumulative knowledge. But since, unlike science, philosophy is not cumulative knowledge; when it pretends to be a science people see right through that pretence.

11. For more details, see my "A Philosopher Looks at Quantum Mechanics (Again)," *British Journal for the Philosophy of Science* 56, no. 2 (2005): 615–34.

12. For a simple statement and proof of Bell's Theorem see: http://www4.ncsu.edu/ unity/lockers/users/f/felder/public/kenny/papers/bell.html.

13. For an account of Weinberg's views, see Yemima Ben-Menahem, *Conventionalism* (Cambridge: Cambridge University Press, 2006), 93–94. Although Weinberg claims that his version of general relativity (which employs a flat space-time) is equivalent to Einstein's version (with its curved space-time), this is not the case because the topologies of space are different in the Einstein version and the Weinberg version. The three-dimensional spatial cross-sections of the universe were finite epsilon seconds after the Big Bang, and while inflation may have produced a situation in which the curvature of 3D space became undetectably small later, a change from a "finite universe" to an "infinite universe" (or from a curved finite cross section to a flat one) would involve a discontinuity in the universe's evolution, which no astrophysicist has postulated. The fact is that Weinberg simply ignores the difference between "undetectably small" and "zero" curvature—a typical verificationist move. In addition, he ignores the fact that, as he has confirmed to me in correspondence, his "flat space-time" version has no description of black holes.

14. Sometimes conceptual crises arise in science such that in order to make progress science has to find a reasonable way to resolve them, and at such times—one example being the time of Newton and the great debate between Newton and Leibniz about whether space is absolute or not, another more recent example being the birth of modern set theory and the debates between Cantor, Frege, Poincaré, Hilbert, Brouwer, and others. Today there are raging debates about string theory (is it physics at all? or metaphysics? etc.). It is characteristic about these crises that both scientists and philosophers contribute to the discussion.

15. I heard this in my undergraduate years at the University of Pennsylvania from my teacher C. West Churchman, who attributed it to his teacher E. A. Singer Jr., who was in turn a student of William James.

16. Cf. Putnam, *The Collapse of the Fact/Value Dichotomy*.

17. I don't think that philosophy can be turned into a science because there are areas of philosophy that are essentially humanistic and I think turning the humanities into science is a fantasy, and a dangerous fantasy at that. But there are parts of philosophy that overlap with science.

18. Think, for example, of the debates between Clarke and Leibniz concerning the reality or lack of reality of action at a distance and absolute space.

5

WHY SCIENTIFIC REALISM
MAY INVITE RELATIVISM

Carol Rovane

There is a widespread and fairly longstanding assumption that realism and relativism are mutually opposed doctrines. When Kant first introduced the doctrine of transcendental idealism, he argued that we cannot know things as they are in themselves but only as they appear to us, given our particular forms of sensibility. He thereby left open at least the logical possibility that there might be other kinds of knowers, with other forms of sensibility different from ours, who could not know the things that we know; rather, they could only know the things that appear to them given their particular forms of sensibility, things that we in turn cannot know. If we think of such objects of knowledge as truths rather than things, it is tempting to infer that truth is relative to a kind of knower.[1] In the early twentieth century, Carnap exploited logical positivism to much the same effect. He held that objectivity requires confirmability by experience and that experience can confirm a given existential claim as true or false only if a linguistic framework is in place that provides clear rules for affirming and denying sentences on the basis of experience. If we grant that there is more than one way to set up a linguistic framework, it is again tempting to infer that truth is relative to a linguistic framework. Ever since, arguments for relativism have taken a standard form that follows the same pattern as those invited by Kant and Carnap: reality is first portrayed as dependent on some (broadly speaking) mental condition or other—a form of sensibility, a linguistic framework, a conceptual scheme, a scientific paradigm, a culture, and so on—and then it is inferred that truth is relative to that mental condition. For this reason, the widespread assumption has prevailed without question. Relativists assume that they must deny the mind-independence of reality, a cornerstone of realist doctrine, just as realists assume that, when they affirm the mind-independence of reality, they thereby rule out relativism.

But the widespread assumption is false. Relativists need not reject realism. On the contrary, realism may be the only way to *save* relativism in the wake of Davidson's arguments against it.[2]

The reason this point is so often overlooked—and the widespread assumption has prevailed without question—is not lack of acquaintance with Davidson's work. The reason is that no one has bothered to elucidate the doctrine in a satisfactory manner to begin with. It is fair to say that most discussions of it proceed with only a vague and intuitive sense of its content. This is unfortunate on two counts: first, we lack clarity about what the doctrine actually is, and, second, as a result, we lack much needed critical distance on the widespread assumption. Davidson himself lacked such critical distance. Although he saw both realism and relativism as falling into a common philosophical error, he saw them as mutually opposed doctrines nonetheless.[3]

◎ ◎ ◎

The common error of which Davidson accused both relativism and realism is the scheme-content distinction. If there is a dualism of scheme and content, then we can never have direct epistemic contact with a mind-independent reality (content) but only with our mental representations (scheme). In Davidson's view, this dualism leaves philosophers with two unattractive options: either we affirm mind-independence and accept the skeptical implication that we may fail to know things as they really are, or we deny mind-independence and accept some form of idealism, usually with relativistic implications. I will be arguing that Davidson didn't get matters entirely right. On the one hand, there is room for a kind of realism—scientific realism—that concedes more to the mind-independence of reality than he was prepared to concede, but without bringing skepticism in train. On the other hand, in his own efforts to get away from what he saw as a common error shared by relativists and realists, Davidson fell into another—in my view equally common—error, which is to accept the widespread assumption that realism and relativism are mutually opposed doctrines. Yet even without getting everything right, Davidson's assault on relativism remains the most powerful response to the doctrine on record. We shall see that he didn't leave much scope for a significant relativism. But insofar as he did, it is only because he left more scope for a significant realism than he realized—the realism that animates and informs the scientific enterprise. However, it is easy to miss this last point because of the influence of the widespread assumption.

I want to emphasize that my main concern in this paper is neither exegesis nor criticism of Davidson's philosophical vision. My main concern is to bring out exactly why the widespread assumption is mistaken—why realism, in particular scientific realism, does not rule out relativism. In order to explain this, I first need to elucidate the doctrine of relativism with somewhat greater care than is generally taken in most discussions of it.

AN ELUCIDATION OF THE DOCTRINE OF RELATIVISM

My approach to elucidating the doctrine of relativism will be indirect. I will start by trying to develop what I take to be the most natural first thoughts about what the doctrine is and then showing why these thoughts do not lead to a satisfactory outcome. My own proposal will emerge as a positive lesson drawn from these failures.

When I outlined what has become the standard form of argument for relativism, I loosely characterized the doctrine as saying that truth is relative. This seems natural enough given the name of the doctrine. Yet if we examine the idea of relative truth more closely, we notice that it can't really be the dividing issue between relativists and their opponents. The only clear model we have of relative truth is indexical truth. Sentences containing indexical pronouns like "I," "here," "now," and "this" are not true or false tout court but true or false relative to context of utterance—who is making the utterance, where and when it is made, and what item is being referred to by the speaker there and then. Thus, a sentence like "I am a woman" is not true or false tout court but true or false relative to whether the speaker is a woman. Obviously, those who oppose relativism need not oppose the idea that truth may be relative to context in the way that the truth of indexical sentences is. So relativists must have something else in mind when they claim that truth is relative.

It seems to me that they must have the following in mind: truth is tied to context in such a way that, if truths are tied to different contexts, then they cannot be embraced together. From here on, I shall refer to such truths that cannot be embraced together as *alternatives*. The idea of alternativeness was implicit in my brief descriptions above of the Kantian and Carnapian positions. Kant did not suppose that there is a common vantage point from which the truths that would be accessible to different forms of sensibility can be embraced together; likewise, Carnap did not suppose that there is a common vantage point from which the truths that would be expressible within different linguistic frameworks can be embraced together.[4] The idea of alternativeness is also strongly suggested by much of the descriptive terminology and metaphor associated with the doctrine of relativism—as when it is claimed, for example, that different languages and theories are *incommensurable* or that there are *others* who have their truths which are not for us. I shall take it as my point of departure that, whatever else relativists may be committed to, they are certainly committed to alternativeness.

The primary reason it is pointless to try to elucidate the doctrine of relativism in terms of the idea that truth is relative is that it doesn't help us to make sense of the idea of alternativeness. I can best explain why by returning to the case of indexicals. I've already observed that the relative character of indexical truth is not a basis on which to affirm relativism. I can now elaborate the point in terms of alternativeness: the relative character of indexical truth does not stand in the way of embracing true indexical claims together. On the contrary, it helps to ensure that true indexical claims can be embraced together, even when surface grammatical appearances would suggest they are inconsistent and therefore cannot be embraced together. So for example, if I say "I am a woman" and my husband says "I am not a woman," there is a surface appearance of inconsistency. But because the truth of indexical claims is relative to context, we can remove the appearance of inconsistency by employing the formal maneuver of semantic ascent and explicit relativization of truth to context. Then we can make clear that the sentence "I am a woman" is true relative to a context in which the speaker is a woman, while the sentence "I am not a woman" is true relative to a context in which the speaker is not a woman. Since I am a woman and my husband is not, our claims are consistent after all. I see no reason why this same strategy shouldn't work for all cases of relative truth. For, presumably, there must always be something that relative truth is relative *to*; and this would seem to ensure that the formal maneuver of explicit relativization of truth to context can always be applied, with the result that all such relative truths are consistent and co-tenable and, hence, not alternatives.[5]

If the idea that truth is relative doesn't help us to make sense of the idea of alternativeness, one might wonder whether anything can. For there is a serious difficulty that seems to stand in the way of making sense of alternativeness at all. This difficulty takes the form of a *dilemma*: any two truth-value bearers must be either inconsistent or consistent; if they are inconsistent, then they cannot be embraced together, but they cannot both be true; if they are consistent, then they can both be true, but they can also be embraced together; in neither case do we have alternatives in the sense required by relativism, of truths that cannot be embraced together. Clearly, anyone who aims to elucidate the doctrine of relativism in terms of the idea of alternativeness must find a way out of this dilemma.

In many discussions of relativism, the doctrine is implicitly construed as rejecting the first horn of the dilemma. On this construal, relativists hold that there are, or can be, *irresoluble conflicts*. Here I don't mean that the conflicts are irresoluble *by us*, because we are too ignorant or stupid to figure out which party is right. What I mean is that they are irresoluble in principle because *both sides can be right*. Or, to go back to the terms of the dilemma, inconsistent claims can be equally true.

Although I'm going to argue against this response to the dilemma for alternativeness, there is much to be said in its favor. It attempts to explain what makes alternatives *alternatives*—that is, what prevents us from embracing them together—by exploiting our sense that it would be logically absurd to embrace inconsistent claims together. In addition, it accords well with the widespread assumption that realism and relativism are mutually opposed doctrines, which is something many would see as a virtue. For if we ask why it should ever be supposed that a conflict cannot be resolved in principle, the most natural answer is that there are no objective facts of the matter that could resolve it. Since this is tantamount to denying realism, relativism and realism do appear to be opposed doctrines when we advocate this response to the dilemma for alternativeness. In fact, it will be useful to state the response in terms of antirealism: In domains where antirealism holds, there is no more to the truth than standards of warranted assertibility that hold locally within a given antirealism; because this is so, it cannot be ruled out that a given truth-value bearer might be true by one set of standards while its negation is true by another. The inconsistency could never be resolved unless there were further facts beyond these local standards—which is to say, mind-independent facts—in the light of which at least one of the truth-value bearers was not really true. But such mind-independent facts are not available in domains where antirealism holds; the inconsistent truths that arise within different truth practices are, therefore, viable candidates for the status of alternatives.[6]

Despite its virtues, this response to the dilemma incurs a high cost: it requires the relativist to allow exceptions to the law of noncontradiction. Moreover, it deprives the relativist of our ordinary understanding of why the law holds. In this ordinary understanding, it is our interest in *truth* that speaks against affirming contradictions; that is, it is because we want to affirm only truths and we believe that both halves of a contradiction cannot be true. It is not obvious what other reason, besides our conviction that inconsistent claims cannot be equally true, should stand in the way of embracing inconsistent claims together.

I suppose that anyone who was prepared to allow that the framework of antirealism can throw up counterexamples to the law of noncontradiction might expect that the framework should also provide us with a nonstandard reason to refrain from embracing inconsistent claims together. So I want next to explore whether it does. This exercise will be instructive not because it turns up the needed reason but because it points to a different, less problematic response to the dilemma for alternativeness.

The following reflection might seem to supply the needed reason: in domains where antirealism holds, there is no more to truth than meeting standards of warranted assertibility that hold locally within a given truth practice; it follows that truth itself must be local in those domains. Since inference aims to preserve truth, inferential practice should be local in any domain where truth is local. This reflection makes clear why, in domains where antirealism holds, we ought not to see claims that arise within distinct truth practices as falling under the law of noncontradiction. But it doesn't directly address the point at issue. We wanted to know what reason we can give to refrain from embracing inconsistent claims together when they are equally true. It is one thing to say that such claims are allowable exceptions to the law of noncontradiction; it is quite another to say what *else*, besides that law, might stand in the way of embracing them together. The reason on offer is the locality of truth and inference. But it seems to me that this reason speaks against supposing that such claims stand in any logical relations at all, including the relation of inconsistency itself. For, as far as I can see, there is no more to the reality of a logical relation than its logical significance, by which I mean its normative force as that force is registered in laws of logic. If I'm right, then we ought not to suppose that the relation of inconsistency can hold without its usual normative force, as that force is reflected in the law of noncontradiction. Rather, we ought to suppose that these two things go together: the law of noncontradiction applies everywhere, and only where, inconsistency arises. I conclude that the framework of antirealism doesn't give us a reason to reject the first horn of the dilemma for alternativeness by throwing up inconsistent claims that are equally true. At best, antirealism gives us a reason to doubt whether claims that arise within distinct truth practices can be inconsistent to begin with.

So far, my effort to explore how the framework of antirealism might allow us to reject the first horn of the dilemma for alternativeness has only served to reinforce it—and, for that matter, the other horn as well. All the dilemma really does is exploit the view that I've just been defending, according to which logical relations carry normative force. (Specifically, it affirms that the relation of inconsistency carries the normative force that is registered in the law of noncontradiction and that the relation of consistency carries the normative force that is registered in the law of conjunction.) Yet this doesn't mean that alternativeness can't arise in domains where antirealism holds. To see how it might, all we need to do is complete the reflection I started above, which looked to the locality of truth for a reason not to embrace allegedly inconsistent claims together when they are both true. In the course of that reflection I argued that if truth is local in domains where antirealism holds, then inferential practice should be local, and if inferential practice is local then the logical relations that inferences track should be local too. It follows that claims that arise within distinct truth practices do not stand in any logical relations at all but are *normatively insulated* from one another. Such normative insularity constitutes a *third possibility* that the dilemma for alternativeness overlooks entirely. When claims are normatively insulated from one another, they are neither inconsistent nor consistent. Because they are not inconsistent, the law of noncontradiction doesn't prevent them from being equally true, and because they are not consistent, the law of conjunction doesn't apply. The upshot is precisely what alternativeness requires: truths that cannot be embraced together.

If we construe alternativeness along these lines, as arising when there is normative insularity, then the real dividing issue between relativists and their opponents turns on a normative question: Do logical relations necessarily run everywhere, among all truth-value bearers? Relativists answer no, and their opponents answer yes. We shall soon see that this dividing issue has significant metaphysical and practical implications.

I'll start with the metaphysical implications. If logical relations run everywhere, among all truth-value bearers, then all of the true truth-value bearers should be consistent and conjoinable. This yields a positive metaphysical conception that is very close to the one that Wittgenstein announced at the start of the *Tractatus* when he said that the world is all that is the case—or, as I prefer to say, there is a single, consistent, and complete body of truths. Because these two conceptions are driven mainly by formal considerations, they are neutral with respect to a great many substantive metaphysical issues. Nevertheless, they do take a stand on at least one metaphysical issue: both affirm the *oneness* of reality. That is part of what it means to affirm that the world is *all* that is the case, or that there is a *single and complete* body of truths. I shall call this one-world thesis *unimundialism*. It is thus the oneness of reality that relativists most fundamentally oppose when they affirm alternativeness. For if some truth-value bearers are normatively insulated from one another, then they may be true without being consistent and conjoinable. It follows that there isn't one single and complete body of truths but, rather, many incomplete bodies of truths. This many-worlds thesis—*multimundialism*—is the true metaphysical significance of relativism when it is elucidated along the lines I am proposing.

I said that the normative issue that divides relativists from their opponents has practical as well as metaphysical implications. What I meant is that the unimundial and multimundial views impose different *practical stances* toward inquiry and interpersonal relations.

When we adopt *the unimundial stance*, we can never be entirely indifferent to any truth-value bearer that we might come across. For we recognize that each and every truth-value bearer that we might come across is guaranteed to stand in logical relations to what we already hold to be true, and these logical relations carry a normative force that we are never free to ignore. In particular, there is always a question whether a given truth-value bearer is consistent or inconsistent with what we already hold to be true. If it is consistent, then we may embrace it as true. But if it is inconsistent, then either we must reject it as false, accommodate its truth by revising our prior beliefs so as to make them consistent with it, or suspend belief on it along with the prior beliefs that conflict with it. These same normative constraints carry over to our interpersonal relations. If others' beliefs are consistent with what we already hold to be true, then we may embrace them, but if they are inconsistent then at most one of us can be right. What I want to underscore is this: anyone who believes that these normative constraints hold without exception implicitly embraces the unimundial conception of reality as consisting in a single, consistent, and complete body of truths. This normative commitment of the unimundial stance should not be confused with the goal of actually learning all of the truths. We may reject this goal on many grounds. We might think it does not lie within our power to achieve it.[7] Or even if it does lie within our power to realize it, we might reject it because we think there are other things more worth doing. However, the normative significance of unimundialism would remain in place, for to reject the goal of learning everything would not release us from the normative constraints on inquiry and interpersonal relations that follow upon unimundialism. The only way to avoid these normative constraints is by rejecting the unimundialists' one-world thesis and embracing multimundialism instead.

There are three related ways to describe what the *multimundial stance* involves, and all of them have to do with recognizing normative insularity. First, we must think of our inquiries as circumscribed by *boundaries* in such a way that logical relations hold only within them but not across them. Second, when we come across such boundaries, we must be *epistemically indifferent* to what lies outside them. Such epistemic indifference is very unlike the more usual forms of indifference that we exhibit toward certain truths. Usually, indifference simply reflects a lack of interest, and, usually, there is no bar to *becoming* interested and commencing appropriate inquiries. But the case is different with truths that are normatively insulated from our own. When a truth is normatively insulated from what we hold to be true, then it doesn't stand in any logical relations to our beliefs. Consequently, it can't be conjoined with them, but nor can it be ruled out by them. It lies permanently outside the scope of our inquiries—which is why we are consigned to epistemic indifference. The third way of describing what is involved in adopting the multimundial stance concerns the interpersonal counterpart to such epistemic indifference. If we were to come across subjects whose beliefs are normatively insulated from ours, they would be subjects from whom we cannot learn and with whom we cannot disagree. We would have to regard them as having *their* truths, which are not for *us*, and this would result in a profound sense of *otherness*.

Now that I have elucidated the doctrine of relativism—in its normative, metaphysical, and practical aspects—I am ready to turn in the next section to the task of arguing against the widespread assumption that realism and relativism are mutually opposed doctrines. But let me close this section with a few preliminary observations. It should be obvious that the unimundial view that relativists oppose does not carry any commitment to realism, since idealists like Berkeley can perfectly well endorse the one-world thesis. It may be less obvious that realism doesn't carry a commitment to unimundialism either. In fact, this latter claim is likely to seem implausible. For when we conceive reality as consisting in mind-independent facts, there seems to be a perfectly straightforward way to conceive it as *one*. We can simply think of all the facts as belonging to a *single set*. The trouble is that this way of conceiving the oneness of reality doesn't explicitly rule out the relativists' multimundial stance, which follows upon the recognition of normative insularity. For the idea of a single set that contains all of the mind-independent facts is silent on the question of whether logical relations must hold among all truth-value bearers, and so it doesn't directly entail the normative constraints that go together with the unimundial ideal. This strongly suggests that the issue that divides realists from their opponents, concerning whether reality is mind-independent, is orthogonal to the issue that divides relativists from their opponents, concerning whether reality is one or many.

However, these preliminary observations do not suffice to show that the widespread assumption is false. Let us grant that realism doesn't directly deliver a unimundial conclusion. Still we must ask, what would ever deliver a multimundial conclusion instead? I have to admit that the answer encouraged by my own dialectic so far is that only antirealism could. After all, I arrived at the suggestion to construe alternativeness in terms of normative insularity by exploiting an antirealist conception of reality, in which there is no more to the facts than standards of warranted assertibility that hold locally within a given truth-practice, and in which there is more than one such set of standards. I've made no other suggestion about what else, besides the framework of antirealism, ever would or could give rise to normative insularity. If nothing else could, then the widespread assumption would stand as true. Realism and relativism would indeed be mutually opposed doctrines. Even though realism wouldn't suffice to rule out relativism, it would still be the case that antirealism must serve as a crucial first premise in any argument for relativism, just as we find in the standard form of arguments for relativism that I identified at the start.

But let us not confuse the order of discovery with the order of justification. I did *arrive* at the suggestion to construe alternativeness in terms of normative insularity while exploring how the framework of antirealism might yield relativism. But it is significant that I managed to *formulate* the dividing issue between relativists and their opponents without mentioning the issue that divides realists from antirealists. As we cast about for reasons to take a stand on the former issue, I recommend that we not make any assumptions about how the latter issue might be brought to bear. I recommend that we let the connections between these two issues—the oneness of reality versus its mind-independence—come to light through unbiased investigation.

THE REAL RELATION BETWEEN REALISM AND RELATIVISM

Before proceeding any further, I need to broach a delicate matter concerning who should bear the burden of proof in the debate about relativism. It is more or less standard practice in philosophy to give our commonsense views default status. Since the unimundial view accords much better with common sense than the relativists' multimundial view, it might seem that we ought to accord it default status. However, before doing so, we ought to consider the dialectical consequences. Unimundialism and multimundialism are mutually opposed positions. To accord either of them default status would automatically beg the question against the other. The issue is whether we have a right to beg the question against multimundialism simply on the ground that the opposed unimundial view is favored by common sense. After all, insofar as multimundialism is an internally coherent position, we could in principle beg the question in its favor instead. I realize that this last claim runs counter to the expectations of many philosophers who take it for granted that relativism is not a coherent doctrine. But it seems to me that these expectations are not well placed and are in large part attributable to the insufficiency of philosophical attention to the task of trying to provide an adequate formulation of the doctrine of relativism to begin with. Once it is formulated as multimundialism we cannot dismiss it as incoherent. But nor should we dismiss it simply on the ground that it violates common sense. For it is not necessarily a mark of falsehood that a view violates common sense. Furthermore, it is not as though common sense provides us with a well thought out defense of the unimundial view. It merely takes it for granted that logical relations run everywhere, among all truth-value bearers, and thereby directly begs the question against the opposed multimundial view. It seems to me that the more judicious approach would be to try to do more than, or perhaps I should say do better than, simply begging the question one way or the other. In other words, we should look for supporting reasons that go beyond *merely* begging the question—precisely because it seems feasible to beg the question *either way* if we are determined to do so. Thus my suggestion is that we should start from a position of agnosticism, and regard both views as requiring some positive support that goes beyond merely begging the question against the other view.

It is a strength of Davidson's conceptual-schemes argument that it doesn't merely beg the question against relativism. On the contrary, he tried to make positive sense of the doctrine and expose reasons why we cannot do so. I don't have enough space in this short essay to give a full exposition of his argument or try to defend it in all its details. I can only explain in a very general way how it speaks against the doctrine of relativism as I have elucidated it.[8]

Davidson's argument exploits the same holistic considerations that Quine brought to bear against the analytic-synthetic distinction in "Two Dogmas of Empiricism."[9] There are really two mutually supporting forms of holism at work in the argument. One is a holism of meaning and belief, according to which we cannot clearly distinguish what belongs to the meanings of words from what belongs to the beliefs that guide our uses of them. The implication of meaning-belief holism is that we cannot come to understand another's language without coming to know what they believe about the world. The second holism is a holism of concepts and beliefs, according to which a concept is a position in a cluster of beliefs in which it figures and, correlatively, each individual belief is a position in an overall system of beliefs. It follows from these two forms of holism together that we cannot understand another's language without coming to see them as believing, by and large, what we believe about the world.[10] This is not to say that we must agree on every single point; it is rather to say that intelligible disagreement can take place only against a background of overall agreement on most matters. If this background of overall agreement must be in place when we take ourselves to understand others, then we can never encounter an alternative in the sense I've been discussing in this essay. Others' beliefs will mostly be the same as our own and, when they are not, they will still stand in logical relations to our own. In short, thoroughgoing holism precludes normative insularity.[11]

The natural first thought about how to resist Davidson's conclusion exploits the fact that intelligibility does not require *perfect* agreement but only *by and large* agreement. This seems to leave room for the logical possibility of a spectrum of subjects in which all close neighbors by and large agree but in which more distant subjects agree somewhat less, with the implication that subjects occupying the extreme ends of the spectrum do not hold any beliefs or meanings in common at all. Now the idea of such a failure of intelligibility is not exactly the same as the idea of normative insularity. For it might be possible that if two subjects did fail to find each other intelligible, their beliefs still could be embraced together if true. We can go some distance in imagining this by imagining a third party who could understand them both, and this is a point to which I will return. But at this juncture, it will do to register that, given the Davidsonian view of meaning and belief, such a failure of intelligibility is at least a necessary, if not a sufficient, condition for alternativeness. So if we want to preserve the possibility of alternativeness in the face of his conceptual-schemes argument, we must establish the possibility of such a failure. The most promising way to do this is by trying to imagine a spectrum of intelligibility along the lines I just gave.

Davidson did not think that his view left any room for this possibility. This is something he made clear when he took up the case of "partial translatability"—which is a linguistic description of the conditions that would afford a spectrum of intelligibility. He noted, correctly, that those who stand on one end of such an alleged spectrum would have no reason to think that those who stand on the other end are subjects at all, and he seemed to think that that suffices to show that there can't be such subjects.

I am sure that many philosophers are completely unconvinced by this quick and dismissive response on Davidson's part. I am also sure that the reason why they are not convinced is an underlying commitment on their part to a realist conception of the facts as mind-independent. When we view the facts as mind-independent, we must admit that they might not be knowable by us. If we apply this realist attitude to the case of other minds, then we cannot assume with Davidson that all other minds are necessarily intelligible to us. We must allow that there might be other kinds of knowers whose beliefs and meanings are unintelligible to us, just as the spectrum of intelligibility suggests.

In a minimal way, I have already shown what I promised to show. The widespread assumption is false. Realism and relativism are not mutually opposed doctrines. They cannot be, since one way to save the possibility of alternativeness in the face of Davidson's conceptual-schemes argument is by embracing realism.

But this is not the end of the story. There is much more yet to bring out about how realism might invite relativism. First, I need to explain why Davidson thought that his attack on the scheme-content distinction speaks just as much against realism as against relativism. Then I want to explain how some realist intuitions may be saved in the face of that attack. Only then will we be in a position from which to appreciate why there is a deep—if counterintuitive—connection between multimundialism and scientific realism.

When Davidson discussed realism, he mainly had in mind the sort of realist view that brings skepticism in train because it portrays the facts as radically mind-independent. He saw such skeptical realism as sharing in the same philosophical mistake as relativism, which is to embrace a version of the scheme-content distinction. This is very much part of what he was getting at in the much quoted concluding sentence of his presidential address: "By giving up the duality of scheme and contact we re-establish unmediated touch with the objects whose antics make our beliefs and sentences true." For reasons that I'll bring out below, it would be going too far to insist that every variety of realism is wedded to this duality of scheme and content that deprives us of direct epistemic contact with what we aim to know. But it does seem that the varieties of realism that entail skepticism are so wedded. For it is hard to see how our knowledge can be called into skeptical doubt if it consists in unmediated epistemic contact. Since Davidson saw his argument against relativism as an argument against the scheme-content distinction, it stands to reason that he thought it should also be effective against these varieties of realism.

In the antiskeptical version of his argument, Davidson asked us to imagine an omniscient interpreter who by hypothesis knows all of the truths.[12] He took for granted that such an interpreter would have to use the same methods to know our minds that we use in order to know other minds. From this he took it to follow that we by and large agree with the omniscient interpreter or, equivalently, that our beliefs are by and large true.

As any one can see, realists have resources with which to resist the omniscient interpreter argument. All they need to do is thump the table with their basic intuition that reality is radically mind-independent. The independence in question is independence from *our* minds. This makes room for the possibility that an omniscient interpreter who knew reality would have a mind quite unlike ours. If that were so, then the basis on which she knew our minds might be quite different from the basis on which we take ourselves to know other minds. Perhaps, unlike us, she could have grounds on which to attribute meanings and beliefs that are by and large mistaken by her lights, or in other words, false—thus undermining Davidson's antiskeptical conclusion.

To his credit, Davidson did not rest his entire case against skepticism on the omniscient interpreter argument. He went on to give a much more interesting and ambitious—I'm inclined to say, transcendental—argument against skepticism that rests on considerations having to do with what he called "triangulation."[13] Here again, space does not permit me to discuss his argument in any detail. In this case, I cannot even state the argument. All I can do is state its central claim, which is that knowledge of our own minds is possible only via our knowledge of *other* minds and, more specifically, via knowledge of how others in turn know *our* minds. If this claim is correct, then it is impossible to cleave to skepticism in the face of the omniscient interpreter argument, unless the skeptic is prepared to give up *self*-knowledge as well as knowledge of anything else that she proposes to doubt. This is enough to put the skeptic in a compromising position, since without self-knowledge she cannot claim to know what she is up to in the course of mounting her skeptical arguments.

A large part of the interest of the triangulation argument lies in how it completes and augments Davidson's extended attack on the scheme-content distinction by giving us a better handle on what unmediated epistemic contact is supposed to consist in, as well as some reasons why we should think we have it. After all, this is left somewhat obscure in both the original conceptual schemes argument against relativism and in the omniscient interpreter argument against skepticism. Both of those arguments rely heavily on the so-called *principle of charity*, which emphasizes that in our attempts to know other minds we have no other basis from which to proceed than our own best lights, and this leaves unexplained how or why our own best lights should amount to direct epistemic contact with anything. In contrast, the triangulation argument emphasizes that others don't count as having thoughts at all unless they systematically respond to various aspects of their environment, where this responsiveness constitutes a form of direct epistemic contact. Then the argument simply applies the same picture to ourselves. If we must triangulate through others in order to know our own minds, then we must view ourselves in the same way that they view us, with the result that we ourselves don't count as having thoughts unless we systematically respond to various aspects of our environment, where again this constitutes a form of direct epistemic contact. The philosophical mistake that Davidson saw in the scheme-content distinction is that it deprives us of this picture of the mind-world relation, which, as I've been saying, he took to rule out both relativism and skepticism together.

Even though Davidson officially opposed realism, it is striking that many scientific realists accept something very close to his picture of the mind-world relation. Like him, they see thought as involving systematic responsiveness to an environment and, in that sense, direct epistemic contact with an environment. This is really the heart of any naturalistic, "causal" account of content. When we see content in these naturalistic terms we get much the same result that Davidson gets through Quine's principle of charity: it is in the nature of belief to be true, at least by and large. The main difference between scientific realists and Davidson is that they do not share his transcendental ambition to refute skepticism. They have no stake in showing that various skeptical scenarios are incoherent or impossible but are content to set aside those scenarios as improbable in the light of our current theories. Yet they are closer to skeptics than to Davidson in one respect: they see the mind-independence of reality as requiring us to acknowledge certain epistemic limitations. Only it is not *error* that they primarily worry about; it is, rather, *ignorance*. While they are prepared to allow that there is much that we *do* know, they also want to insist that there are many things we do *not* know and, moreover, *cannot* know because we are unfit to know them. Thus, scientific realists would regard it as a form of hubris to presume that everything should be epistemically accessible to human intelligence. They would also regard it as a form of hubris to presume that human intelligence is the only form that intelligence can take. By this route, the epistemic modesty of scientific realism returns us to the same possibility that was raised by the spectrum of intelligibility: for all we know, there might be other kinds of knowers who are so unlike us that we cannot understand them.

However, when scientific realists raise this possibility, they do so in a spirit that is strikingly different from the spirit in which I raised it earlier. Then I had raised it in the same spirit that skeptical possibilities are typically raised against our claims to knowledge, where the rule is to resist a conclusion on any ground, no matter how far-fetched, so long as that ground identifies something which is logically possible. But scientific realists are not interested in challenging our claims to knowledge. They want to exploit our knowledge in order to probe *real* possibilities, by which I mean possibilities that are left open by our best theories. It is in this spirit that they may affirm the possibility of other kinds of knowers whose beliefs are not intelligible to us, even in the face of Davidson's arguments. If anyone doubts whether the possibility of knowers unlike us is indeed supported by our current best theories, they need only to consider Chomsky's conception of cognitive abilities. In his view, abilities essentially limit as they enable—for by enabling us to do one thing they prevent us from doing something else. This means that a cognitive problem that is soluble by one kind of knower might be insoluble by another kind of knower with different cognitive abilities. In such cases, these knowers could not possibly solve the problem of knowing each other—that is, they would not find each other intelligible. For, given the models of knowledge of other minds that I've been working with in this essay—Davidson's interpretive model and the more naturalistic variant—it is impossible to know another's mind without knowing what it knows.

As I pointed out earlier, to establish the possibility of a failure of intelligibility falls short of establishing the possibility of alternatives in the sense required by relativism. In addition, we would have to establish that there is no further point of view from which the beliefs of different kinds of knowers could be apprehended together, and embraced together if true. This is a heavy burden of proof for relativists to bear. But on the other hand, I have insisted that unimundialists also bear a substantial burden of proof. For we ought not to presume, without supporting argument, that all of the truths that could in principle be known by all of the different kinds of knowers can in principle be conjoined. If Davidson had been right, then unimundialists would have been spared this particular burden of proof. For when he denied that there can be failures of intelligibility, he was claiming that there is only one kind of knower. In contrast, the realist conception of the facts as mind-independent forces us to concede the possibility of knowers unlike us. It also imposes on us a certain epistemological aspiration that is characteristic of scientific inquiry. This is the aspiration to a *view from nowhere* that transcends the limits of our distinctively human vantage point, so as to reach a more "objective" view of reality—the view of things as they are in themselves that Kant thought impossible.[14] But if scientific realists want to rule out relativism they need to establish something else—not a view from nowhere but a *view from everywhere*. This would be the view of a super-knower who could know, and embrace together, all of the truths that could be known by every other kind of knower, regardless of whether such knowers could know one another. Unless we can establish the possibility of such a view from everywhere, we cannot rule out the possibility of alternativeness—the possibility that some truths cannot be embraced together.[15]

It has been my argument in this essay that the realist conception of facts as mind-independent makes the task of establishing the unimundial view that relativists reject harder, not easier. Not only does it impose a substantial burden of proof on unimundialists, of showing that there could be a view from everywhere, but in addition, our current best theories about the nature of cognitive abilities cast doubt on whether there could be such a view from everywhere. For if cognitive abilities carry limitations, it hard to see how they could be conjoined together in a single knower. And if cognitive abilities themselves cannot be conjoined in a single knower, it is hard to see how the different kinds of knowledge they afford could be embraced together from a single point of view. Nevertheless, this is what realists must show if they want to complete Davidson's conceptual schemes argument, so as to rule out the very idea of an alternative.

NOTES

1. Of course, Kant posited forms of understanding as well as forms of sensibility. But although he allowed that we can conceive forms of sensibility different from our own, he did not allow that we can conceive forms of understanding different from our own. If we agree with him on this point, then we shall have to see the threat of relativism that emerges from his transcendental idealism in the way I described it in the text, as tied to the variability of conditions of sensibility. However, we shall see further on that realists have good reason to depart from Kant's vision of understanding and allow that conditions of understanding might also conceivably vary.

2. See Donald Davidson's presidential address to the American Philosophical Association, "On the Very Idea of a Conceptual Scheme," reprinted in Donald Davidson, *Essays on Truth and Interpretation* (Oxford: Oxford University Press, 1984), 183–98.

3. For an illuminating account of the themes of relativism and realism, see Akeel Bilgrami, "Realism and Relativism," *Philosophical Issues* 12 (2002): 1–25. Bilgrami argues that once we understand Davidson's arguments against relativism, we can appreciate that realism need not be at odds with relativism. But he fails to acknowledge the extent to which Davidson himself remained in the grip of the widespread assumption. Furthermore, his own discussion—like Davidson's and most other discussions—is marred by the usual failure to provide a proper elucidation of the doctrine of relativism to begin with. These shortcomings of Bilgrami's article are

perhaps understandable given that its real focus and interest lie in his positive proposal of a pragmatist conception of truth that is neither realist nor relativist. But they nevertheless reflect a certain neglectful tendency among philosophers in their discussions of relativism.

4. This is not to say that we could not *understand* two different linguistic frameworks; it is rather to say that there is no one point of view from which they could be evaluated together for truth because they would be governed by different logical and semantic rules. The root of the problem for Carnap is that there are no objective grounds on which to adjudicate questions about which logical and semantic rules to adopt.

5. It may seem that this proposal to portray relativism as carrying a commitment to the idea that there are alternatives should be compatible with the proposal I just rejected, which portrays the doctrine as carrying a commitment to the idea that truth is relative. For I've just suggested that if some truths are alternatives, in the sense that they cannot be embraced together, this would most likely be caused by the way in which those truths are tied to contexts; prima facie, it would seem reasonable to suppose that whenever truths are tied to contexts, then they can appropriately be thought of as *relative* to the contexts to which they are tied. In this way of thinking, the main difference between the indexical case and the cases of relative truth that would pose a threat of relativism would be the choice of contextual parameter to which truth is relativized. While indexical truths are relative to contexts of utterance, other truths might be relative to different evaluative or epistemic contexts—as when truth is relativized to different standards of taste, conduct, or warrant. There is a growing literature in which positions are labeled "relativist" just in case they portray truth as relative to such other parameters. Two prominent examples are Max Kobel, *Truth Without Objectivity* (New York: Routledge, 2002); and John MacFarlane, "Making Sense of Relative Truth," *Proceedings of the Aristotelian Society* 105 (2005): 321–39. However, as interesting as these accounts of relative truth are, they do not automatically provide for alternativeness in the sense that I think is required for relativism. For in general, the very act of specifying how a given truth is relative to a given context makes that truth accessible from outside that context, and this opens a prospect for embracing truths together even when they are relative to different contexts. This shows that we cannot expect to capture the idea of alternativeness in terms of the idea that truth is relative, at least not without adding some extra conditions and qualifications. In particular, we would need to show that, in some cases, when different truths are tied to different contexts, there is no further context from which someone could represent those different truths as relative to the different contexts to which they are tied, and thereby apprehend and embrace them together. But if something like this could be shown, it seems to me that the deeper point wouldn't be that the truths in question are *relative* to the different contexts to which they are tied; the deeper point would be that *there is no single context from which they could all be represented as relative to the different contexts to which they are tied.* In other words,

the deeper point would be that certain contexts are neither mutually accessible nor accessible together from some third context. Although this deeper point may not be entirely ruled out by the idea that truth is relative, it is certainly not automatically provided for by that idea. In fact, it seems to me that the idea would tend to work *against* the deeper point rather than *in favor* of it. For when we portray truths as relative to contexts, we do so from a position outside those contexts, and as I've said, this opens a prospect for apprehending and embracing truths together even when they are relative to different contexts. For further discussion of this issue, see section 1 of my "How to Formulate Relativism," in *Mind, Meaning, and Knowledge: Themes from the Philosophy of Crispin Wright*, ed. A. Coliva (Oxford: Oxford University Press, forthcoming).

For good general discussions of why recent accounts of relative truth are unlikely to help us either to formulate or defend the doctrine of relativism, see Crispin Wright, "Relativism About Truth Itself: Haphazard Thoughts About the Very Idea," in *Relative Truth*, ed. M. Garcia-Carpintero and M. Kolbel (Oxford: Oxford University Press, 2008), 157–58; and Paul Boghossian, "What Is Relativism?" in *Realism and Truth*, ed. P. Greenough and M. Lynch (Oxford: Oxford University Press, 2006), 13–37. I should register, however, that these discussions do not proceed from my own starting point—that is, they do not take it for granted that relativists are committed to the existence of alternatives as I have defined them, simply as truths that cannot be embraced together.

6. I am drawing heavily on Crispin Wright's recent account of what he calls "True Relativism." However, the view that I discuss in the text is not the view he actually defends. For although he does see relativism as tied to the possibility of irresoluble inconsistency, he does not think that relativists can flatly assert that there is such irresoluble inconsistency without getting into logical difficulties. This leads him to develop an extremely subtle and ingenious account of relativism that makes a surprising connection with the problem of vagueness. It would go beyond my purposes in this paper to discuss this novel proposal. I merely want to focus on a very common picture of relativism that is evident in other, much less sophisticated philosophical discussions. The picture arises mainly in the domain of value, where the charge of relativism is often levied against an ethical or political position precisely because the position does not guarantee that all conflicts can be resolved. Although the label "antirealist" is not always used in these ethical and political discussions, the reason it is supposed that such conflicts may be irresoluble is, nevertheless, implicitly antirealist. The thought is that there are no objective, or mind-independent, facts of the matter available to resolve the ethical and political conflicts in question. To see how Wright has proposed to cope with the logical difficulties that attend this intuitive conception of relativism, see his "On Being in a Quandary: Relativism, Vagueness, Logical Revisionism," *Mind* 110 (2001): 45–90. For an extended critical discussion of this intuitive conception see my "How to Formulate Relativism."

7. It might be thought that the goal of learning everything is unachievable by us because of formal difficulties that would attend any totalizing conception of all of the truths. I have nothing very informed or interesting to say about that. But it seems to me that if there are such formal difficulties, we can safely set them aside when we focus on the practical significance of unimundialism. Then we can embrace unimundialism as a *regulative ideal* that governs our inquiries and interpersonal relations, by imposing the normative constraints I just described in the text, safe in the knowledge that we shall never be in a position to contemplate the totality of all of the truths.

8. See Davidson, "On the Very Idea of a Conceptual Scheme." His argument does not directly address the issue of unimundialism versus multimundialism. Nor does it directly attack the idea of alternativeness, construed in terms of normative insularity. The official target throughout is the scheme-content distinction. All the same, he implicitly held that relativism involves a commitment to alternativeness. For he explicitly formulated the doctrine in terms of the idea that there may be distinct conceptual schemes that are true but not translatable, and this idea could not generate a *plurality* of conceptual schemes unless such nontranslatable truths were *not conjoinable*—which is to say, alternatives. Furthermore, it seems to me that he was not far from construing alternativeness in terms of normative insularity. For he did not assume that alternative conceptual schemes must conflict with one another, and if they don't conflict, I see no other reason why they shouldn't be conjoinable than that they fail to stand in logical relations at all. Yet it cannot be said that Davidson actually recognized the possibility of such normative insularity.

9. W. V. O. Quine, *From a Logical Point of View* (Cambridge, Mass.: Harvard University Press, 1953).

10. As I've said, I don't have the space to try to defend the considerations that Davidson brought to bear against relativism. This means, in particular, that I don't have the space to defend his holistic picture of meaning and belief, or its implication that we all by and large agree. Yet I cannot resist trying to address, however briefly, a few of the most obvious difficulties with it.

Some philosophers object that there is no argument for holism. This is probably true. So let me register that my own conviction about it derives from contemplation of examples that illustrate it and my sense that I can proliferate such examples indefinitely.

Some philosophers complain that holism has absurd implications, such as that learning a language or acquiring concepts would be impossible because we could not grasp any part of a language or system of beliefs without already possessing the whole in which it figures, and that two people cannot share any single belief without agreeing on everything. To these complaints I can only answer that concept possession cannot be all or nothing on a holistic view but must be a matter of degree—something I go on to observe in the text.

11. When holism isn't thoroughgoing, it may not preclude normative insularity, and this is one way in which arguments for relativism have gotten off the ground in philosophy of science. If we think of the meaning of a scientific term as deriving from its place in a given scientific theory, and if we think of scientific theories as self-contained wholes, then the same terms can never figure in different theories. (From a syntactic point of view it might appear that the same terms figure in different theories, but they can't really be the same terms if they don't carry the same meanings.) This makes it hard to see how there could be any points of logical contact between different theories. If such a failure of logical contact between theories were to preclude our conjoining them, then it would amount to normative insularity. In any event, the sort of logical isolation that would arise if scientific theories were self-contained holistic structures is one of the targets of Davidson's conceptual-schemes argument. He saw holism as being so thoroughgoing as to ensure logical contact between all beliefs.

12. See Donald Davidson, "The Method of Truth in Metaphysics," in *Essays on Truth and Interpretation* (Oxford: Oxford University Press, 1984), 199–214.

13. See Donald Davidson, "Three Varieties of Knowledge," in *Subjective, Intersubjective, Objective* (Oxford: Oxford University Press, 2001), 205–37.

14. The phrase is, of course, Thomas Nagel's, from *The View from Nowhere* (Oxford: Oxford University Press, 1987).

15. This complements my point about relative truth in note 5. What I said there is that the mere claim that truth is relative doesn't suffice to establish the possibility of alternatives in the sense that I'm claiming is required by relativism. The reason is that the mere idea that truth is relative doesn't suffice to rule out the intelligibility of a view from everywhere.

PART III

PHILOSOPHY AND THE HUMAN SCIENCES

6

TAKING THE HUMAN SCIENCES SERIOUSLY

David Macarthur

> *It is an unjustified leap to say that [good and right] . . . are*
> *not as real, objective, and non-relative as any other part of the*
> *natural world. The temptation to make this leap comes partly*
> *from the great hold of natural science models on our entire*
> *enterprise of self-understanding in the sciences of human life.*
> —Charles Taylor, *Sources of the Self*

A major problem in contemporary analytic metaphysics, one that has a precedent at the end of the nineteenth century in the context of continental philosophy, is the apparent irreconcilability of the natural and the normative. This metaphysical problem can be approached by asking how to place normative phenomena in a natural scientific world—call it *the placement problem*.[1] This problem strikes many naturalistic philosophers today as inevitable and unavoidable. The aim of this paper is to show that this impression is mistaken, even for one committed to Scientific Naturalism.

The placement problem is not a problem that just any scientific naturalist must address. It only arises for a certain questionably narrow version of that doctrine, that is, an orthodox understanding of Scientific Naturalism, which is defined solely in terms of the *natural* sciences. From an enlightened or Broad Scientific Naturalism that takes the human sciences seriously (in a sense to be explained below), the placement problem need not arise. I want to show that its sources are not based on the most plausible understanding of Scientific Naturalism available but only on an understanding of that doctrine that presupposes metaphysical commitments that a scientific naturalist of all people has good reason to abandon.

I argue that the central tenets of Scientific Naturalism can be more or less liberally interpreted and that there is considerable pressure within current scientific naturalist orthodoxy to move to a more liberal position that can, in principle, accept the legitimacy and autonomy of the explanations of the human sciences such as psychology, sociology, and economics.[2] Insofar as these explanations require positing or presupposing semantically and onto-logically irreducible normative truths, there need be no problem in the scientific acknowledgment of the legitimacy of normative phenomena like reasons, meanings, and values if properly construed.

I shall further argue that Scientific Naturalism itself is best understood as a *normative* doctrine where the normativity in question is irreducible rational normativity so that the scientific naturalist accommodation of irreducible normative phenomena—paradigmatically, reasons—is not just one problem among others but an unavoidable matter of self-consistency.[3]

I conclude by showing the there remains a residual problem about the relation between the natural and the normative. It is related to the widely recognized hermeneutic insight that our understanding of normative phenomena is not the same as, and is not exhausted by, the kinds of understanding provided by *any* of the sciences, natural or human, insofar as these are concerned with various kinds of nonnormative explanations, for example, nomological, causal, or statistical.

BASIC NATURALISM

Let us begin by attempting to articulate a minimal set of commitments shared by many prominent contemporary positions that fly the flag of "naturalism," taking this term in as broad a sense as possible. Under this heading, then, let us include the reductive versions of Scientific Naturalism of David Armstrong, Michael Devitt, Fred Dretske, and David Papineau,[4] as well as various nonscientific (or liberal) naturalisms such as Jennifer Hornsby's "naïve naturalism," John McDowell's "naturalism of second nature," Peter Strawson's "catholic naturalism," and Barry Stroud's "expansive naturalism."[5] We might think of what is common to these and many other contemporary versions of naturalism as defining a position that we might call *Basic Naturalism*. It can be characterized by the following three commitments:

1. *Anti-Supernaturalism*: A rejection of any commitment to the supernatural whether in the form of supernatural entities (e.g., God, Platonic Ideas) or supernatural faculties of mind (e.g., Cartesian transparency, mystical intuition).

2. *Human beings are part of nature and can be properly studied by the sciences.*
    There can be what Hume called a "science of man," a doctrine that leaves
    open the question whether scientific understanding can provide a complete
    understanding of the human.
3. *"The naturalist is one who has respect for the conclusions of natural science."*[6]

Obviously, a commitment to Basic Naturalism leaves plenty of room for disagreement about many things, such as what to include in the category of the
supernatural, what we are to understand by "the sciences" or a "scientific"
account of the human, and what constraints on philosophy flow from having
"respect" for the natural sciences. I will return to these issues below.

Given the connection between Basic Naturalism and science one might
think that it is a form of *Scientific* Naturalism, but I shall reserve the latter
title for a set of doctrines that give a more preeminent role to science within
philosophy. Let us say that Scientific Naturalism is a commitment to *both* of
the following claims:[7]

Ontological claim: The only things that there are in the world are those things
    that are presupposed or posited by the successful sciences.[8]
Methodological claim: The only genuine and irreducible form of knowledge or
    understanding is that resulting from the methods of inquiry of the success-
    ful sciences.[9]

At this point I am deliberately leaving it vague what to count under the
category of "the sciences." Success is a matter of predictive power and explanatory fruitfulness. These two naturalist doctrines, the ontological and the
methodological, constitute the standard background assumptions in analytic
philosophy, although the ontological claim tends to take precedence in contemporary analytic metaphysics. It is typically interpreted as a commitment
to only those things that are the objects of successful scientific inquiry—
although even Quine saw that physics could not avoid a commitment to the
existence of numbers on the grounds that explanations in physics presuppose
them.[10]

A wholly scientific ontology is a precondition for projects of "naturalization" that are typically conceived as semantic projects aiming to reduce or
explain the concepts in some supposedly problematic area of discourse (e.g.,
intentional, moral, mathematical, or modal discourse) in favor of scientifically kosher concepts. Standard options within this philosophical project
include reductionism, eliminativism, instrumentalism, and nonfactualism.
What drives the program of naturalization is the assumption that there is a

stark divergence between what Sellars called the scientific and the manifest images of the world. If the world is nothing but the world-as-posited-by-the-sciences then we confront the problem of how to "place" items that appear in the more expansive manifest image in the restrictive world that science has supposedly revealed to us.[11]

It should be clear that different types of Scientific Naturalism are possible depending on the range of inquiries accorded the status of being successful and irreducible sciences–something I shall speak of in terms of which sciences are taken *seriously*. Let us define three positions out of a spectrum of possibilities: *Extreme Scientific Naturalism* treats physics as the only science worth taking seriously; *Narrow Scientific Naturalism* takes this attitude to the natural sciences as a whole; and *Broad Scientific Naturalism*, beyond the natural sciences, takes at least some human sciences—those that are pulling their explanatory weight—seriously, too. As we shall see, there is good reason to think of this last version of the doctrine as the default position, the one that should be preferred in light of the failure of past reductionist dreams: of all the sciences to physics; and of the human sciences to the natural sciences.

The narrowness of the orthodox understanding of the scientific image is a consequence of assuming that the only successful and irreducible sciences are *natural* sciences, including physics, chemistry, biology, behavioral psychology, and the like. According to this understanding, the discoveries of the human sciences, such as sociology, are seen as illegitimate or too immature to be taken seriously or they are supposed to be reducible in principle to the discoveries of the natural sciences. Consequently, the human sciences play no autonomous role in the orthodox understanding of naturalism, that is, in what it is we take to be *really* real and what we take to *really* know or understand.

THE NATURAL AND THE NORMATIVE

Let us turn to consider the widely held sense that there is a tension between the naturalistic conception of nature and core normative phenomena such as reasons. Normative phenomena are those concerning evaluative claims and how things should or ought to be and they come in many different kinds. For example, there are rules of etiquette that tell us how we ought to conduct ourselves in company if we want to be seen in a certain light (say, as civilized or well brought up). Such rules are clearly optional and relative to a particular social milieu. Or there are rules in games such as tennis, which are hypothetical in the sense that if one wants to play tennis then one must abide by the rules. But, of course, one need not play tennis so, again, whether we are constrained by these rules is optional or hypothetical or conditional. Rules of both these sorts pose no obvious trouble for Scientific Naturalism since they can plausibly be treated in terms of individual or social psychology and social conventions. Whence, then, the long history of a sense of a fundamental tension between the natural and the normative, which dates back at least to Kant and was the central issue animating the *Methodenstreit* of the 1880s and 1890s in the German-speaking world?[12]

Plausibly, the primary conception of the tension arises from the confrontation between the orthodox understanding of Scientific Naturalism and basic forms of normativity that are, arguably, indispensable, irreducible (to social conventions or natural facts), and pervasive aspects of human thought and talk. Reasons, meanings, and values are good examples of the relevant forms of normativity. The fact that such normative items do not appear in the scientific image of the world, narrowly conceived, gives rise to the placement problem. How do reasons, meanings, and values "fit" into the natural world? How are we to understand the role of our concepts for such normative phenomena?

LIBERAL NATURALISM

For some time there has been a movement within contemporary naturalism—those committed to the tenets of Basic Naturalism—toward more and more liberal interpretations of its commitments. I shall call this *a process of liberalization*. As we will see, to a significant extent this turns on a growing appreciation within the camp of Scientific Naturalism of the range and diversity of actual scientific practices themselves. Of course, there is a further question whether Scientific Naturalism of *any* stripe is the most plausible and promising interpretation of Basic Naturalism. An important challenge here is possibility of a Liberal Naturalism distinct from both Scientific Naturalism and Supernaturalism. John McDowell's "naturalism of second nature" provides the best known example.[13]

In McDowell's account this new form of naturalism invokes a richer conception of nature than that provided by the scientific image, no matter how broadly construed, one that makes room for sui generis normative phenomena such as reasons.[14] It does not deny that there are sciences of the human but it does deny that such sciences can provide an understanding of the space of reasons and values that an acculturated rational human being typically becomes responsive to through maturation, learning a language, and enculturation.[15] Liberal Naturalism, in McDowell's conception of it, promises a setting in which it is possible to recognize the normative as both natural (in the sense of nonsupernatural) and yet nonscientific (in the sense that normative items do not fall under laws of nature).

For the purposes of this paper I want to leave aside the important question of the development of a viable form of Liberal Naturalism—although I do want to provide an important reason to favor Liberal Naturalism over any form of Scientific Naturalism.[16] The main focus of the paper will be on the question, What is the most plausible form of Scientific Naturalism? Answering this question involves considering a stage of liberalization *internal* to Scientific Naturalism, one that comes *before* the stage of liberalization leading to Liberal Naturalism. This prior conceptual move has been largely overlooked in the current debate between Scientific Naturalism and Liberal Naturalism.[17] But it has important implications both for how to conduct this debate and for our attitude to the shortcomings of Scientific Naturalism. My purpose is to show that a commitment to Scientific Naturalism per se need not give rise to the placement problem—although, as we shall see, it inevitably gives rise to problems about our *understanding* of autonomous forms of normativity.

THE FIRST STEP IN THE LIBERALIZATION OF SCIENTIFIC NATURALISM

More than half a century ago there was an influential ambition shared by a considerable number of philosophers to reduce all sciences to physics.[18] This was associated with an extremely narrow interpretation of the commitments of Basic Naturalism, which we may call Extreme Scientific Naturalism, according to which the only irreducible and legitimate science is physics itself. In this interpretation naturalism coincides with strict physicalism, the view that the only entities there are in the world are those posited by (current or idealized) physics.

Although this position still exerts some influence, most analytic philosophers today do not share its extremely restrictive conception of science and the scientific image of the world.[19] A growing consensus now sees biology as providing another distinct paradigm of science and scientific explanation. That is, many philosophers are happy to admit the irreducibility of biology and biological taxonomy to physics and physical taxonomy. Evolutionary biology, for example, aims to provide a causal history for a highly specific sequence of actual historical events—say, the evolution of such and such organisms under such and such circumstances—not laws for a possible sequence. Indeed, the same is true of physics itself. Cosmology, for example, aims to describe the actual development of the universe, another highly specific sequence of historical events. And, to give another biological example,

Mendelian genetics involves predictions by way of statistically discovered patterns of phenotypic variations in populations of biological entities. Furthermore, teleosematicists in the philosophy of biology have long accepted the irreducibilty of biological functions to the entities of physics or chemistry.[20] The point is that biology provides examples of genuine scientific explanations and predictions that do not aim to explain or predict in terms of general laws of nature. The recognition of the failure of the deductive-nomological conception of science, a conception built around generalizing the case of an idealized physics, and the recognition of biology as an autonomous science form what I shall call the first step in the process of liberalizing the philosophical conception of the sciences within a naturalist outlook.

It is important to see that this move is largely motivated by one of naturalism's central features, its opposition to metaphysics. Metaphysics is that traditional a priori subject that tries to explain, or explain away, the "appearances" of things in terms of some idea of an underlying "reality" that is supposed to be explanatorily fundamental, that is, absolute or necessary or fixed. One of the main attractions of modern naturalism has always been its promise to provide a genuine alternative to the long and pervasive history of philosophy inspired by, or grounded in, the metaphysical doctrines associated with theism.[21] For this reason anti-Supernaturalism is a central tenet of Basic Naturalism, an opposition to *any* philosophical appeal to a supernatural entity or power (such as Descartes's appeal to a good God to explain our epistemological access to the world). Apart from God and his powers, naturalism has also abandoned other supernatural entities, including Platonic forms, Kantian noumena, and the immaterial minds of Descartes, on much the same grounds, namely, that such items are explanatorily otiose. They do not bear the right sort of relations to observation and experiment.[22]

This is not to say that there are not scientific naturalists who think of themselves as practicing metaphysics, for example, Frank Jackson, David Armstrong, and David Lewis. Such philosophers understand Scientific Naturalism as being opposed to certain traditional or outmoded forms of metaphysics but not to metaphysics as such. Their hope is that metaphysics can be made suitably "scientific." What that might mean and how likely it is to succeed I shall not endeavor to say. But it is important to recognize a prima facie tension between the aspiration to metaphysics and the aspiration to make philosophy scientific. The important point for present purposes is simply that there is an antimetaphysical tendency internal to Scientific Naturalism that involves putting weight on a posteriori discoveries and scientific explanations and denying or severely narrowing the scope of a priori discoveries and purely rational explanations.

At this point it is worth briefly reviewing some of the more significant metaphysical ideas that served as ideological props for the once-popular Extreme Scientific Naturalism. Here are three examples:

1. *(Strict) physicalism.*[23] This is the metaphysical doctrine that all that exists in the world are the posits of physics as it stands, or perhaps an imagined future physics. It has usually been defended by supposing that everything not posited by physics can be reduced to (combinations of) the entities posited by physics. It is important to see that this sort of physicalism is not a claim internal to physics itself. Physics might presuppose that the posits of its successful explanations exist, but it does not imply that nothing else exists. The claim that physics, and only physics, provides a complete inventory of what there is in the world is obviously a metaphysical claim. Of course, the naturalist's allegiance ought not to be to some metaphysical (non- or prescientific) conception of ontology but to whatever ontological commitments our best scientific explanations themselves suggest. The very failure of the programs to reduce all sciences to physics implies that naturalist ontology must reflect whatever scientific explanations are successful, whether or not they lie outside the domain of physics.

2. *Causal fundamentalism.* This is the metaphysical idea that the world has a single causal structure. It goes with the view that it is the business of science (and, ultimately, of physics) to describe this order, typically in terms of laws of nature. Apart from allegiance to strict physicalism, one of the chief reasons that philosophers have accepted this view is, as Papineau puts it, "The widespread acceptance of the doctrine now known as the 'causal closure' or the 'causal completeness' of the physical realm, according to which all physical effects can be accounted for by basic physical causes (where 'physical' can be understood as referring to some list of fundamental forces)."[24]

But it is simply a mistake to think the causal closure thesis implies causal fundamentalism. What tends to get overlooked is that the same event might have more than one complete causal history.[25] Some of the most exciting recent work on causation suggests that we see causation as having close conceptual ties to explanation and explanatory contexts. Such an account of causation poses a significant challenge to causal fundamentalism by lending strong support to the doctrine of causal pluralism. If causal talk is always understood relative to a background explanatory context, the fact that there are different levels of explanation implies that there can be a plurality of complete causal explanations for the same event.[26]

It is also worth noting that causal fundamentalism is not obviously compatible with contemporary physics. For instance, it is unclear whether fundamental physics posits causal relations at all since the mathematical equations of quantum mechanics are symmetrical whereas the causal relation, at least as we ordinarily understand it, is asymmetrical, implying an arrow of time; that is, the cause typically comes *before* the effect.[27]

3. *The unity of science.* The metaphysical idée fixe that the sciences must constitute an ultimate unity has been subjected to intense criticism, but it is still widely current as witnessed by this remark by the renowned biologist, Richard Lewontin: "Historians of science, epistemologists, and, when they are in a contemplative mood, natural scientists picture science as having a single mode and form."[28] In philosophy the idea of the unity of science is, for the most part, expressed in two ways: *reductionism*, the thought that all the sciences reduce to physics, or *the deductive-nomological conception of science*, the idea that all legitimate sciences aim to provide explanations by way of subsuming phenomena under laws of nature together with a statement of initial conditions. Of course, these views often go together in the thought that all of the laws discovered by the various sciences ultimately reduce to the laws of physics. We have already commented on the failure of physicalist reductionism. And the deductive-nomological conception of science is open to the objection that, as we have seen, other forms of scientific explanations exist (even in physics itself!). To give another example, the human sciences involve explanatorily fruitful generalizations, which are distinguishable from laws of nature—even supposing these are ceteris paribus laws—on the grounds that they may admit too many exceptions and only hold over quite limited domains.[29]

THE SECOND STEP IN THE LIBERALIZATION OF SCIENTIFIC NATURALISM

The first step of liberalization in our understanding of Scientific Naturalism can be explained on the basis of a growing realization that Extreme Scientific Naturalism is itself not a part of science but is really based on a dubious and dispensable metaphysical account of science. The claim that physics is the only legitimate and irreducible science is plausibly interpreted as a metaphysical claim based on an a priori demand for unity, a unity that the plurality of actual scientific practice obviously does not corroborate. Or, if one regards the unity of the sciences as an hypothesis, then it should surely be abandoned on the basis of the notorious failure of past attempts at reductionism in the philosophy of science.[30] As Fodor put it, "Reductionism . . . flies in the face of the facts about the scientific institution."[31]

For the scientific naturalist there are important pluralist consequences of taking the first step of liberalization. The default positions for the resulting Narrow Scientific Naturalism are *ontological pluralism* since, for example, some biological kinds will now be irreducible elements of autonomous biological explanations, and *methodological pluralism* because biology involves providing functional and historical explanations of particular relevant phenomena (e.g., the adaptation of a particular creature to its specific environment, the development of a specific organ, such as the human eye) that are distinct from explanations from covering laws.

Once one has advanced to the first step in the process of liberalizing Scientific Naturalism then one cannot fail to ask, why give such a preeminent place to the so-called *natural* sciences in the interpretation of Basic Naturalism? If we keep in mind that the narrow scientific naturalist is *already* committed to pluralism in both ontology and method, then a commitment to Scientific Naturalism by itself provides no reason for denying that the human sciences, too, might have their own ontological commitments and distinct methods of inquiry. To continue to hold on to strict physicalism, causal fundamentalism, and the deductive-nomological conception of science would be to fly in the face of the naturalistic spirit insofar as these are metaphysical commitments that clash with a naturalistic account of actual scientific practice. Naturalism is surely committed to a naturalistic account of science.

This last point is important but often overlooked. A naturalist of all people ought to be concerned with the actual empirical reality of the sciences and their implications rather than preconceived ideas or ideological constructions of metaphysics into which, it is supposed, science *must* fit. The questionably metaphysical character of both Extreme and Narrow Scientific Naturalism is apparent when we consider that their conceptions of science fail to do justice to the variability and plurality of actual scientific inquiries, practices, and commitments, or what Dupré calls "the disunity of the sciences."[32]

In particular, the explanatory successes of at least some human sciences (say, economics) and the growth in the number of human sciences in the past century owe nothing to a nonexistent reduction of the human sciences to physics. Moreover, nothing in our general concept of science blocks regarding the human sciences as legitimate or irreducible sciences. In a naturalistic or nonmetaphysical conception, the sciences have no fixed essence, that is, no fixed set of necessary or sufficient conditions.[33] The concept "science" is more realistically seen as a family-resemblance concept that does not appear to admit of any interesting unification in terms of the objects of scientific inquiry or of scientific methodology.[34] Human sciences such as economics, sociology, and anthropology earn their right to the title of "science" on the grounds that they do not involve supernatural posits, admit empirical evidence, and provide fruitful explanatorily generalizations, even if these are only local.

Despite the push toward taking the human sciences seriously, the influence of old metaphysical ideas still exerts a curious influence over debates about the form and viability of Scientific Naturalism. John McDowell, for example, although contesting any form of Scientific Naturalism, retains a misguided monistic conception of scientific explanation in terms of subsumption under laws of nature, one according to which legitimate social sciences differ from physics only to the extent of involving laws with richer ceteris paribus clauses.[35] And another opponent of Scientific Naturalism, Charles Taylor, tends to treat all the natural sciences as articulating an overarching mechanistic view of the universe that clearly does not conform to the worldview of modern physics or modern biology.[36] Furthermore, although many narrow scientific naturalists have given up strict physicalism and scientific reductionism, the doctrine of causal fundamentalism is still widely influential.[37] As a consequence it is still a popular misconception that the failure of the human sciences to discover laws shows either their illegitimacy or their infancy.

CONSEQUENCES OF TAKING THE HUMAN SCIENCES SERIOUSLY

I have argued that the human sciences have not been taken seriously by Extreme or Narrow Scientific Naturalism because of metaphysical commitments that scientific naturalists in particular have good reason to give up. Paradoxically, orthodox Scientific Naturalism has not been naturalistic enough by its own lights! If we apply the naturalist methodology to naturalism itself by respecting the actual plurality of explanatorily useful scientific practices then we remove needless metaphysical obstacles to acknowledging the human sciences as at least potentially legitimate and autonomous sciences. Proceeding in this way we arrive at the claim *that Broad Scientific Naturalism is the default position for Scientific Naturalism* since it is the position at which we arrive by taking the second step in the liberalization of our interpretation of Scientific Naturalism. As we have seen, this represents a continuation of the earlier liberalization of Scientific Naturalism from an extremely reductive, physics-based paradigm of science to a pluralistic conception that includes biology as an additional paradigm of legitimate and autonomous science.

It is important to see that taking the human sciences seriously in this way does not depend on settling any of the well-known disputes about whether there is a philosophically significant line to draw between the natural and human sciences or, if there is, how to draw it. The question of a dividing line between the *Naturwissenschaften* and the *Geisteswissenschaften* seems to assume that there is some important unity that characterizes all of the sciences that fall on each side of this line, say, that the former involves *Erklären* (explanation by way of laws) while the latter involves *Verstehen* (hermeneutic understanding from the "inside"). However, from the position developed here, there is as much reason to think that there are significant differences *within* the category of natural science, and *within* the category of human science, as there are *between* the natural and the human sciences. This vitally important point is gradually gaining credence. For example, in a late paper Kuhn expresses uncertainty "whether [differences between the natural and human sciences] are principled or merely a consequence of the relative states of development of the two sets of fields."[38] The differences between the human sciences (and those between the natural sciences) might be just as principled, or just as important, as the differences between the natural and human sciences.

THE NATURAL AND THE NORMATIVE AGAIN

From the perspective of Broad Scientific Naturalism, the philosophical issues look very different. The scientific image of the natural and human sciences together is prima facie much closer to the manifest image of the world. Insofar as social scientific explanations involve recognizing, say, irreducible values, meanings, and reasons, then the move to a Broad Scientific Naturalism has the important consequence that a great many of the currently fashionable naturalization projects would lose their motivation since they simply take for granted a narrow conception of the scientific image. Most importantly, the placement problem that these projects are designed to answer simply lapses. Since the human sciences prima facie recognize a realm of reason-governed, meaningful, and valuable human actions, artifacts, and institutions, then we can simply admit into our scientific ontology many items, including many irreducibly normative items that Narrow and Extreme Scientific Naturalism threaten with elimination or reduction.

But do social scientific explanations require the recognition of values, meanings, and reasons in their own right? It has often been supposed that scientific methodology demands that we do not include normative notions within the content of explanations. Following Weber, it has become a commonplace to say that the discoveries of science are, or ought to be, "value-free."[39] There is an element of truth in this claim insofar as no science investigates reasons or values or meanings understood as abstract standards or ideals. There is no current science of what is a good reason for what or what the objective values or meanings of things are since these are not causal matters. And to suppose that there is a possible future science of these things is to beg the question by assuming that we can, in principle, reduce these normative notions to nonnormative notions. But even if we admit that there is no science of norms understood as abstract standards or ideals, the social sciences can obviously recognize people's intentional states *about* reasons, values, or meanings—say, their acceptance of them or their beliefs about them—since these psychological states are clearly causal in nature. For example, it might be that many people's belief that slavery was unjust helped bring about the end of slavery in America. One way to capture the content of such beliefs as they figure in a social scientific explanation is to suppose there are truths about the injustice of slavery that we have come to appreciate by way of the right sort of reasoning.

Another reason to think the claim of value freedom is an overstatement is that the idea of what entities and objects are recognized by the sciences ought not to be limited to only those things that are the objects of scientific study or that are posited by scientific explanations. There is an important class of abstract (or noncausal) entities that are not studied by the sciences and so do not appear in the content of scientific explanations but that, nevertheless, are *presupposed* by them. This is one of the most important lessons of the recent work of Hilary Putnam in his ongoing attempt to challenge the orthodox picture of a fact/value dichotomy.[40] Scientific explanations presuppose scientific rationality, which, in turn, presupposes the existence of ineliminable epistemic or cognitive values such as simplicity, Occam's razor, fruitfulness, testability, universality, coherence, reasonableness, and so on.[41] In making this point Putnam is really extending Quine's well-known indispensability argument that scientific rationality presupposes such abstract entities as numbers or sets to the case of cognitive values.

Nonetheless, even if the placement problem lapses, there remains a tension of sorts between the natural and the normative. Values, reasons, and meanings are not *fully* accounted for in terms of scientific explanations since when they are understood as standards or ideals they will not appear as part of the content of scientific explanations.[42] To say that one proposition provides a good reason to accept another, for example, is not a causal or historical or functional claim; it is a normative claim that can only be (provisionally) settled by thinking about the relevant propositions in the right kind of way.[43] So such normative notions, although presupposed by scientific inquiry, are not objects of such inquiry. But they must figure in the scientific image of the world nonetheless.

SCIENTIFIC NATURALISM
IS A NORMATIVE DOCTRINE

In the previous section it was pointed out that the naturalist suspicion of irreducible normativity is predicated on extreme or narrow interpretations of Basic Naturalism that the scientific naturalist has good reason to abandon since his own view is committed to adopting a naturalistic conception of the sciences. The best expression of Basic Naturalism's antimetaphysical spirit—growing out of its respect for the sciences as they are actually practiced—is a Broad Scientific Naturalism that recognizes a richer (or somewhat "enchanted") conception of the scientific image. For this version of Scientific Naturalism, normative phenomena, far from being spooky or "queer,"[44] are both involved in and presupposed by scientific inquiry and explanation, even if they are not direct objects of scientific investigation and explanation. Thus the old conception of a strict dualism of the natural and the normative is revealed to be a metaphysical prejudice rather than a matter of what the sciences, realistically conceived, show us.

But for all its virtues even Broad Scientific Naturalism faces an acute epistemological embarrassment. Such a naturalism is committed to the completeness of scientific knowledge and understanding and it is plausible, as we have seen, that there is no way of explaining irreducibly normative notions in terms of scientific laws, causes, or statistical correlations. This problem is particularly acute because, although many of its adherents suppose otherwise, Scientific Naturalism is itself a *normative* doctrine.[45] Consequently, it is really a claim about how philosophy *should* be conducted, or what it *should* admit, from the rational point of view given the great success of modern science and its implications. That is, we can more perspicuously represent the two doctrines that make up Broad Scientific Naturalism as follows:

Normative ontological claim: One rationally *ought* to admit the actual existence of only those things that are recognized by the successful natural and human sciences.

Normative methodological claim: One rationally *ought* to admit that the only genuine knowledge or understanding we have is that provided by the successful explanations of the natural and human sciences.

In both cases the statement of the naturalist thesis presupposes a notion of rational normativity that even naturalists are inclined to admit is both indispensable and irreducible and for which we have no plausible scientific story to tell.[46]

CONCLUSION

It has been argued that the placement problem lapses from the perspective of a Broad Scientific Naturalism. Although many suppose otherwise, irreducible normative phenomena can figure in the scientific image, suitably liberalized. In that sense there is no dualism of the natural and the normative. But, in another sense, a tension remains. As there is, as far as we can see, no possibility of providing a scientific explanation of core normative items such as reasons, then Broad Scientific Naturalism is committed to the existence of things for which it can give no complete naturalistic understanding. Consequently, what understanding we have of, say, rational normativity is properly called *nonscientific* understanding. This is particularly embarrassing since, plausibly, Scientific Naturalism itself is expressed in terms of two claims about what is rationally normative. I conclude that orthodox narrow scientific naturalists have good reason to become broad scientific naturalists and that the latter have good reason to abandon their position for a Liberal Naturalism that allows for the possibility of nonscientific understanding and explanation.

NOTES

I have benefited from audiences at Macquarie University and Università Roma Tre and from comments by Eric Oberheim on an earlier draft.

1. I am aware that this is a restrictive employment of the expression "placement problem." Similar problems arise with respect to intentional entities, mathematical entities, and so on. But for present purposes I shall retain the focus on normative entities.

2. For dialectical purposes I shall restrict the focus to Scientific Naturalism, leaving the question of Liberal Naturalism aside.

3. A scientific naturalist is, of course, committed to the view that the methods of science are the right tools for understanding normative phenomena.

4. David Armstrong, "Naturalism, Materialism, and First Philosophy," in *The Nature of Mind* (St. Lucia: University of Queensland Press, 1980), 149–65; Michael Devitt, *Realism and Truth* (Princeton, N.J.: Princeton University Press, 1984); Fred Dretske, *Naturalizing the Mind* (Cambridge, Mass.: MIT Press, 1995); David Papineau, *Philosophical Naturalism* (Oxford: Blackwell, 1993).

5. Jennifer Hornsby, *Simple Mindedness: A Defence of Naïve Naturalism in the Philosophy of Mind* (Cambridge, Mass.: Harvard University Press, 1997); John McDowell, *Mind and World* (Cambridge, Mass.: Harvard University Press, 1994); P. F. Strawson, *Skepticism and Naturalism: Some Varieties* (New York: Columbia University Press, 1985); Barry Stroud, "The Charm of Naturalism," reprinted in

Naturalism in Question, ed. Mario De Caro and David Macarthur (Cambridge, Mass.: Harvard University Press, 1994), 21–35.

6. John Dewey, "Anti-Naturalism in Extremis," *Naturalism and the Human Spirit*, ed. Y. H. Krikorian (New York: Columbia University Press, 1944), 2.

7. Of course, one could hold one of these doctrines and not the other, but that would be a minority position within contemporary philosophy.

8. It is being assumed that the sciences presuppose the possibility of things that we do not yet know about given the current stage of scientific development.

9. The sciences also presuppose fallibility, the possibility that the current sciences will be superannuated by future sciences.

10. See, e.g., W. V. Quine, *Theories and Things* (Cambridge, Mass.: The Belknap Press of Harvard University Press, 1981), chap. 1. It is important to distinguish matters of ontological commitment from the metaphysical subject of Ontology. Quine is contributing to the philosophical elucidation of scientific standards of ontological commitment. This should be distinguished from the metaphysical subject called "Ontology," which is a theory of existence that purports to make necessary or explanatorily basic claims about existents and their categories from the perspective of a priori reason.

11. For a representative statement of the problem see Frank Jackson, *From Metaphysics to Ethics* (Oxford: Oxford University Press, 1994), chap. 1.

12. The *Methodenstreit* originated in the context of a dispute within the German-speaking world between two views of the methods of economics. The historicist Gustav von Schmoller argued against Carl Menger that a complete science of human behavior based purely on reason and the natural sciences was impossible.

13. McDowell, *Mind and Word*, 86.

14. David Macarthur and Mario De Caro, eds., *Naturalism in Question* (Cambridge, Mass.: Harvard University Press, 2004), presents papers dedicated to criticizing orthodox Scientific Naturalism and defending the philosophical promise of a more liberal conception of naturalism.

15. Since McDowell thinks of science in terms of nomological explanations and there are no natural laws of normative items, he takes it that these items fall outside the purview of science.

16. I defend McDowell's version of naturalism from the charge of Supernaturalism in David Macarthur, "Naturalizing the Human or Humanizing Nature: Science, Nature, and the Supernatural," *Erkenntnis* 61 (2004): 29–51.

17. Opponents of McDowell's view include Jerry Fodor, "Encounters with Trees," *London Review of Books* 17, no. 8 (1995): 10–11, and Crispin Wright, "Human Nature?" in *Reading McDowell: On Mind and World*, ed. Nicholas Smith (London: Routledge 1995), 140–74. For his replies see John McDowell, "Responses," in *Reading McDowell: On Mind and World*, ed. Nicholas Smith (London: Routledge, 1995), 269–306.

18. It involved reducing all natural scientific laws to the laws of physics. For a nice expression of this view see Hilary Putnam and Paul Oppenheim, "The Unity of Sci-

ence as a Working Hypothesis," in *Concepts, Theories, and the Mind-body Problem*, ed. Herbert Feigl, Michael Scriven, and Grover Maxwell, Minnesota Studies in the Philosophy of Science 2 (Minneapolis: University of Minnesota Press, 1958), 3–36.

19. The most popular form of physicalism today, at least amongst analytic philosophers, is nonreductive physicalism.

20. See, e.g., Ruth Millikan, *Language, Thought, and Other Biological Categories* (Cambridge, Mass.: MIT Press, 1984).

21. Here it is relevant to mention Auguste Comte's idea that human understanding has passed through three stages of development: theological, metaphysical, and then "positive" (that is, the stage of nomological science).

22. One of the interesting paradoxes about naturalism is the way it has appealed to a metaphysics of experience, empiricism (or "positivism"), to undermine other transcendent forms of metaphysics. The empiricist tradition tends to picture science as really a matter of discovering lawlike relations at the level of the empirically observable. And it has historically been associated with phenomenalist and antirealist understandings of science (e.g. Berkeley, Mill, Mach, Russell). With regard to the human sciences that deal with the meanings, intentions, and interpretations of actions, texts, artifacts, symbols, institutions, and so on, the tendency has been to suppose that in so far as such items are not properly empirical, they are not the subject matter of any science. However, this is clearly too restrictive a conception of the observable. Meanings, for example, are public phenomena expressed and observable in action and language. There is an important difference, for example, between seeing a sentence in one's native tongue in which the sense is manifest and seeing a sentence in a foreign language (that, let us say, one has no acquaintance with) as just a series of marks. The values of things, too, can be experienced.

23. I use the term "strict" to distinguish this form of physicalism from a weaker version defined in terms of nonreductive supervenience, for the latter doctrine allows for the reality of nonphysical entities so long as they supervene on the physical.

24. David Papineau, "Naturalism," in *Stanford Encyclopedia of Philosophy*, ed. E. Zalta, 2007, http://plato.stanford.edu/entries/naturalism.

25. This is not the same thing as overdetermination since each complete causal history will involve a different context of explanation. For example, an act of waving might be fully causally explained in neurophysiological terms and also fully causally explained at the intentional level in terms of reasons for so acting.

26. Cf. Peter Menzies, "The Causal Efficacy of Mental States," in *Physicalism and Mental Causation*, ed. Sven Walter and H. D. Heckmann (Charlottesville, Va.: Imprint, 2003), 195–223; James Woodward, "Explanation and Invariance in the Special Sciences," *British Journal for the Philosophy of Science* 51 (2000): 197–254.

27. See Bertrand Russell, "On the Notion of Cause," *Mysticism and Logic* (London: Longmans, Green, 1918), for skepticism about whether modern physics involves the ordinary notion of cause.

28. Richard Lewontin, "Facts and the Factitious in Natural Science," *Critical Inquiry* 18 (1991): 141.

29. "Other widely used generalizations in the special sciences have very narrow scope in comparison with paradigmatic laws, hold only over restricted spatio-temporal regions, and lack explicit theoretical integration" (James Woodward, "Scientific Explanation," in *Stanford Encyclopedia of Philosophy*, ed. E. Zalta, 2007, http://plato.stanford.edu/entries/scientific-explanation/).

30. For example, see John Dupré, *The Disorder of Things: Metaphysical Foundation of the Disunity of Science* (Cambridge, Mass.: Harvard University Press, 1993) on the irreducibility of biology to physics.

31. Jerry Fodor, "Special Sciences" (1974), reprinted in *Readings in the Philosophy of Social Science*, ed. Michael Martin and L. C. McIntyre (Cambridge, Mass.: MIT Press), 439.

32. Dupré, *The Disorder of Things*.

33. This point was famously argued by Kuhn and Feyerabend against various attempts to solve the problem of the demarcation of science from nonscience, including Popper's appeal to falsificationism and the logical positivists' appeal to verificationism. See also Hilary Putnam, "The Idea of Science," in *Words and Life* (Cambridge, Mass.: Harvard University Press, 1995), 481–91; and Richard Rorty, "Is Science a Natural Kind?" in *Objectivity, Relativism, and Truth: Collected Papers*, Vol. 1 (Cambridge: Cambridge University Press,1991), 46–62.

34. Scientific method can be fruitfully thought about, following Dewey, as a kind of democratic experimentalism which involves a set of quasi-moral virtues of inquiry (e.g. openness to criticism, reasonableness) and a fallibilist, experimental attitude to knowledge.

35. McDowell, *Mind and Word*.

36. Charles Taylor, *Philosophical Papers*, 2 vols. (Cambridge: Cambridge University Press, 1985).

37. Abandoning type-type reduction of laws of one science to another still allows one to continue to think that each instance of a higher-level causal law is token-identical with an instance of a physical law. See Papineau, "Naturalism."

38. T. S. Kuhn, "The Natural and Human Sciences," in *The Interpretive Turn*, ed. J. F. Bohman, D. R. Hiley, and Richard Shusterman (Ithaca, N.Y.: Cornell University Press, 1991), 22.

39. Richard Miller, "Fact and Method in the Social Sciences," in *The Philosophy of Science*, ed. Richard Boyd, Philip Gasper, and J. D. Stout (Cambridge, Mass.: MIT Press, 1991): "That the social sciences should be value free is the closest thing to a methodological dogma in Anglo-American social science" (774).

40. See, e.g., Hilary Putnam, *The Collapse of the Fact/Value Dichotomy* (Cambridge, Mass.: Harvard University Press, 2002).

41. Of course, it also presupposes the meaningfulness of scientific claims.

42. We might think of this as an insight of the hermeneutic tradition.

43. With regard to practical reason, Thomas Scanlon in his "Metaphysics and Morals" (reprinted in this volume) writes: "It is . . . plausible (in a way that it is not in the case of beliefs about the physical world) to suppose that these are matters that

we can discover the truth about simply by thinking about them in the right way. It may not be easy to say more exactly what this way is, and the answer clearly depends on the concepts involved. In some cases it may involve the right kind of sequential reasoning, in others, perhaps, the right kind of mental picturing, in others, carefully focusing our attention on the relevant features of the ideas in question and being certain that we have distinguished them from other, irrelevant factors. But as long as there is some way of doing this correctly, there is a difference between getting it wrong and getting it right" (175).

44. John Mackie, *Ethics: Inventing Right and Wrong* (New York: Penguin, 1977).
45. This is rarely admitted, although Mark Colyvan is one prominent exception; see his "Naturalizing Normativity," in *Conceptual Analysis and Philosophical Naturalism*, ed. David Braddon-Mitchell and Robert Nola (Cambridge, Mass.: MIT Press, 2008), chap. 8.
46. See, e.g., Fodor, "Special Sciences."

7

REASONS AND CAUSES REVISITED

Peter Menzies

What kind of theory is folk psychology? In particular, what kind of theory is intentional psychology—that part of folk psychology concerned with mental states with intentional content, such as beliefs, desires, intentions, fears, hopes, and so on? These are the questions this essay addresses. Of course, they presuppose the thesis that intentional psychology is a theory, a thesis that is much contested between so-called theory theorists and simulation theorists.[1] However, I shall not be able to enter into the debate between theory theorists and simulation theorists in this essay. Rather, I shall simply assume that folk psychology is a theory of some kind so that I can focus on the issue of what kind of theory it is. As I see things, there are two broad traditions of thought about this issue.

I shall call the first tradition *Scientific Naturalism*. According to this view, intentional psychology is much like a scientific theory, as described by the deductive-nomological model of scientific theories. In this instance the theory is used to predict and explain observable human behavior on the basis of posited intentional states. The theory consists of a set of contingent laws describing the internal relations among perceptual experiences, intentional states, and actions. (The folk may not be able to articulate the laws in any detail—that is the job of the philosopher or cognitive psychologist. Nonetheless, they have tacit knowledge of the laws in the same sense in which they have tacit knowledge of the grammatical rules of their language.) The folk appeal to these laws to make causal judgments about intentional states. It is a crucial feature of intentional states, in Scientific Naturalism's picture, that they enter into causal relations with perceptual experiences, other intentional states, and actions because scientific naturalists believe that an intentional state is individuated in terms of its profile of typical causes and effects. In other words, they believe that what distinguishes a given type of intentional state from others is the pattern of typical causes and effects that type of intentional state exhibits. Indeed, some scientific naturalists go so far as to assert an identity between intentional states and causal roles.

As its name is meant to suggest, Scientific Naturalism's view of intentional states gets its rationale and motivation from the naturalistic project of trying to locate intentionality within the physical realm: if intentional states are individuated in terms of their causal relations, and causal relations are physical relations, then it is reasonable to think that intentional states can be located in the physical order of things. But it is crucial to this project that intentional psychology should be essentially descriptive or nonnormative in character. If the generalizations that make up intentional psychology do not describe real causal transitions in thought and action but rather prescribe in normative fashion how people should make certain transitions in thought and action, then the naturalistic project of locating intentionality within physical reality would become significantly more problematic.

This is a very rough sketch of a very general view, which I believe is held by a large number of contemporary philosophers of mind. Perhaps not all of them would accept every part of what I just sketched. Nonetheless, I would argue that among many influential philosophers of mind, D. M. Armstrong, Jerry Fodor, David Lewis, and the early Hilary Putnam held something like this view of intentional psychology.[2]

The other tradition of thought about intentional psychology has a longer history, going back at least to Kant.[3] I shall call this tradition *Kantian rationalism*. According to this view, intentional psychology is not a scientific theory at all but rather a theory of interpretation based on a set of normative principles that prescribe how people should reason and behave if they are rational. Kantian rationalists see the generalizations that license predictions and explanations of human behavior in a very different light from scientific naturalists: rather than being contingent laws that are known a posteriori, they are conceptual truths that are known a priori. Most significantly, these generalizations are essentially normative: they state how rational subjects should move from perceptual experiences to intentional states, from some intentional states to others, and from intentional states to action.[4] The rationality of such transitions can be demonstrated by explicit articulation in the form of arguments. This is straightforward in the case of theoretical reasoning: the rational subject moves, for example, from the belief that if his flight has been announced, it will soon depart and the belief that his flight has been announced to the belief that his flight will depart soon. The arguments in the case of practical reasoning are more complicated because they involve intentional states like desires with motivational force and because they result in action rather than belief. Nonetheless, patterns of rational transitions between intentional states and action are regimented in action theory and, in more sophisticated guise, in decision theory. Whether the reasoning is theoretical or practical, the rationality of the arguments and the transitions in intentional states they mirror depend crucially on the semantic content of the intentional states involved. For instance, it is rational for a subject to move from his belief that his flight is about to depart and his desire to catch the flight to the action of running toward the departure gate because the specific contents of his beliefs and desires rationalize this specific action.

By contrast to scientific naturalists, Kantian rationalists think that intentional states belong to the space of reasons rather than the space of causes, to use Sellars's terminology.[5] In other words, we predict and explain human behavior by considering how intentional states rationalize rather than cause other intentional states and actions: the important predictive and explanatory relations are structured by the reason-giving rather than the causal relation. In keeping with their demotion of causation from a central place in their framework, Kantian rationalists tend to be skeptical about the project of trying to locate intentionality in the physical world. In voicing their skepticism on this score, however, they typically reject suggestions that their account of intentionality within the space of reasons must be seen as positing some supernatural realm. Instead, they often present themselves as operating with a more generous conception of naturalism and a more inclusive picture of the natural world that extends beyond the purely physical.

Again this is a very rough sketch of a very general view, which I believe is endorsed by a sizeable number of contemporary philosophers of mind. Kantian rationalists are perhaps outnumbered by the scientific naturalists, but they still represent a significant strand of thinking about intentional psychology in contemporary philosophy of mind. Among the Kantian rationalists about intentional psychology, I would include Robert Brandon, John McDowell, and the later Hilary Putnam.[6]

Traditionally, philosophers of mind have identified themselves as belonging to one or another of these traditions of thinking about intentional psychology. Indeed, many (but not all) have seen the positions as incompatible, though explicit arguments for their incompatibility are surprisingly rare in contemporary discussions. Nonetheless, one argument has had a long—and sometimes subterranean—influence on thinking about this matter. The *logical connection argument*, first formulated by Wittgensteinians in the 1950s and 1960s, purports to establish that the generalizations that license the predictions and explanations of intentional psychology have a very different character from the empirical laws that back causal prediction and explanation.[7] Thus, it is argued that the generalizations of folk psychology are, first, conceptual truths that are known a priori and, second, normative or prescriptive in character; and that these two characteristics make them ill suited to act in the role of empirical laws supporting causal predictions and explanations. Though this argument has been much discussed on and off for forty years, I do not believe that its relevance to the issue of the compatibility of the Scientific Naturalism and Kantian rationalism about intentional psychology has been properly appreciated. I set out the argument in more detail in the next section of this essay, explaining why some scientific naturalist responses to the argument have misdiagnosed the source of the argument's apparent cogency.

This is not say that I agree with the conclusion of the logical connection argument. On the contrary, I wish to argue in this essay for a compatibilist position about intentional psychology: we appeal to intentional states to predict and explain behavior because they both rationalize and cause behavior. The proper domain of intentional psychology is both the space of reasons and the space of causes. Of course, compatibilist views are well known in the literature. Donald Davidson famously argued for the thesis that reasons are causes in "Actions, Reasons, and Causes" and "Mental Events."[8] I shall outline Davidson's compatibilism, explaining how it offers a two-tiered approach to intentional psychology. For it implies that intentional psychology really employs two sets of generalizations: one set of norm-based generalizations covering the role of intentional states in rationalizing explanations of behavior, and a second set of generalizations expressing empirical laws that cover the role of intentional states in causal explanations. As I shall argue, this two-tiered approach is unsatisfactory because intentional psychology seems to employ only one set of generalizations to provide both rationalizing and causal explanations.

By contrast with Davidson's compatibilism, I shall advance a one-tier approach to intentional psychology in the remaining sections the essay. I shall argue that only one set of generalizations is needed to support the predictions and explanations of intentional psychology and that these generalizations have very much the character imputed to them by the logical connection argument. However, contradicting the conclusion of the logical connection argument, I shall argue that the conceptual and normative character of these generalizations does not undermine their capacity to support causal explanations. Philosophers of mind of both the scientific naturalist and Kantian rationalist traditions, have mistakenly thought this to be the case because they have been wedded to the deductive-nomological account of theories and of theoretical explanation. In place of the deductive-nomological account, I shall advance a model-based conception of intentional psychology and a model-based account of the way in which prediction and causal explanation takes place within intentional psychology. Intentional psychology should be seen, I claim, as employing idealized models of rational agents as the basis for both rationalizing and causal explanations. Despite the fact that the relations on which two kinds of explanations rest—the reason-giving relation and the causal relation—are very different, there is a principled reason that they coincide in extension for the most part in the explanations of intentional psychology.

THE LOGICAL CONNECTION ARGUMENT

In setting out the logical connection argument, my aim is not to provide a historically accurate reconstruction of the argument developed by Wittgensteinians in the mid-twentieth century. Rather, it is to capture two different strands of thought that have persuaded Kantian rationalists that the generalizations of intentional psychology are ill suited to support causal prediction and explanation.

The version of the logical connection argument that I shall discuss starts from the assumption that the generalizations of intentional psychology have two distinctive characteristics. First, when the generalizations are stated in a form that gives them a chance of being true, they seem to be conceptual truths. Consider, for example, the following generalization governing the connection between beliefs and desires, on the one hand, and actions, on the other: an agent who desires X; believes that action A will lead to the satisfaction of this desire; and, after an evaluation of all other desires, forms the all-things-considered judgment that A is the best available option, will in fact do A. This generalization is false, as shown by any case of weakness of will. For example, I may want to get to Bologna in time for the start of the conference, believe that catching the morning flight from Rome will enable me to get to Bologna in time for the start of the conference, and form the all-things-considered judgment that this is the best option for me, and yet I may fail to catch the flight because of weakness of will, as I indulge my passion for sightseeing in Rome. No matter what further conditions are added to the antecedent of this schematic generalization, the resulting generalization need not be true of every agent. This suggests that the generalization should not be seen as an empirical generalization about actual agents. Rather, it is better seen, I think, as a definition about how a rational agent acts on the basis of certain intentional states. Read in this way, the generalization remains true regardless of the existence of irrational agents who suffer from weakness of will. For it is simply a definitional truth about rational agents that any such agent who desires X; believes that action A will lead to the satisfaction of this desire; and, after evaluating her other desires, forms the all-things-considered judgment that A is the best option, will in fact do A. Anyone who did not act in this way would simply not count as completely rational by definition.

The second distinctive feature of the generalizations of intentional psychology is that they can be expressed synonymously using evaluative expressions such as "should." For example, an alternative formulation of the above generalization that preserves its meaning is: if an agent desires X, believes that action A will lead to the satisfaction of X, and after considering all her other relevant desires forms the all-things-considered judgment that A is the best option, then she *should*—from the point of view of rationality—do A. This formulation reveals the second important feature of such generalizations: namely, that their content is normative or prescriptive. It is no simple matter to explain the nature or the basis of this normativity. Indeed, it may well be the case that different generalizations—for example, those governing theoretical reasoning and those governing practical reasoning—express very different kinds of norms. In any case, I shall not attempt here to offer a systematic account of their normativity. Nevertheless, when I claim that the generalizations of intentional psychology are essentially normative, I mean simply that such generalizations are most informatively expressed in terms of evaluative terms like "should" and that, when agents fail to reason or act in accordance with these generalizations, they lay themselves open to criticism and correction.

Scientific naturalists have noted the special character of the generalizations of intentional psychology, especially their logical or conceptual character, but given a different account of it. For example, Braddon-Mitchell and Jackson explain the conceptual character of these generalizations as arising from the fact that intentional psychology implicitly defines intentional states in terms of their causal roles.[9] In view of this fact, it is not surprising that generalizations linking beliefs and desires with actions are conceptual truths since it is part of the meaning of expressions referring to these beliefs and desires that they give rise to certain actions. According to this explanation, these generalizations are conceptual truths not because they are definitions of how a rational agent should reason and act but because they follow from the implicit definitions of intentional states. A problem facing this explanation is that it implies that it should be completely tautological to explain an agent's behavior in terms of her beliefs and desires, whereas it is often informative to do so. In response to this problem, Braddon-Mitchell and Jackson argue that intentional psychology implicitly defines an intentional state not as a role state—as the state of having some lower-level state occupy the defining causal role—but rather as a realizer state—the lower-level physical state that occupies the causal role. Hence, an agent's beliefs and desires can causally explain her behavior because they are physical realizer states that are nontrivially linked to behavior.

This account of the conceptual character of the generalizations of intentional psychology is far from satisfying. In the first place, it involves a conceptual doublethink. It starts by claiming that intentional states are causal role states in order to explain how the generalizations of intentional psychology are conceptual in character, and then it shifts to claim that they are causal realizer states to accommodate the fact that causal explanations of behavior in terms of intentional states can be informative. Second, the account of the conceptual character of the generalizations does not make any mention of the essential role of the concept of rationality, and so it predicts that these generalizations will stand as conceptual truths even when they are read without restrictive quantification over rational agents. But this is clearly not the case. The generalizations of intentional psychology are simply false when they are not restricted to rational agents: actual agents sometimes fail to act in accord with them simply because they are not completely rational. Third, when the generalizations are read as restricted to rational agents, their status as conceptual truths does not seem to be affected by whether one accepts or rejects the causal role analysis of intentional states. For example, suppose that one were to accept an alternative understanding of intentional states—say, a dualist conception of intentional states as irreducible sui generis states. It would, nonetheless, still be reasonable to accept a generalization like the one described above as a conceptual truth. For its status as a conceptual truth derives not from the specific content of the concepts of intentional states, taken by themselves, but from the interconnections between these concepts and the concept of rationality.

In conclusion to this part of my discussion, let us grant that the generalizations of intentional psychology have their special character—as conceptual and normative truths—because they are definitions of how rational agents should reason and act. On the basis of some fairly standard assumptions, it would seem to follow that they are not the kinds of empirical laws that are required to back causal predictions and explanations. Does this mean that the Kantian rationalist is correct in thinking that when intentional states explain behavior, they do so by virtue of rationalizing rather than causing the behavior? Before I turn to answer this question, I want to consider Donald Davidson's response to this question.

TWO-TIERED COMPATIBILISM

Davidson famously argued that explanations of behavior in terms of intentional states are rationalizing explanations, but such rationalizing explanations are nonetheless a species of causal explanation.[10]

His 1968 article "Actions, Reasons, and Causes" was his initial statement of the compatibilist view that an intentional state that is the reason for some behavior is also its cause. Central to Davidson's argument is a conception of causation according to which causation is an extensional relation between events conceived of as concrete particulars. When a causal relation holds between events, it does so no matter how the events are described. He allows that a rationalizing explanation of action will typically refer to an intentional state under a description that is conceptually linked to the description of the action. However, since events are different from their descriptions and since causation relates events and not their descriptions, this fact is compatible with the existence of a causal relation between the relevant events. Davidson also holds that causation is nomological in character, which is to say that where a causal relation holds between events, the events must have descriptions that instantiate a true strict law. In the case of the causal relations that hold between an agent's intentional state and his action, Davidson claims that these descriptions will not be intentional but almost certainly be neural or physical in character. So we have the following basic picture: one event can rationalize and also cause another event, but it will rationalize the second event only if these events have appropriate conceptually linked intentional descriptions and it will cause the second event only if they have physical descriptions that instantiate a strict physical law.

Davidson elaborates his compatibilist view in his 1970 article "Mental Events." In particular, he supports his claim that physical laws cover the causal relations between intentional states and actions with an argument to the effect that there are no laws governing the interaction of the psychological with the physical. His argument starts from the assumption that every causal relation requires the existence of a strict or exceptionless covering law. His argument then runs that psychophysical laws by their nature cannot be strict since the psychological and physical realms are governed by different constitutive principles, with the psychological realm being governed by a constitutive principle of rationality that does not apply to the physical realm. If there were strict psychophysical laws, it would be possible to formulate definitional or nomological necessary or sufficient conditions for the psychological in terms of the physical and vice versa, but these conditions would conflict with the constitutive principles of each realm: "There cannot be tight connections between the realms if each is to retain allegiance to its proper source of evidence."[11] It follows from the nonexistence of strict psychophysical laws that if intentional states causally interact with physical states, there must be strict physical laws covering these causal interactions. But this means that the intentional states have physical descriptions that instantiate these laws, and this suffices to show that the intentional states are physical. The resulting position is what Davidson calls anomalous monism: the intentional is not reducible definitionally or nomologically to the physical, but every intentional event is nonetheless describable in physical terms and therefore is a physical event.

Davidson's compatibilism is an instance of what I shall call two-tier compatibilism. For his position is committed to the idea that there are two tiers of generalizations operating in intentional psychology. First, there are the rough-and-ready generalizations that we cite in support of claims that intentional states rationalize actions. Davidson allows that these are conceptual truths holding by virtue of normative principles of rationality. Second, there are strict laws that cover the causal relations between intentional states and actions. These are perhaps unknown to common sense but are in principle available to physics: they are empirical and established by usual inductive methods. The obvious benefit of this two-tier structure is that it allows Davidson to reconcile the distinctive conceptual and normative character of rationalizing explanations with the claim that they point to underlying contingent causal relations.

However, as well as having this benefit, Davidson's two-tiered compatibilism has a significant cost. The position implicitly involves the postulate that whenever two events have intentional descriptions according to which one rationalizes the other, they also have physical descriptions that instantiate a strict physical law. Or, alternatively, whenever an intentional event rationalizes some behavior, an underlying causal relation holds between the relevant events. Davidson simply assumes this postulate without giving any reason to believe it is true. But why should the rationalizing relation almost always coincide in extension with the causal relation (setting aside cases of irrationality)? Why should a conceptually linked pair of rationalizing descriptions almost always match up with a pair of descriptions that instantiate a strict physical law? Is it not a remarkable fact that the generalizations at the two tiers line up with each other in the required way? Without some justification for the postulate, Davidson's position simply effects the reconciliation required of compatibilism by stealth rather than honest toil.

There is another difficulty for Davidson's position that arises from its two-tiered structure. This difficulty is roughly that intentional psychology does not seem to draw the distinction implied by Davidson's framework between the properties of events by virtue of which one event rationalizes another and the properties of events by virtue of which one causes another. Let us assume, for the moment, that it makes sense to introduce the following notions into Davidson's framework: an event c *rationalizes* another event e *by virtue of their having the properties F and G, respectively,* if and only if there are descriptions of the two events referring to the properties F and G under which they instantiate a rationalizing generalization; and an event c *causes* another event e *by virtue of their having the properties F and G* if and only if there are descriptions of the events referring to the properties F and G, respectively, under which they instantiate a strict law. Then Davidson's two-tiered compatibilism implies that the properties by virtue of which one event rationalizes another are different from the properties by virtue of which the first event causes the second. But this goes against the ways of intentional psychology. When we explain an agent's behavior by saying, for example, that he is running toward a departure gate because he wants to catch his flight and he believes that it is about to depart from that gate, we assume that the explanation that gives the reason for his running can double as an explanation that gives its cause. And we do so because we implicitly assume that the properties of the events by virtue of which one rationalizes the other are the same properties by virtue of which the first event causes the second. (In this example, the properties of the first event are its involving a person's wanting to catch the flight while believing it is about to depart and the property of the second event is that it involves a person's running to the departure gate.)

That intentional psychology makes this assumption about the coincidence of the rationalizing and causally efficacious properties is best shown by considering cases in which the assumption is violated. Cases of deviant causation, of which the following is a typical example, often involve a violation of this assumption. A waitress wishes to embarrass her employer by dropping a pile of glasses in front of some guests. She forms the intention to drop the glasses. However, her audacity at forming the intention so unnerves her that she drops the glasses.[12] A natural diagnosis of why this example is anomalous is that the property that gives the reason for her dropping the glasses—being an intention to embarrass her employer—is not the property that causes her to drop the glasses—being unnerved by the audacity of the intention. This diagnosis enables one to distinguish such anomalous cases from standard cases in which the intentional properties of mental events are both reason-giving and causally efficacious with respect to the intentional properties of actions. Of course, this is a point emphasized by orthodox causal theories of action.[13] But this explanation of the difference between standard and non-standard cases is not available to Davidson's compatibilism precisely because of its two-tiered structure. In terms of his framework, there are no grounds for differentiating deviant causation cases from other cases in terms of the lack of coincidence between the reason-giving and the causally efficacious properties because all cases exhibit this lack of coincidence.

This difficulty that I am raising for Davidson's position is related to the well-known objection that his anomalous monism is a form of epiphenomenalism because it makes mental properties causally inefficacious.[14] In response to this objection, Davidson has denied that it makes sense within his theory of causation to talk of one event causing another by virtue of their properties: "For me, it is events that have causes and effects. Given this extensionalist view of causal relations, it makes no literal sense, as I remarked above, to speak of an event causing something as mental, or by virtue of its mental properties, or as described in one way or another."[15] As I do not have the space to go into these issues in depth, I simply say here that I find the arguments of Davidson's critics persuasive in showing that there is, in fact, no inconsistency between Davidson's account of causation as an extensional relation and the talk of one event causing another by virtue of their properties.[16] As McLaughlin points out, Davidson's own example of an extensional relation illustrates this point. The relation of one object weighing more than another is a perfectly extensional relation. Nevertheless, it is consistent with its extensionality to say that an object a weighs more than another object b by virtue of these objects' properties, namely their having the weights W_a and W_b. McLaughlin argues that Davidson thinks the opposite is true in the case of causation because he mistakenly thinks that "by virtue of F, the event c causes event e" means the same thing as "c's being F causes e." But in fact these claims have different implications since the first implies that the event c is the cause whereas the second implies that the state of affairs of c's being F is the cause. In any case, it is very plausible to think in general terms that if a causal relation holds between a pair of events c and e but not between another pair of events c' and e', this is not a difference of brute fact, but rather a difference between the properties of the events c and e, on the one hand, and the properties of c' and e', on the other hand. (I would also add here that causal relations hold by virtue of relations between events as well as properties of events.)

Even though Davidson has disputed the intelligibility within his framework of the idea of causal relations holding by virtue of properties of events, he has also conceded that it may nonetheless be possible for him to appeal to the concept of one property being causally relevant to another.[17] I mention this because it may be thought that Davidson's views on this issue may help to bring his position into line with intentional psychology to the extent that it takes the intentional properties of mental events to be causally relevant respect to actions. In the article "Thinking Causes," Davidson suggests that he could consistently adopt Fodor's suggestion that mental properties are causally relevant if descriptions of these properties appear in nonstrict psychophysical laws.[18] He also suggests in the same article that mental properties can be said to make a difference to actions in the sense that they supervene on physical properties that are linked by strict laws with the physical properties of the actions.[19] Whether these explications are successful or not, it is clear that the notion of causal relevance they are explicating is weaker than the notion of causal efficacy; that is, the notion of full-blooded causation that Davidson thinks holds between events when their descriptions instantiate strict laws. For the two-tiered structure of Davidson's compatibilism necessitates that the only genuine causal efficacy occurs where there are strict covering laws. Whatever the merits of the notion of causal relevance that Davidson seeks to explicate, it does not do justice to intentional psychology, which requires, plausibly in my view, that intentional properties are causally efficacious with respect to actions, in no lesser sense than physical properties are.

In conclusion, even when Davidson's views are modified to take account of the role played by properties in causation, we can see the defects of his two-tiered compatibilism. For this position is essentially committed to the idea that there are two very different kinds of generalizations having different functions. One kind of generalization has the function of specifying the properties that rationalize actions, where these properties may be causally relevant, in a weak sense, to the actions. The other kind of generalization has the function of specifying the properties that are genuinely causally efficacious with respect to actions. But I take this picture to be in fundamental tension with the way things are viewed in intentional psychology. This recognizes but one kind of generalization with the dual function of specifying properties that both rationalize actions and are causally efficacious with respect to them.

INTENTIONAL PSYCHOLOGY AS A MODEL

In the view that I am advancing, intentional psychology is a theory that deploys a set of conceptual, normative generalizations in explanations that are jointly rationalizing and causal in character. This view contradicts the orthodox view of the matter that is influenced by the deductive-nomological model of causal explanatory theories. The deductive-nomological model says that a theory that is in the business of providing causal explanations and predictions must contain strict, empirically confirmable laws. These laws license predictions and causal explanations about particular events by implying statements about them when conjoined with statements about other events.[20] This is the roughest outline of the model, but it is enough to suggest how the model has influenced the thinking of both scientific naturalists and Kantian rationalists: both camps have thought that intentional psychology must conform to the deductive-nomological model of a theory if it is a causal-explanatory theory. However, there are other models of the way that causal-explanatory theories operate. I shall describe one of these models in this section as an instructive corrective to the distorting influence of the deductive-nomological model.

In looking at this alternative approach, a useful place to start is to consider the work of philosophers of science who have criticized the unrealistic conception of laws fostered by the deductive-nomological conception of theories. Under this conception, the laws of a scientific theory are supposed to be universal generalizations that are true without exception. However, several philosophers of science have argued that the laws used in actual successful scientific theories fail to meet this exacting requirement and are known to do so.[21] For example, even before the advent of relativity and quantum theory, Giere observes, it was known that no two bodies in the universe exactly satisfied Newton's laws of motion and law of gravitation. The only possible way two bodies could precisely satisfy Newton's laws would be if they were the only two bodies in the universe, or if they existed in a perfectly uniform gravitational field. But the existence of numerous other bodies with their own gravitational influences rules out the first possibility, and inhomogeneities in the distribution of matter in the universe rule out the second possibility. These familiar facts made it obvious that Newton's laws could not be viewed as literally true descriptions of the behavior of real-world systems. Many similar examples from a broad range of sciences have been cited to illustrate the same point. On the basis of such considerations, Giere suggests that the laws cited in scientific theories should not be seen as describing exceptionless regularities about real-world systems; rather, they should be seen as defining the kinds of idealized systems that are central elements of scientific models.

Many contemporary philosophers of science have adopted a model-based approach to understanding scientific theories and the role of laws within them. While all these philosophers take the notion of a model to be central to scientific theorizing, they differ slightly in the way they understand the notion. Some exponents of the semantic view of scientific theories identify models with the kind of set-theoretic structures that provide semantic interpretations within a logical system.[22] Other exponents identify a model with a state-space structure.[23] Yet other philosophers advancing a model-based approach regard models as abstract but nonmathematical in character. Giere describes models as abstract entities that are the creations of the human imagination. Models, so construed, seem to be fictional entities that are constructed to provide simplified and idealized representations of actual complex systems. He says that models can be presented in many different ways—in terms of a diagram, a set of equations, or an informal description. But these modes of presentation are different from the model itself. It will prove to be convenient in our subsequent discussion to adopt Giere's way of understanding models, which is, I believe, close to one important scientific understanding of the concept.

As I have said, in Giere's model-based approach, the laws of a theory are taken to be elements in stipulative definitions of models. For example, Giere construes the laws of motion and the law of gravitation in Newtonian mechanics not as descriptions of exceptionless regularities in real-world systems but as principles that define what it is for a system to be a Newtonian gravitational system. Such a system is simply stipulated to be one that satisfies these laws precisely. Of course, classical mechanics defines more than just one kind of system. Corresponding to the different force functions, there are different kinds of systems. For example, adding a linear restoring function instead of an inverse square function to the laws of motion yields a general model for a simple harmonic oscillator. This general model can be applied to specific kinds of systems: a pendulum with a small amplitude, a mass hanging from a spring, a diatomic atom, and so on. So classical mechanics is best understood as a general explanatory scheme for generating many different models, all constructed from a common toolkit of explanatory entities. What the physicist acquires in learning classical mechanics is a facility for using this common explanatory scheme to construct models in application to particular problem situations.

Model-based approaches to scientific theories such as Giere's emphasize the role of the human imagination in the construction of models. But the identification of theories with families of models does not mean that theories lack empirical import. On the contrary: models are supposed to be devices for representing the world and real-world systems. How do models represent the world? Some exponents of the model-based approach argue that models represent the world by virtue of their isomorphism,[24] or partial isomorphism,[25] with real-world systems. On the other hand, Giere and Teller argue that many scientific disciplines regularly appeal to highly idealized models that are not even partially isomorphic with real-world systems.[26] They advocate that the important representational relation between model and target system is similarity: models can represent a system by being similar to it in certain respects and to certain degrees. Giere and Teller acknowledge that representation by virtue of similarity will be a trivial matter unless the relevant respects and degrees of similarity are specified precisely. But they point out that a specification of these is a very context-sensitive matter, depending on the specific problem situation and its larger scientific context.

The picture of theoretical representation that results from this approach is very different from the picture expressed in the deductive-nomological model, which says that a theory represents the world by virtue of its laws being true or false of the world. In contrast, the model-based approach says that a theory's representational capacities lie in its associated models, which, as nonlinguistic entities, cannot be true or false of the world. Nonetheless, models are no less representational than maps, graphs, and physical scale models. The crucial point is that something counts as a representation because it is used for representational purposes. The specific way in which a model is supposed to represent a real-world system is implicit in its application to the system. For example, the application of the model of a two body gravitational Newtonian system to an actual system, say the Earth-Moon system, can be expressed in an hypothesis to the effect that the actual system is an instance of a gravitational Newtonian system. Giere calls these *theoretical hypotheses*. He stresses that when properly formulated, they specify the relevant respects and degrees of similarity between model and target system that are intended. The hypothesis of the example should say, for instance, that positions and velocities of the Earth and the Moon are very close to those of a two-body gravitational Newtonian system. Such hypotheses are truth evaluable. So in summary, Giere's model-based approach says that a theory is best identified not just with a family of models but also with the theoretical hypotheses associated with its particular applications.

Returning now to intentional psychology, I wish to suggest that intentional psychology is best understood as a model. More precisely, the suggestion is that when ordinary people engage in predicting and explaining a person's behavior, they do so by bringing a model to bear on that person's behavior. If their aim is predict the person's behavior, the model is used to generate an output in the form of a decision or action from an input about known beliefs and desires. Alternatively, if their aim is to explain the person's behavior, the model is used to construct a plausible etiology for a known decision or action.

On this suggestion, the generalizations of intentional psychology have a limited role. They do not describe regularities in the behavior of real agents, but stipulatively define the model of a rational agent. Such a rational agent is as idealized as a perfectly homogenous gravitational field or a pendulum operating in a frictionless environment or any other fictional theoretical entity. Such entities are needed in models because real-world systems are too complex to describe in perfectly accurate detail. Given the nature of models, the generalizations that define them do not have to be universal generalizations that are true without exception. The generalizations merely have to specify simplified templates outlining how a rational agent would behave under various highly idealized circumstances.

I have suggested that ordinary people understand intentional psychology in terms of a model about rational agents. But it may be more realistic to see intentional psychology as involving a family of models rather than a single model, somewhat in the same way that a scientific theory like classical mechanics involves a family of models built on a common explanatory scheme. Perhaps what the folk first grasp in learning intentional psychology is a basic general explanatory scheme that can have many specific instances. They learn the basic picture of the contents of normal minds (beliefs, memories, desires, wishes, and fears) and how these contents normally interact; they learn how a rational agent acquires beliefs by perception and inference and combines these beliefs with desires to make decisions that result in actions. But as their grasp of intentional psychology becomes more sophisticated they learn how to construct specific models for particular interpretative tasks. They learn how to assemble from a common toolkit of intentional elements a detailed explanation of how a rational agent would behave under increasingly complex combinations of intentional states. So, for example, they may develop a sophisticated explanatory model of weakness of will according to which a rational agent's normal deliberative processes are subverted by a powerful desire or impulse. But if something like this is the case, the more sophisticated models would all be variations on a general model based on basic belief-desire psychology.

The idea that intentional psychology is to be understood as involving a model or a family of models is not entirely new. Several other philosophers have recognized the appeal of this idea and sought to use it to explain some distinctive features of intentional psychology. Heidi Maibom and Peter Godfrey-Smith have argued that folk psychology is a family of models, and they also take as the starting point for their arguments Giere's version of the model-based approach to scientific theorizing.[27] Their theses are slightly broader than the one I am advancing, as they concern folk psychology as a whole, so that their models include nonintentional elements such as sensations, emotions, moods, and personality traits that strictly fall outside the domain of intentional psychology. Nonetheless, their papers make some excellent points about the explanatory power of the model-based approach, which will prove a useful springboard for my discussion in the next section.

The central focus of Maibom's "The Mindreader and the Scientists" is the problem of how ordinary people can know folk psychology. If one assumes that folk psychology is a theory that consists of universal generalizations, it is implausible that the folk have ordinary knowledge of folk psychology; ordinary people are not able to articulate explicitly any universal generalizations that are both true and nontrivial. Indeed even philosophers of mind have trouble doing this. In response to this point, it is often said that the folk have tacit knowledge of universal generalizations in the same way that Chomskian psycholinguists say they have tacit knowledge of language grammars. However, Maibom argues that the ordinary people's knowledge of folk psychology does not conform to the any of standard accounts of tacit knowledge worked out in psycholinguistics. In her view, the solution to the problem is to abandon the orthodox view about the content of folk psychology. If one sees folk psychology as about models rather than universal generalizations, one can suppose that people have ordinary, everyday knowledge of folk psychology and one no longer needs to resort to appeals to special, tacit knowledge. Her suggestion is that knowledge of folk psychology consists in the ability to construct idealizing models and the ability to apply these models to the behavior of real agents on the basis of theoretical hypotheses. The picture that Maibom sketches is very similar to the one that I wish to sketch, though her emphasis on the broader subject of folk psychology means that the models go beyond simple models of rational intentional psychology that I posit.

The focus of Godfrey Smith's "Folk Psychology as a Model" is somewhat different from the focus of Maibom's paper and of this paper. He proposes a model-based approach to folk psychology not so much as an application of the semantic view of theories but as an illustration of a distinctive style of scientific theorizing. Scientists deploy this style, he says, when they are ignorant about the detailed workings of very complex system. The strategy involves constructing a range of models about the workings of the system in ways that are noncommittal about their applications. Folk psychology is, he believes, a prime target for this strategy of theorizing. He also compares the way in which scientists and ordinary people construct families of models on the basis of common schematic patterns and common repertoires of explanatory elements. He likens the ordinary person's grasp of folk psychology to a scientist's facility for creating a range of models according to a general schematic pattern. As well as the multiplicity of models, Godfrey Smith also emphasizes the multiplicity of possible applications of models to real-life systems. He adopts Giere's view that the important representational relation between models and reality is similarity: a model is applied to a real-world system by a theoretical hypothesis, or "construal" in his terminology, that states that the system is like the model in certain respects and to certain degrees. Instead of seeing this as a defect of Giere's model-based approach, he sees it as a virtue, especially in connection with the particular scientific strategy that he has in mind. For what is distinctive about this strategy is that models are investigated with scant regard to how they are applied to real-world systems. Likewise, the folk are sometimes interested, he stresses, in considering folk psychological models of mind without paying attention to their construals.

With variation in models and variation in construals possible, Godfrey Smith claims that many ascriptions of mental state are semantically indeterminate and even lack truth value. While I do not think there is as much semantic indeterminacy as he does, some indeterminacy must be allowed. Indeed, the kind of indeterminacy that is allowed on Giere's model-based approach is crucial if a defense is to be mounted against an objection that accounts of intentional psychology emphasizing rationality are overly sophisticated. This objection takes the form of claiming that human infants and nonhuman animals are said to have beliefs and desires, even though they fail to possess the full-fledged rationality of adult humans. The model-based approach that I am advancing can readily answer this objection if it is permitted to say that the kinds of models that we apply to infants and nonhuman animals are very simple models centered on basic belief-desire psychology or that in applying uniform models to them we employ relaxed standards of similarity.

With this brief introduction to the model-based approach to intentional psychology, I turn to consider how it may be used to formulate a one-tiered compatibilism that provides a unified treatment of rationalizing and causal explanations.

ONE-TIER COMPATIBILISM

I started by considering a version of the logical connection argument that purports to show that the generalizations of intentional psychology have a special conceptual, normative character that renders them incapable of supporting predictions and causal explanations. However, in the last section I suggested that advocates of this argument have been overly influenced by the deductive-nomological model of the way theories license predictions and causal explanations. I claimed that if a model-based approach to scientific theories is adopted in place of the deductive-nomological approach, the conceptual and normative character of the generalizations pose no special difficulty regarding empirical prediction and causal explanation. In this section, I shall try to substantiate this claim in more detail as it applies to a model-based approach to intentional psychology. I shall also explain how such a model-based approach can deliver a form of one-tier compatibilism about reasons and causes.

As we saw earlier, the generalizations of intentional psychology, when read as restricted in scope to rational agents, have an a priori, conceptual character. If read as generalizations about all agents, rational and irrational, they are readily falsified. But, as generalizations about the behavior of rational agents, they are truistic. A model-based approach to intentional psychology can accommodate such truistic generalizations by construing them as principles defining models of rational agency. These principles, as enunciated by the folk, may be very rough and ready, but that is to be expected of the highly simplified mental models the folk employ.

As we saw in the last section, Giere's model-based approach to scientific theories explains how such principles, construed as definitional truths, are nonetheless capable of supporting empirical predictions and causal explanations. There is a simple logical point here that is worth pausing to consider. If one considers the way in which a model is applied to make a prediction about a particular case, one can capture the structure of the reasoning in the form of an explicit argument. Consider the following simple illustration of an application of a model from the point of view of Giere's approach.

> A two-body Newtonian gravitational system is one that precisely satisfies Newton's laws of motion and law of gravitation.
> The Earth-Moon system is very similar to a two-body Newtonian gravitational system in which the bodies have certain specified masses and certain specified distance apart.
> Therefore, the positions and velocities of the Earth-Moon system should approximate those of this two-body Newtonian gravitational system.

The simple logical point this argument illustrates is that even where the first premise is a definitional truth, the fact that the second premise is a contingent theoretical hypothesis means that the conclusion is also a contingent thesis. This is an enormously simplified representation of the structure of the application of a model, but it serves to make this logical point evident. (This point does not depend on Giere's construal of a theoretical hypothesis as a statement of similarity between model and target system: even if the hypothesis were to state an isomorphism, the same point about the contingency of the hypothesis would apply.)

The application of a model of rational agency to a particular case has a similar argumentative structure. Let us consider how a simple model of rational agency might be applied to the example used earlier to illustrate weakness of will. Let us suppose that the model is characterized by a number of generalizations including the generalization that if a rational agent judges that it is better, all things considered, to catch the morning flight, he catches the morning flight. Then this model can be used to make an empirical prediction about the target agent's behavior:

> If a rational agent judges that it is better all things considered, to catch the morning flight, he catches the morning flight.
> PM is a rational agent, or is close to being one in respect of basic deliberative psychology.
> Therefore, if PM judges that it is better, all things considered, to catch the morning flight, he will catch the morning flight.

Again, notwithstanding the presence of a definitional first premise, the argument yields a contingent empirical conclusion because the second premise is a contingent theoretical hypothesis. As it turns out, the prediction expressed by the conclusion is false in the particular example: prone as I am to weakness of will when it comes to sightseeing in Rome, I fail to catch the flight to Bologna even when I think it is best thing to do, all things considered, because I am not a completely rational agent. Again, the logic of the situation is clear: the definitional status of the generalizations of intentional psychology does not stop them from issuing in genuine empirical predictions.

One of the virtues of the deductive-nomological approach to scientific theories is that it yields a symmetrical account of prediction and causal explanation: a theory licenses a prediction or causal explanation about a particular event only when appropriate deductive relations hold between laws and statements about events, the only difference between prediction and causal explanation being that prediction takes place before the given event occurs and causal explanation after the event has occurred. By contrast, the model-based approach to theories leaves open the possibility of asymmetrical treatment of these things. As it happens, philosophers of science who have advocated a model-based approach to scientific theories have not discussed in any depth the way causal explanation is to be treated in this approach. On the other hand, there is an extensive literature, especially in the social and bio-medical sciences, that attempts to describe how causal reasoning and causal explanation work within models. Much of this literature uses the technical framework of structural equations. Judea Pearl's book *Causality* gives a state-of-the-art presentation of the framework.[28] Woodward and Hitchcock have made the basics of this framework accessible to philosophers.[29] I shall use some highly simplified parts of this framework to suggest in a brief outline how causal explanation might work in the models of intentional psychology.

A good place to start this outline is to describe more precisely how the models of intentional psychology might be represented within the structural equations framework. A characteristic feature of such models is that they are couched in terms of variables describing the intentional states and actions of a rational agent. Such a variable will more often than not be a simple binary variable, specifying whether the agent is in a certain intentional state or has performed an intentional action. Such a model typically does not employ variables describing nonintentional states such as neural states of the agent that do not bear on rational processes of thought and behavior. Another characteristic feature of the models, as I have already remarked, is that they involve generalizations about the intentional profiles of rational agents as they evolve over time. These generalizations are best understood, I believe, as generalized counterfactuals about what a rational agent would do under various hypothetical circumstances. Many of these generalizations have a high degree of invariance in that they accurately describe the behavior of rational agents under a wide variety of background conditions. As Woodward has emphasized, such invariance is characteristic of generalizations used in causal explanation.[30] In the framework under consideration, these general-izations are represented as structural equations. A structural equation is sim-ply an equation of the form

$$Y = f(X_1, \ldots, X_n),$$

where $X_1, \ldots, X_n$ list all the variables whose values play a role in determining the values of Y. Such a structural equation encodes a set of counterfactuals of the form "If it were the case that $X_1 = x_1, \ldots, X_n = x_n$, then it would be the case that $Y = f(x_1, \ldots, x_n)$" with one such counterfactual for each possible combi-nation of values for the variables $X_1, \ldots, X_n$.

A model can be more precisely specified as an ordered pair <V, E>, where V is the set of variables used to characterize the system in question, and E is a set of structural equations governing the system. There is a structural equation for each variable of the model, which appears on the left-hand side of the equation. The form of the equation depends on whether it is an exogenous (that is, one whose values are determined by factors outside the model) or endogenous variable (that is, one whose values are determined by factors within the model). If it is an exogenous variable, the equation states its actual value. If it is an endogenous variable, the equation states its value as a function of the values of the other variables that play a role in its causal determination.

A causal model and its set of structural equations can be depicted in a graphical representation. The variables form the nodes of the graph, and these nodes are connected by directed edges according to the following rule: an edge is drawn from X to Y if and only if the values of X play a role in determining the values of Y. In this case, X is said to be a *parent* of Y. An exogenous variable is without any parent in the graph for its model.

It will be useful to consider a particular example to illustrate this formal description of a model. Consider, for instance, the example involving deviant causation that was considered earlier as a problem for Davidson's two-tiered compatibilism. Recall that in the example a waitress intends to drop some plates noisily to embarrass her employer, but she is so unnerved by these thoughts that she inadvertently drops the plates noisily. A model of the situation will be somewhat untypical of intentional models in that it will employ variables for nonintentional states as well as intentional states of the agent. A simple model of the situation can be described as follows :

$I = 0$ if agent does not intend to drop glass
$I = 1$ if agent intends to drop glass noisily
$I = 2$ if agent intends to drop glass not too noisily

$N = 0$ if agent is not unnerved
$N = 1$ if agent is unnerved

$D = 0$ if agent does not drop glass
$D = 1$ if agent drops glass noisily
$D = 2$ if agent drops glass not too noisily

Its structural equations are:

$I = 1$
$N = 0$ if $I = 0$; and 1 otherwise
$D = 0$ if $I = 0$ and $N = 0$; and 1 otherwise

The graph for this model is represented in figure 7.1.

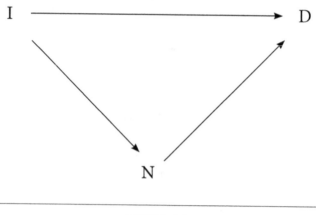

FIGURE 7.1

How does causal explanation fit into this picture? It is reasonable to suppose that causally explaining some event within a model involves citing the
event that counts as the cause of the given event in the model. (I shall assume
that it is possible to represent an event's occurrence or nonoccurrence in this
framework in terms of a variable taking appropriate values.) But when does a
model count one event as the cause of another? Various accounts have been
given of this matter, but I will, for expository convenience, settle on the following simplification of an account proposed by Woodward in *A Theory of
Causal Explanation*:

> X = x is *a cause* of Y = y in a model if and only if there is at least one route from
> X to Y in the graph for the model such that changing the value of the X will
> change the value of Y, given that the other direct causes of Y that are not on this
> route have been fixed at their actual values.

In the simple model of the example, we can see that the event of the waitress's
intending to drop the glass noisily (represented by I = 1) does not count as a
cause of her dropping it noisily (represented by D = 1) on this account. For
changing the value of I will not affect the value of D, when the actual value of
the variable N is held fixed. In other words, with the fact that the agent was
unnerved held fixed, a change in her intentions does not result in a change in
her actions. It appears then that her intentions do not control her actions in

the given circumstances, so that it is reasonable to conclude that her action is not intentional.

Much more needs to be done to motivate and justify this account of causal explanation. The important point, however, that I would like to emphasize is that this kind of account of causal explanation, whatever its precise details, is the kind that supports a one-tier form of compatibilism. Recall that a one-tier form of compatibilism, which I wish to propose in place of a two-tier form of the kind advocated by Davidson, says that the properties by virtue of which one event rationalizes another are the same properties as those by virtue of which the first event causes the second. I claim that our explanatory practices in intentional psychology support one-tier compatibilism and that the present framework captures this feature of our practice. To see this, consider a modification of the example of deviant causation just discussed. Let us suppose that the waitress forms the intention, as before, to embarrass her boss by noisily dropping the plates but is confident enough not to be unnerved by this intention. (The structural equations in the corresponding model will perhaps omit the variable N.) Within a simple model representing this new scenario, it is reasonable to infer that her intention both rationalizes and causes her action of noisily dropping the plates. Moreover, the model will vindicate the thesis that the property by virtue of which her intention provides a reason for her action—its being an intention with a specific content— is the same property as that by virtue of which her intention causes the action. Causal explanation in the structural equations framework is a relatively straightforward matter: the properties that are cited as causally efficacious do not have to be obscure properties mentioned in fundamental physical laws but may be intentional properties mentioned in the rough-and-ready generalizations of intentional psychology.

I have focused so far on explaining how a model-based approach to intentional psychology explains prediction and causal explanation in a way that vindicates one-tier compatibilism. But the original observation that motivated the logical connection argument was that the generalizations of intentional psychology have not only a conceptual, a priori character but also a normative character. How is this normative character to be accounted for in the model-based approach to intentional psychology? One initially attractive thought is that the idealization involved in the models of intentional psychology must play a role in the account. The models of intentional psychology concern highly idealized rational agents whose behavior often diverges from the behavior of real agents. Perhaps this idealized behavior can represent the way real agents should behave. There is something in this thought, but it is far from being sufficient by itself to capture the normative character of the generalizations of intentional psychology. Let us review the relevant feature of models, as I have represented them. I have taken the counterparts of the generalizations of intentional psychology within models to be structural equations. These encode counterfactuals that simply tell us what an ideal rational agent *would* do under various circumstances. They do not tell us what an agent *should* do in these circumstances. As Kripke has argued forcefully, the first kind of descriptive counterfactual is very different in content from the second kind of prescriptive counterfactual.[31] But it is the second kind of prescriptive counterfactual that is expressed by the generalizations of intentional psychology.

All the same, the fact that the models employed in intentional psychology represent the behavior of idealized rational agents is, I think, one part of the account of the normativity of the generalizations of intentional psychology. Another part of the account is, I think, that the idealizing models of intentional psychology have two separate functions—a descriptive function and a prescriptive function—and each of these functions is associated with a distinctive direction of fit in our practices of relating models to real systems. In their descriptive function, models are used, naturally enough, to represent and describe the way real agents behave. This function is central to the use of models in prediction and causal explanation. This function is evidenced in the applications of models to reality that presuppose a direction of fit from model to real system: the models must fit the real systems in the sense that where there is divergence between them, the model stands in need of correction. But models can also have a prescriptive function: as representations of the behavior of idealized rational agents they tell us how one should reason and behave if one is rational. This second function is evidenced in those applications of models to reality that presuppose the opposite direction of fit from real system to model: the real agent must fit the model in the sense that, where there is divergence between model and real agent, we take the real agent to stand in need of correction. (Both sorts of application require auxiliary assumptions, which may be rejected where model and real system diverge.)

Of course, no single application of a model to a real system can realize both functions at the same time. But it is possible, and indeed common, for one and the same model to be applied in its two different roles to the same real agent. On many occasions we deploy a model of a rational agent in its prescriptive role to tell us how a real agent should act in a given set of circumstances and then employ it in its descriptive role to predict how she will act in those circumstances. Models, by virtue of their idealized character have a Janus-faced nature that enables their use in description and prescription.

That is one of the fundamental insights that a model-based approach can bring to the explanation of the intentional psychology.

Much more needs to be said about the contrast between models' descriptive and prescriptive functions and their associated directions of fit. In particular, more needs to be said about how an application of a model to a real system has a certain direction of fit. I do not, however, have the space here to explore these issues further. My intention is to provide the barest outlines of an account of how the normativity of the generalizations of intentional psychology can be captured within a model-based approach. My suggestion is that the normativity of such generalizations is captured within this approach by the idea that the models of rational agents defined by these generalizations have a prescriptive role, as is evidenced by applications presupposing a distinctive direction of fit from real agents to models. No doubt there is a great deal more that could be said about the normative character of the generalizations of intentional psychology. But my programmatic hope is that these further aspects of their normativity can be systematically accounted for in terms of the prescriptive function of models.

CONCLUSION

The starting point for my discussion was the opposition between Scientific Naturalism and Kantian rationalism over the kind of theory that intentional psychology is. Scientific Naturalism takes it to be a theory that individuates intentional states in terms of the causal relation, whereas Kantian rationalism takes it to be a theory that individuates intentional states in terms of the reason-giving relation. Since these relations are very different in nature, it would appear that both sides cannot be right about the nature of intentional states. Indeed, the logical connection argument purports to show that Scientific Naturalism is incorrect on the grounds that the generalizations of intentional psychology have a special character that precludes them from supporting empirical predictions and causal explanations.

I have tried to argue that the logical connection argument is not successful, as it is based on the unsatisfactory deductive-nomological model of how a causal explanatory theory works. If one adopts an alternative model-based approach to theories and applies it to intentional psychology, one can see that even though the generalizations of intentional psychology have the character imputed to them by the logical connection argument, they can nevertheless support empirical predictions and causal explanation. Moreover, model-based account of intentional psychology has no need to posit two tiers of generalizations, creating an artificial dualism between rationalizing explanations and causal explanations. A model-based approach supports, I argue, a one-tier compatibilism that vindicates the commonsense view that the properties that underpin reason-giving relations between intentional states and actions also underpin causal relations between them.

Of course, the model-based approach to intentional psychology admits that the reason-giving relation is a fundamentally different kind of relation from the causal relation—one is conceptual and a priori, and the other, contingent and a posteriori. Nonetheless, the approach explains how these different relations overlap in extension for the most part if cases of irrationality are set aside: in cases of rational reasoning and action, the intentional states that provide the reason for some other intentional state or action are also its causes. The explanation of this overlap consists in the fact that we apply the same model to give the rationalizing explanation as we apply to give the causal explanation. The harmony of the two kinds of explanation is no remarkable coincidence: it follows straightforwardly from the commonality of model. The model has, to be sure, slightly different roles in the two kinds of application. It has a prescriptive role with a certain direction of fit in its application to rationalizing explanation and a descriptive role with the opposite direction of fit in its application to causal explanation. Nonetheless, the striking fact that these two kinds of explanation harmonize so happily itself receives an explanation.

NOTES

1. Martin Davies and Terry Stone, eds., *Folk Psychology: The Theory of Mind Debate* (Oxford: Blackwell, 1995); Shaun Nichols and Stephen Stich, *Mindreading: An Integrated Account of Pretence, Self-Awareness, and Understanding of Other Minds* (New York: Oxford University Press, 2003).

2. David Armstrong, *A Materialist Theory of Mind* (London: Routledge and Kegan Paul 1968); Jerry Fodor, *Psychosemantics* (Cambridge, Mass.: MIT Press, 1987); David Lewis, "An Argument for the Identity Theory," *Journal of Philosophy* 63 (1966): 17–25; Hilary Putnam, "The Nature of Mental States," reprinted in *Mind, Language, and Reality* (Cambridge: Cambridge University Press, 1975), 429–40.

3. See Gary Hatfield, *The Natural and the Normative* (Cambridge, Mass.: MIT Press, 1990); Grant Gillett, "Actions, Causes, and Mental Ascriptions," in *Objections to Physicalism*, ed. Howard Robinson (Oxford: Clarendon Press, 1993), 81–100.

4. Nick Zangwill, "The Normativity of the Mental," *Philosophical Explorations* 8 (2005): 1–19.

5. Wilfrid Sellars, *Empiricism and the Philosophy of Mind* (Cambridge, Mass.: Harvard University Press, 1956).

6. Robert Brandon, *Making It Explicit* (Cambridge, Mass.: Harvard University Press, 1994); John McDowell, *Mind and World* (Cambridge, Mass.: Harvard University Press, 1994); McDowell, "Naturalism in the Philosophy of Mind," in *Naturalism in Question*, ed. Mario De Caro and David Macarthur (Cambridge, Mass.: Harvard University Press), 91–105; and the later Hilary Putnam, "Why Reason Cannot Be Naturalized," in *Realism and Reason* (Cambridge: Cambridge University Press, 1983), 229–47.

7. Stuart Hampshire, *Thought and Action* (London: Chatto and Windus, 1959); Antony Kenny, *Action, Emotion, and Will* (London: Routledge and Kegan Paul, 1963); Abraham Melden, *Action* (London: Routledge and Kegan Paul, 1961); Norman Malcolm, "The Conceivability of Mechanism," *The Philosophical Review* 77 (1968): 45–72.

8. Donald Davidson, "Actions, Reasons, and Causes," *Journal of Philosophy* 60 (1963): 685–700; Davidson, "Mental Events," in *Experience and Theory*, ed. Lawrence Foster and J.W. Swanson (Amherst: University of Massachusetts Press, 1970), 79–101.

9. David Braddon Mitchell and Frank Jackson, *The Philosophy of Mind and Cognition* (Oxford: Blackwell, 1996).

10. Donald Davidson, "Causal Relations," *The Journal of Philosophy* 64 (1967): 691–703; Davidson, "Mental Events."

11. Davidson, "Mental Events," 222. For this reconstruction of Davidson's argument, see Jaegwon Kim, "Psychophysical Laws," in *Actions and Events: Perspectives on the Philosophy of Donald Davidson*, ed. Ernest Lepore and Brian McLaughlin (Oxford: Oxford University Press, 1985), 369–86.

12. Kathleen Lennon, *Explaining Human Action* (London: Duckworth, 1990).

13. See, for example, John Bishop, *A Causal Theory of Action* (Cambridge: Cambridge University Press, 1989).

14. See Jaegwon Kim, "The Myth of Non-Reductive Materialism," *Proceedings of the American Philosophical Association* 63 (1989): 31–47; Jerry Fodor, "Making Mind Matter More," *Philosophical Topics* 17 (1989): 59–80; Fred Dretske, "Reasons and Causes," *Philosophical Perspectives* 3 (1989): 1–15.

15. Donald Davidson, "Thinking Causes," in *Mental Causation*, ed. John Heil and Alfred Mele (Oxford: Clarendon Press, 1993), 13.

16. Jaegwon Kim, "Can Supervenience and 'Non-Strict Laws' Save Anomalous Monism?" in *Mental Causation*, ed. John Heil and Alfred Mele (Oxford: Clarendon Press, 1993), 19–26; Brian McLaughlin, "Davidson's Thinking Causes," in *Mental Causation*, ed. John Heil and Alfred Mele (Oxford: Clarendon Press, 1993), 41–50.

17. Davidson, "Thinking Causes."

18. Ibid., 10.

19. Ibid., 14.

20. C. G. Hempel, *Aspects of Scientific Explanation and Other Essays in the Philosophy of Science* (New York: Free Press, 1965); Ernest Nagel, *The Structure of Science: Problems in the Logic of Scientific Explanation* (New York: Harcourt, Brace and World, 1961).

21. Nancy Cartwright, *How the Laws of Physics Lie* (Oxford: Oxford University Press, 1983); Ronald Giere, *Explaining Science: A Cognitive Approach* (Chicago: University of Chicago Press, 1988); Giere, *Science Without Laws* (Chicago: University of Chicago Press, 1999).

22. Patrick Suppes, "A Comparison of the Meaning and Uses of Models in Mathematics and the Empirical Sciences," *Synthese* 12 (1960): 287–301.

23. Bas van Fraassen, *The Scientific Image* (Oxford: Oxford University Press 1980); Frederick Suppe, *The Semantic View of Theories and Scientific Realism* (Urbana: University of Illinois Press, 1989).

24. Van Fraassen, *The Scientific Image*.

25. Newton Da Costa and Steven French, *Science and Partial Truth: A Unitary Approach to Models and Scientific Reasoning* (Oxford: Oxford University Press, 2003).

26. Giere, *Explaining Science*; Paul Teller, "Twilight of the Perfect Model," *Erkenntnis* 55 (2001): 393–415.

27. Heidi Maibom "The Mindreader and the Scientists," *Mind and Language* 18 (2003): 296–315; Peter Godfrey Smith, "Folk Psychology as a Model," *Philosophers' Imprint* 5, no.6 (2005): www.philosophersimprint.org/005006.

28. Judea Pearl, *Causality, Models, Reasoning, and Inference* (Cambridge: Cambridge University Press, 2000).

29. James Woodward, *A Theory of Causal Explanation* (New York: Oxford University Press, 2003); Christopher Hitchcock, "The Intransitivity of Causation Revealed in Equations and Graphs," *Journal of Philosophy* 98 (2001): 273–99.

30. Woodward, *Making Things Happen*.

31. Saul Kripke, *Wittgenstein on Rules and Private Languages* (Oxford: Blackwell, 1982).

PART IV

META-ETHICS AND NORMATIVITY

8

METAPHYSICS AND MORALS

T. M. Scanlon

Judgments about right and wrong and, more generally, judgments about reasons for action, seem, on the surface, to claim to state truths. They obey the principles of standard propositional and quantificational logic, and satisfy (at least most of) the other "platitudes" about truth enumerated by Crispin Wright and others.[1] Moreover, some of these judgments seem to be true, rather than false, if anything is. It would clearly be wrong for me to present a paper that was in fact written by someone else, and in light of this I have good reason not to do so (even though in some respects it might have been better if I did.) In addition, I find it difficult to resist saying that I believe that these things are so.

I thus find myself strongly drawn to a cognitivist understanding of moral and practical judgments. They strike me as the kind of things that can be true, and their acceptance seems to be a matter of belief. But strong arguments against accepting these appearances have been offered from several quarters. What I want to do this afternoon is to consider some of these arguments and to try to identify and assess my reasons for resisting them. There is, of course, a vast literature on these topics, rich in detail and in the variety of positions delineated in it. I will not do justice to this richness in the present paper. This will be an overview, with brief and selective examinations of a few arguments and positions.

When I just said that some moral judgments seem to be clearly true rather than false, what I was claiming is that they are supported by our "ordinary criteria" for answering such questions. Standard ways of arguing about our obligations to one another clearly support these conclusions. So one thing that those who deny that moral judgments should be understood as straightforwardly true may be claiming is that if moral judgments are understood in this way then they claim, or presuppose, more than these "ordinary criteria" can deliver. We might understand moral terms, know how to apply them, and be confident in their applicability to certain situations; yet they could still fail to be true because they fail to "describe the world."

Here I think there is some ambiguity in what is meant by "the world." On its most expansive reading, "the world" might be taken to be simply the reflection of those sentences that are true, when judged by the criteria appropriate to sentences of their type. If, according to the best moral criteria, it would be wrong of me to deliver on this occasion a paper written by someone else, then on this expansive reading it is, trivially, a "truth about the world" that this would be wrong. So if it is claimed that moral judgments do not state truths *because* if taken as true they would not "describe the world," this claim must be understood in terms of some more restrictive idea of "the world." Commonly, I think, this more restrictive sense is "the world as described by science" or "the natural world" where this is understood to include all and only physical and (insofar as they may be different) psychological facts. Thus it is sometimes said that there is a problem about how to fit moral truths into a scientific view of the world. It is, however, not clear what this problem is.

The thought might be that moral judgments endorsed by ordinary moral criteria, understood literally as true, involve claims about the natural world that conflict with those of our best science. Some kinds of (non-moral) claims do have this problem. Believers in witches, ghosts, and demonic possession, may have "ordinary criteria" for telling when these phenomena are present. Yet their claims fail to describe the world accurately, and are literally false. This is because these claims involve claims about events occurring in space and time, and about which things cause or are caused by them. These causal claims are false according to the proper criteria for assessing causal claims.

Do moral judgments, similarly, make claims that conflict with those of science? Moral judgments are often *based on* claims about the occurrence of events in time and space and about causal relations between them. Whether an action is wrong may depend, for example, on whether it causes injury. But these claims are to be assessed according to ordinary empirical criteria, not special moral ones. Moral judgments themselves, I would say, make no *causal* claims beyond these.

It might be replied, however, that even if moral judgments themselves make no special causal claims, claims to know truths about moral matters would involve or presuppose special causal claims. The idea would be that if moral judgments express truths, then in arriving at our opinions about moral matters we must be responding to the causal powers of special moral properties in the world. But moral knowledge need not be understood on this perceptual model. There is good reason to understand perceptual beliefs, and other beliefs about the physical world, in this way. It is, after all, part of their content that they are about how things are with physical things that exist at a distance from us. How could we be reliably in touch with such things except by being causally affected by them? But moral beliefs, mathematical beliefs, and other beliefs about abstract matters are different. They do not purport to be about things located apart from us in space, or anywhere else. It is therefore plausible (in a way that it is not in the case of beliefs about the physical world) to suppose that these are matters that we can discover the truth about simply by thinking about them in the right way. It may not be easy to say more exactly what this way is, and the answer clearly depends on the concepts involved. In some cases it may involve the right kind of sequential reasoning, in others, perhaps, the right kind of mental picturing, in others, carefully focusing our attention on the relevant features of the ideas in question and being certain that we have distinguished them from other, irrelevant factors. But as long as there is some way of doing this correctly, there is a difference between getting it wrong and getting it right.

Suppose I am correct that neither moral judgments, understood as true, nor our claims to know such truths involve claims about the natural world that conflict with those of science. Taking moral judgments to be true still might be seen as "incompatible with a scientific view of the world" just because they do *not* make claims about things that exist in space and time and the causal relations between them. But this does not seem to me to be a genuine incompatibility. Science may claim to be a complete account of the occurrence of events in the spatio-temporal world and of the causal relations between them. But accepting this claim does not mean accepting that there are no true statements about anything other than what science deals with.

If science does not claim this, perhaps the best metaphysics does. Metaphysics may make claims about what "the world" is like that go beyond those of science (whether or not it allows that there is more to the world than what science deals with.) So it may be that what moral judgments would conflict with if they were understood as being true is not science but metaphysics. It might be said that moral judgments, understood as making truth claims, would involve metaphysical claims that go beyond what is guaranteed merely by their correctness according to ordinary moral criteria: to claim that they are true would be to claim for them an "intrinsic metaphysical *grativas*."[2] But I do not see why this need be the case. Moral judgments have sometimes been understood as based on a teleological conception of nature in which the world, and some kinds of objects within it, are taken to have certain assigned goals or purposes. But moral judgments as I understand them do not depend on such a view of the world, nor do they need the support of other metaphysical claims.[3]

But this lack of metaphysical import on the part of moral judgments might be held against them. If moral judgments do not make, or need the support of, metaphysical or scientific claims, then what kind of claims "about the world" do they make? Why isn't it idle to take them as true? The answer is that moral judgments are not idle because they make claims about what we have reason to do, and this is something of importance to us, as rational beings. The kind of *gravitas* that they require is thus not metaphysical but normative. This gives rise to at least two problems.

The first is a problem about the adequacy of ordinary moral criteria to deliver the normative significance that moral judgments claim for themselves. To claim that it would be wrong to act in a certain way is to claim that so acting would violate standards that people have strong, normally conclusive, reasons to abide by. If, in deploying our "ordinary criteria" for deciding whether an action is wrong, we are just determining whether that action meets certain standards, there remains the question what reason people have to take these standards seriously. Indeed, it may be asked whether there in fact are any standards that have the authority that moral judgments are normally taken to claim. This question cannot be answered simply by showing that certain actions are wrong according to our "ordinary moral criteria," that is to say, according to the standards we generally accept, since it is a question about the authority of those standards. It seems to follow that these criteria alone do not suffice to establish that judgments about moral wrongness are *true*. To show that they are we would need to meet this challenge by showing that these criteria have the authority claimed for them.

I believe that this challenge can be met. But even if it can, a second, deeper problem remains. Insofar as moral judgments involve claims that we have reasons for acting in certain ways, it may be said that they are not the kind of thing that can be true or false because claims about reasons are not the kind of thing that can be true or false. It seems to me that this is the central issue and that most of the problems that are thought to beset a cognitivist understanding of moral judgments are inherited from, or at least shared by, a cognitivist understanding of practical judgments more generally. Let me turn then, from the question of whether moral judgments can be understood as straightforwardly true to the question of whether judgments about the reasons we have can be so understood.

To begin again at the beginning: I am strongly drawn toward a cognitivist understanding of judgments about the reasons we have. These judgments obey the usual principles of propositional and quantificational logic, and (at least most of) the other platitudes about truth. Moreover, some of these judgments seem clearly to be true rather than false. The fact that jumping into the audience in the middle of a lecture would lead to serious injury and embarrassment is a reason for me not to do this. Some might claim that this is a reason only because I have certain desires. But those who hold that this is so and those who deny it agree that I in fact have such a reason. Why not say, then, that the claim that I have this reason is true?

Objections to saying this are of two related kinds. The first holds that judgments about reasons are not true because there is no fact in the world that they describe. The second holds that if we took judgments about reasons merely to describe such facts then we would be unable to account for the practical significance of these judgments. I will comment briefly on the first of these objections, and spend the remainder of my lecture discussing the second.

I said above that moral judgments as I understand them do not conflict with the claims of science, nor do they need the support of, nor could they be supported in a plausible way by, metaphysical claims. The same seems to be true of judgments about reasons. What may lie behind the worry that there are no "facts" that true normative judgments could capture is not this metaphysical worry but two different ones. First, in applying some of our concepts—in identifying objects as chairs, for example—we may be said to be applying relatively clear criteria implicit in our concepts. In the case of reasons, however, it is unclear what these criteria are. Taking C to be a reason for A, it might be said, is not a matter of applying criteria implicit in our understanding of the relation "counts in favor of," but simply a matter of reacting to C in a certain way. People can react to the same things in different ways, and there is no basis for counting one of these ways as uniquely correct. I will refer to this as the problem of insufficiently determinate content (or, for short, the problem of content.)

Second, given the practical significance of recognizing a consideration as a reason, it may seem that the reaction in question is not properly understood as classifying that consideration as one of a certain kind, but rather a reaction of some other, less cognitive sort. This practical significance is often taken to be a matter of motivational force. Since I believe it is better understood in normative terms, I will refer to this problem as the problem of normative significance.

I believe that the problem of content and the problem of normative significance are widely regarded as providing the strongest support for reluctance to say that judgments about reasons are true in any substantial sense, despite their apparently assertoric form. These problems may be stated in metaphysical terms. The problem of normative significance might be put by saying that the world does not contain facts with 'oughts' built into them. But in my view neither problem is fundamentally a metaphysical one.

The problem of content arises, rather, from the view that persistent first-order disagreement about reasons is quite legitimate, because the concepts in question provide insufficient structure for us to tell which way of thinking about the reasons we have is thinking about it "in the right way." It is when our understanding of the question gives us insufficient basis for saying why *our* way of assessing matters is the right one that we may have to fall back on a claim "just to see it," which may be characterized as an appeal to "intuition" in a problematic quasi-perceptual sense.

This problem raises difficult issues about which I will confine myself here to two brief remarks. The first is that in the case of practical reasoning, the problem of content can be made to seem more serious than it is by considering it as a problem about reasons in general. It is quite true that the bare concept of a reason—a consideration that counts in favor of some attitude—provides us with very little guidance about how to go about deciding what reasons we have. But it is unrealistic to expect more structure at this level of abstraction. Questions about reasons take on determinate structure only when they are more specific questions about reasons of particular kinds, for particular things (such as the examples I have given about my reasons for not jumping into the audience, or breaking into song, in the middle of a lecture.) Even at this level there remains room for disagreement in some cases, but in most cases of this kind it no longer seems to be just a matter of "how one reacts."

Second, it should be borne in mind that an understanding of the question being addressed can lend support to one's judgment that a certain consideration is a reason in other ways than by providing substantive grounds for the truth of that claim. An understanding of the concepts involved (for example, of what the consideration in question is supposed to be a reason for) may determine standards of relevance. By helping to identify the kinds of distinctions that need to be made in order to be thinking about the question in the right way, and hence about the kind of errors we might be making, this can give us grounds for confidence that we are not making such errors. Experience in thinking carefully and reflectively about such matters can also provide such support. The fact that a judgment is "intuitive" in the sense being discussed therefore does not mean that it is a guess, or a hunch, or "just a feeling."

Let me turn now to the problem of normative significance, and hence to what I earlier called the second line of objection to construing judgments about reasons as true or false. This is the objection that if we take judgments about reasons as things that can be true, and understand the acceptance of such a judgment as a belief, we will be unable to account for the practical significance of these judgments.

The difficulty here is not a problem about how to fit the idea of a reason or the relation "counting in favor of" into a scientific world view. Our beliefs about reasons, if they are beliefs, need not be caused by interaction with this relation. Nor need this relation be causally active in producing actions. It is an agent's *acceptance* of a judgment about the reasons he or she has that does this. Such acceptance, whether it amounts to belief or not, is a psychological state, and hence the kind of thing that figures in ordinary psychological explanations. The state of accepting a judgment, or having a belief, is just as "naturalistic" as having a desire. Desires may, phenomenologically, present themselves more as "urges" or "tugs" than other intentional states such as beliefs or judgments do. But an explanation of behavior that appeals to factors which have this phenomenological character is not for that reason closer to being a causal explanation. The experience of judging something to be true is a psychological phenomenon just as much as feeling a "desire." Both are, I presume, occurrences with some causal basis. How these experiential states (feeling desire or conviction), and the corresponding enduring states (such as belief) are related to a genuinely causal account of what goes on in us is a deep problem. But a desire (or the feeling of an urge) is not, any more than a belief is, an awareness of a cause. So if there is a problem, it must lie elsewhere than in the demands of a scientific outlook.

The problem is often put in terms of motivation. The acceptance of a judgment about one's reasons for action cannot be a belief because, it is said, beliefs are not the kind of things that, by themselves, can move one to act. Only desires can do that.[4] I have doubts about the idea of motivation that this statement of the problem appeals to. It seems to me to be troublingly ambiguous between a causal notion and a normative one. But I want to set aside this formulation of the problem. Whether or not there is a problem about how beliefs can motivate, I believe that there is a parallel problem (at least an apparent one) about the requirements of rationality. In what follows I want to describe this problem, discuss its anti-cognitivist thrust, and consider a possible solution to it. I believe that this solution, if it is one, would bring with it a solution to the problem considered in its motivational form.

The problem I have I mind is this. If a person judges that she has conclusive reason to do X at t, then two things follow. First, insofar as she does not abandon or forget this judgment, she is irrational if she does not intend to do X at t.[5] Second, the fact that she holds this judgment about her reasons can explain her intending to do X at t, and her so acting. The problem is to explain this connection—to explain how a *belief* could rationally require a certain intention to act, and how it can explain this intention and this action.[6]

Before addressing the problem, I want to say something about the idea of rationality in terms of which I am understanding it. We should, I believe, distinguish between two kinds of normative claims. The first kind are claims about the reasons that people have—about what counts in favor of doing what, about what counts in favor of believing what, and about what counts in favor of having other attitudes. The claim that the embarrassment that would result from my singing "The Hills Are Alive with the Sound of Music" right now counts as a reason not to do so is a true claim of this kind.

But not all normative claims are direct substantive claims about reasons. In particular, some claims about what it would be irrational for someone to do are not claims about the reasons that person has. For example, if a person believes that p, then it would be irrational for him to refuse to rely on p as a premise in further reasoning, and to reject arguments because they rely on it. To say this is not to say that the person has good reason to accept such arguments. Perhaps what he has most reason to do is to give up his belief that p. The claim is only that *as long as he believes that p*, it is irrational of him to refuse to accept such arguments. Similar connections hold in practical reasoning: if a person has a certain end, E, and believes that A would advance E, then it is irrational of her not to take this as a reason for doing A. This is not to say that she has any reason to do A. Perhaps what she has most reason to do is to abandon E, or to change her mind about whether doing A would promote it. But as long as she does not abandon E, and believes that doing A would promote E, it is irrational for her to deny that she has any reason to do A. Normative claims of this kind involve claims about what a person must, if she is not irrational, treat as a reason, but they make no claims about whether this actually *is* a reason.

I will call claims about rationality of this second kind *structural* claims, to distinguish them from *substantive* claims about what is a reason for what.[7] I call these claims structural because they are claims about the relations between an agent's attitudes that must hold insofar as he or she is rational, and because the kind of irrationality involved is a matter of conflict between these attitudes. (For example: a matter of believing something but failing to give it the role in further thinking that believing it involves, or having an end, but failing to give it the role in further thinking that is involved in having it as one's end.) The structural requirements of rationality are, it might be said, requirements that are constitutive of the rational attitudes involved.[8]

In earlier work, I have advocated restricting the term 'irrational' to instances of what I am here calling structural irrationality.[9] I am not relying on that restriction here. My present thesis is just that some claims about rationality are of this kind, and in particular that the claim I have mentioned above, about the connection between judging oneself to have conclusive reason to do X at t and forming the intention to do X at t, is such a claim. It is "structurally irrational" to fail to form an intention do to what one judges oneself to have conclusive reason to do. It is also, I would say, structurally irrational to continue to intend to do what one judges oneself to have conclusive reason not to do.

The question is how this rational requirement is to be explained if a judgment about the strength of the reasons one has is a *belief*. How can having a belief that something is the case make it irrational for someone not to form a certain intention and act on it? Having a belief can, as I have already said, make a difference to what one must do if one is not irrational. If I believe that p, then I cannot, without being irrational, refuse to recognize p as a valid premise in further reasoning, or form beliefs that I know to be incompatible with p. These requirements are, it might be said, partially constitutive of belief.

In particular, if judging that I have conclusive reason to do X now is a matter of having a certain belief, and I have that belief, then I must be willing, if I am not to be irrational, to rely on this claim in further reasoning about what reasons I have. But it does not seem to be constitutive of *belief* that if I believe that I have conclusive reason to do X now, I must form the intention to so act.

The idea that there can seem to be a problem here may, I believe, help to explain why there seems to Christine Korsgaard to be a problem with what she calls a substantive realist understanding of judgments about reasons.[10] In *The Sources of Normativity* Korsgaard tends to state this as a problem about *moral* realism, which partially blunts its force. Faced with someone who accepts it as a fact that it would be wrong to act in a certain way but asks why he should do it, Korsgaard sees the (substantive) moral realist as simply insisting that it is obligatory to so act. Her charge that this is unresponsive foot-stomping has merit. It makes sense, as she says, to ask why moral requirements are something we have reason to care about.

But Korsgaard believes that the same problem arises for a (substantive) realist about reasons—that is to say, for a view according to which judging oneself to have a reason to do X is a matter of having a certain belief. Here it is less clear what she is claiming. Suppose a person believes that he has conclusive reason to do X at t. How can this fall short of what is required? What is lacking does not seem to be a reason. A person cannot coherently say "Yes, I see that C is a conclusive reason to do X, but what reason do I have to do it?" The problem might be one about motivation. The problem might be that someone might believe that he or she had conclusive reason to do X, but fail to be at all moved by this. This can certainly happen, due to depression, lassitude, or simple irrationality. But the fact that practical realism allows for this does not seem to be an objection to it. A view that ruled out irrationality would be too strong, as Korsgaard herself observes.

I suggest that the problem Korsgaard sees for (substantive) practical realism is better understood in terms of the distinction between substantive and structural normative claims. Suppose you advance some substantive normative claim about the reasons I have. You might say, for example, that the fact that the APA is meeting in Philadelphia is a reason for me to go there. I might ask, "Why is that a reason?" If you offer some further substantive claim in reply, I can again ask for a further reason, and as long as you keep offering further substantive claims I can keep replying in this way. If you stop at some point and say "It just is a reason!" then, Korsgaard might say, you are engaging in unresponsive foot stomping.

What can stop this regress according to Korsgaard is not a claim about what is a reason but a claim about what I must, if I am not irrational, *treat* as a reason. This would be what I am calling a structural normative claim. For example, Kantians like Korsgaard believe that insofar as one is engaged in practical reasoning at all, one must see one's rational nature (and that of others) as an end in itself. It is not that one sees these things as reason-providing *in order* to be engaged in practical reasoning, or that one must take Kant's argument as a *reason* for taking these things to be reasons. Rather, the claim is just that one *will* take them to be reasons insofar as one is rational. This is what I have been calling a structural normative claim rather than a substantive one.

The point is clearest in what Korsgaard says about instrumental rationality.[11] She writes that a realist is unable to account for even instrumental rationality. If the fact that an action's instrumentality to her end constitutes a reason for her to do it is just another fact, which the agent believes, then "[f]or all we can see, an agent may be indifferent" to this fact. So put, this may sound like a problem about motivation. This interpretation is encouraged when Korsgaard puts the point by saying that on a realist interpretation the principle of instrumental reason "fails to meet the internalism requirement," since internalism is generally understood to be a view about motivation. But she goes on to make clear that the problem she sees for realism is not that it cannot explain how an agent could be motivated to take the means to his ends, but rather that it cannot explain why an agent *must* be so motivated. I take her to mean that it cannot explain why being so motivated is a requirement of rationality.

Korsgaard's larger aim seems to be to show that all valid normative claims can be grounded in this way, in claims of structural normativity about what we must to do if we are not irrational. This strategy is appealing in part because it seems to offer a solution to the problem of content. The requirements of rational agency provide a framework which structures our thinking about what reasons we have, and provides criteria of the kind that seem to be lacking when we address brute questions of substantive normativity on their own. This is what makes it plausible to call Korsgaard a constructivist about reasons.

Her main device for deriving more substance from the bare structure of rational agency is the idea of a practical identity. Insofar as I see myself as having a certain identity, I must, on pain of structural irrationality, see certain things as reasons. But even if this is accepted, there remains the question what reason one has to adopt and maintain any particular identity. I therefore do not believe that this project of grounding substantive normativity in structural normativity can succeed. It seems to me that there always remain substantive normative questions about what is a reason for what which must be faced, and answered, directly. This is why I am not a Kantian, or a neo-Kantian of Korsgaard's sort.

I do, however, also believe that there are valid claims of structural normativity. For example, as I have said, I believe that it is irrational to fail to form an intention to do what one judges oneself to have conclusive reason to do. The question I want now to address is how this can be so if accepting such judgments about reasons is just having a certain kind of belief. It is worth noting that there is no similar problem about explaining structural connections between intentions and other intentions. If I intend to do X at t and do not change my mind about this, then it is irrational for me not to try to do X at t, and irrational for me not to see myself as having reason to do other things that are necessary to my doing this, such as refraining from other incompatible plans of action. These requirements are part of what it is to intend something. They are, one might say, constitutive of intending, just as a readiness to use what one believes as a premise in further reasoning is constitutive of believing. But giving a belief a special content (making it a belief about the reasons I have) does not turn it into an intention.

It might seem from what I have just said that the problem I am calling attention to concerns the rational links between theoretical and practical reasoning. But this is so only if "practical" is understood in a very broad sense to include all judgments about reasons, including reasons for belief. If I judge that I have conclusive reason to believe that p, then I am irrational if, continuing to hold this judgment, I fail to believe that p.[12] But since this requirement depends on the content of the belief in question (the fact that it is a belief about reasons) it is not captured by the "constitutive" requirements I have mentioned above, such as the requirement that someone who believes that p should be ready to rely on arguments that employ p as a premise. So the problem I have been discussing about reasons for action seems to arise as well about reasons for belief. I will return to this in a moment.

In the case of reasons for action one might conclude, and I think many have concluded, that the only way to account for the intrapersonal rational significance of judgments about reasons is to interpret the acceptance of such judgments as something other than a belief. For example, if judging that C is a reason for A is a matter of adopting a certain policy—a policy of reasoning in a certain way, for example—then it would be easy to explain the irrationality of accepting such a judgment but not reasoning in accord with the policy it involves. This would be "constitutive" of judgments about reasons, just as a readiness to use p as a premise in further reasoning is constitutive of believing that p.

Given this, why should judgments about reasons be expressed in the form of assertions, which would normally be understood as expressions of belief? Several factors pull us in this direction. First, it makes sense to reason hypothetically about reasons. We can ask ourselves what reasons we would have if our situation were different in certain ways. This is what gives rise to the well-known Frege-Geach problem of embeddings.

Second, we can reach conclusions about what other people have reason to do, or to believe, but we can't, at least in a literal sense, adopt policies for them, although we can make judgments about what policies they have reason to adopt. In fact, since all conclusions about what we have reason to do, or to believe, are conclusions about what someone in a certain situation has reason to do or believe, they are all, implicitly, judgments about what others have reason to believe, insofar as their situations are similar to ours. When you form the belief that p, or the intention to A, you do something that only you can do. But when you and I think about whether C is good reason to do A, or good reason to believe p, we are thinking about the same question, just as when you and I are thinking about whether the sky is blue.

This is how things seem, and it draws me, at least, toward cognitivist understanding of judgments about reasons. The question is whether, in the face of the intrapersonal rational significance of judgments about reasons we must conclude that this is an illusion, and that these judgments must be understood in some other way. What are the possibilities? And how do they fare as accounts of interpersonal normative discourse?

In a recent article, Allan Gibbard suggests that when I form a judgment about what you have reason to do I am "deciding what to do for the case in which I am you" (or in your position).[13] This seems forced to me. Also, the possibility of irrationality makes it implausible. It seems to me that people can (irrationally) believe that they themselves have conclusive reason to do something but fail to decide to do it. It therefore seems implausible to iden-tify the judgment about reasons (whether one's own or someone else's) with such a decision. The solution Gibbard proposed earlier in *Wise Choices, Apt Feelings*, that a judgment that C is a reason for A expresses one's acceptance of a norm telling us to count C in favor of A, is more appealing in this score.[14] But even this does not seem to capture the full content of disagreement about reasons.

There is, I believe, an explanation for this. Expressivism arises, initially, as an attempt to explain the special intrapersonal normative significance (or, as expressivists might to put it, the special motivational force) of practical judg-ments. In my terms, it arises as way to explain phenomena of structural nor-mativity. But *inter*personal disagreement is mainly about claims of *substan-tive* normativity. Our disagreements are generally about what is a reason for what, such as whether revenge is a good reason to take a certain action in certain circumstances. Disagreements of this kind can arise in many different contexts: when one of us is giving the other advice, or trying to persuade him that he should or should not do something, or when one of us is trying to justify what he has done, or when we are discussing what some third party has reason to do.

The nature of these interpersonal interactions is quite different. But what we are disagreeing about is in fundamental respects the same in each case: whether a certain consideration is or is not a reason for some attitude. The answer to this question is, I am drawn toward saying, something that is prop-erly expressed in the assertoric mode, is capable of being true, and can be the object of belief. It is sometimes suggested that there is something inappropri-ate, even offensive about this way of putting things. The suggestion is that to assert, in such a disagreement, that one's own claim is *true*, is to claim an inappropriate kind of authority.[15] But insofar as this is so, it is equally true (equally a case of unhelpful foot-stomping) when we are disagreeing about some matter of empirical fact, such as who won the 1948 World Series. And even when it is unclear whether there are any criteria for settling our disagreement about reasons, there is nothing inappropriate about taking us to be offering contrasting opinions about a matter of fact that we are both inquiring into. So the charge of inappropriate claim to authority is unfair as a general objection to cognitivism about reasons.

On the other side, it might be charged, against the expressivist, that it is inappropriate, in such a case of disagreement, to be issuing commands, or to be making a decision for the other party. But this charge would be equally unfair. In expressing my acceptance of a norm I may be only explaining, in the mode of advice, what I myself would do. A more serious problem is that these ways of understanding what is being claimed fit the case of advice giving better than the other cases I have mentioned, such as offering a defense of one's own actions.

What all this brings out is the fact that when we are disagreeing about a normative question, such as whether revenge is a good reason for taking a certain action, the assertoric or expressivist character of the judgment we are disagreeing *about* is not central. When I defend my action by saying that revenge was a good reason for doing what I did, the normative claim I am making on you is that you *should accept* this view of reasons. Put in the simplest expressivist terms, I am not saying "Take revenge in circumstances like this" but rather "Accept that revenge is a good reason for such actions." The normative appeal I am making is thus not expressed in the first-order judgment we are disagreeing about but (to put it again in expressivist terms) in the higher-order norms about norm acceptance that, I am claiming, support accepting this norm.[16] In this respect, our disagreement is like a disagreement about a question of empirical fact, in which each of us is claiming that there are good reasons to support his view. As an expressivist might describe such a case, we are each expressing our acceptance of norms (in this case epistemic ones) which, we each claim, favor accepting the view we are advancing. In the interpersonal case, then, the normative character of the view we are advancing, or the expressivist character of the way this view is expressed, is not doing any particular work. Even in expressivist terms, normative disagreement looks much like disagreement in belief.

Let me return, then to the intrapersonal case where, as I have said, the appeal of expressivism seems strongest. Is there any way in which the special intrapersonal significance of judgments about reasons might be accounted for other than by interpreting them as something other than the assertoric statements they appear to be? In conclusion, I want to suggest one possible answer, which emerges once we view the problem as one about rationality rather than about motivation.

Recall that, as I indicated above, there is a problem about the significance of judgments about reasons for belief as well as about the significance of judgments about reasons for action. In the case of belief, it seems to me natural to say that it is constitutive of a belief (the one that is to be formed or modified) that it should be responsive to one's judgment about reasons for it. A belief that p is the sort of thing that should be modified if one judges that there is good reason to conclude that p is false, and should be formed if the agent judges there to be conclusive reason for concluding that p is true.[17] This is not to say that a belief that p commonly arises via a prior judgment that p is true. Many, perhaps even most beliefs do not arise in this way, perceptual beliefs being the most obvious examples.

My claim is only that belief is a judgment-sensitive attitude—a kind of attitude that will, insofar as we are rational, be responsive to our judgments about relevant reasons when we have such judgments. It is the judgment-sensitivity of belief, I would argue, that accounts for the significance, for our beliefs, of our judgments about reasons, and thus makes it unnecessary to construe those judgments themselves as something other than beliefs in order to account for their intrapersonal rational significance. This feature of belief also makes it unnecessary to appeal to a special form of motivation, analogous to a desire, to explain how judgments about reasons can lead to changes in belief. If an agent is rational then her beliefs just will be responsive to her other judgments in this way.

The same thing, I would argue, is true of intentions. Although an intention to do some action need not arise from a prior judgment that there is good reason to so act, intention is a judgment-sensitive attitude. If I judge myself to have conclusive reason to do A, then insofar as I am not irrational I will intend to do A, and if I judge myself to have conclusive reason not to do A, then insofar as I am not irrational I will not so intend. These are claims of *structural* rationality. They do not depend on whether I do in fact have reason to do A. As in the case of belief, however, these facts about structural rationality explain the special intrapersonal rational significance of judgments about reasons for action, and they do so without supposing that those judgments are anything other than beliefs.

The temptation to think that the explanation for these connections must lie in the non-cognitive character of judgments about reasons arises, I believe, from viewing the problem as one of motivation. Insofar as the capacity to motivate is seen as a kind of causal power, it must be attributed to the state that comes first—the desire, or the acceptance of a judgment about reasons—as a power to produce, and to explain, the state that follows. This temporal asymmetry is removed when we view the problem as a question about rationality—about how the acceptance of a judgment about reasons can make it (structurally) irrational to have or not to have another state (a belief, or an intention). Seeing the problem in this way allows us to take a more holistic view. The irrationality of having a certain combination of attitudes can be explained in terms of the relations between those attitudes—the fact that the second is the kind of attitude that must, insofar as the agent is rational, be responsive to the first.

Moreover, once we have this response to the question of rationality, we also have an answer to the corresponding question of motivation. Rationality is, as I have been stressing, a normative idea. But to claim that someone is a rational agent is also in part to claim that that he is so constituted that *in general* though not, of course, invariably, his attitudes fit the patterns required by rationality. They must do so with a certain regularity in order for it to make sense to attribute to him judgments about reasons and beliefs and intentions of the sort that are, ideally, responsive to such judgments. So, in particular, to claim that someone is a rational agent is to claim that he is so constituted that when he judges himself to have conclusive reason to do something he generally responds to this judgment by forming the intention to do it (and, insofar as he is not irrational and does not change this intention, he so acts.) Moreover, these things do not just happen, but also seem to him, in the light of his judgment, to be justified.

The normative aspect of the idea of rationality may, however, raise questions about the nature of what I am calling claims of structural normativity. If what is being claimed is that certain beings ought to behave in certain ways (that, for example, the attitudes of rational beings ought to conform to certain principles of rationality) then it may be asked what the basis of these 'oughts' is supposed to be. Worse, it may seem that practical reason, and hence morality, as I am describing them, do after all have troubling metaphysical presuppositions: namely, that by presupposing the existence of rational agents in this sense they involve a teleological conception of nature that is incompatible with a modern, scientific view of the world.

To allay this worry, I need to consider in more detail the kind of normativity involved in the ideas of rationality and irrationality as I have been employing them. From the point of view of the deliberating agent this normativity is self-effacing. A rational person who believes that p does not accept arguments relying on p as a premise because she sees this as required by some principle of rationality to which she must conform. Nor does she do it "in order not be irrational." Rather, she will be willing to rely on p as a premise because this is simply a part of what it is to believe p. Similarly, a person who believes that doing A would advance some end of hers, will see not this as counting in favor of A because doing so is required by a principle, or because she must do this in order to avoid irrationality. Rather, seeing this as a reason for A is part of what is involved for her in having the end in question.

Taking a more "external" view, however, we can say that a person who does not do these things is irrational.[18] This is a normative judgment; irrationality is a kind of fault. What kind of normativity is involved in such a judgment? One plausible idea is that it is a judgment of functional deficiency, of the same kind that as a judgment that a carburetor, or a kidney, is deficient because it does not operate in the appropriate way.[19] Such judgments are not incompatible with a scientific view of the world. External judgments of this kind are not, however, doing the main work in my account. My argument turns on claims of the first kind, about how rational agents will see things and how they will, as a matter of fact, normally respond.

Even if this is response is accepted, the explanation I have offered of the intrapersonal significance of judgments about reasons may seem to put a great deal of weight on the idea of a reason and the idea of rationality. I have tried to defend the idea that the fact that some consideration is a reason (for an action or a belief) can itself be an object of belief. I have not offered an explanation of the content of such beliefs (of the idea of something's being a reason) except by describing how such judgments are linked with other states, such as beliefs and intentions, by requirements of rationality.[20] I have appealed to these requirements to explain how judgments about reasons, even if they are beliefs, can have special intrapersonal normative significance. Here it may seem that I am building a lot into the idea of rationality. So I should stress that I am relying on this idea only to explain claims of what I have called structural normativity. I have not undertaken to build into the idea of rationality, or to derive from it, substantive claims about what is a reason for what. As I have said, I do not believe that this can plausibly be done.

NOTES

Presidential Address, APA Eastern Division, December 20, 2003. An early version of this paper was presented to the Philosophy Faculty at Oxford and to the Moral Sciences Club in Cambridge. I am grateful to members of both audiences for their comments and questions. I am also indebted to Derek Parfit for comments on a later draft and to Thomas Kelly and Scott Sturgeon for helpful conversation.

1. By Crispin Wright in Ch. 2, sec. 1, and Ch. 3, sec. 1, of his *Truth and Objectivity* (Cambridge, MA: Harvard University Press, 1992.) See also his "Truth in Ethics," in B. Hooker, ed., *Truth in Ethics* (Oxford: Blackwell, 1996), 7–8. I say "at least most of" because there is some disagreement about the range of platitudes that need to be satisfied in order for a kind of sentence to be "truth apt." See, for example, Frank Jackson, Graham Oppy, and Michael Smith, "Minimalism and Truth Aptness," *Mind* 103 (1994): 287–302. They argue that these include the platitude that "a sentence counts as truth apt only if it can [barring certain problems about length, etc.] be used to give the content of a belief." (294) It follows that sentences of a certain kind are truth apt only if they satisfy platitudes about belief. One of these is that beliefs are not conceptually linked to motivation and action. So if there is such a link between accepting an ethical judgment and being moved to act accordingly, then, if this platitude is accepted, sentences expressing ethical judgments cannot be the contents of beliefs, and hence are not truth apt. In the last part of this lecture I will try to show how beliefs are in fact linked to motivation and action in a way that derails this argument. I should say that Jackson *et al* also hold that ethical sentences are truth apt. (299)

2. I take the term from Crispin Wright, "Truth in Ethics," 5. Wright is not endorsing this idea.

3. I will consider, at the end of this lecture, the question of whether we have really escaped a more specific version of this problem.

4. The point is familiar from Hume, *A Treatise of Human Nature*, Book II, Part II, Section III. For a forceful contemporary statement, see Simon Blackburn, *Ruling Passions* (Oxford: Oxford University Press, 1998), esp. 70, 90.

5. I am setting aside here what Parfit has called "state-given" reasons for having an intention (that is, reasons for having an intention to A that are not reasons to A.) What I go on to say could be reformulated to take account of such reasons, but it would then be much more complicated to state.

6. It seems widely agreed that there is a question here. Michael Smith considers a similar question in chapter 5 of *The Moral Problem* (Oxford: Blackwell, 1994). He puts it in terms of desire rather than intention, but I assume he would also accept my formulation. I believe that Christine Korsgaard (a Kantian non-realist) is concerned with a similar problem in *The Sources of Normativity* (Cambridge: Cambridge University Press, 1996). I will say more about her view shortly.

7. What I am here calling claims of structural rationality correspond to what John Broome has called normative requirements. See, for example, his "Reasons," in R. Jay Wallace, Michael Smith, Samuel Scheffler, and Philip Pettit, eds., *Reason and Value: Themes from the Moral Philosophy of Joseph Raz* (Oxford: Oxford University Press, 2004), 28–55. I discuss the content of these requirements in more detail in "Structural Irrationality," which will appear in a volume of essays in

honor of Philip Pettit, edited by Geoffrey Brennan, Robert Goodin, and Michael Smith.

8. This way of putting the matter may have a Kantian ring, and it is indeed inspired in part by some remarks of Chris Korsgaard's in "The Normativity of Instrumental Reason," in Garrett Cullity and Berys Gaut, eds., *Ethics and Practical Reason* (Oxford: Oxford University Press, 1997), 242. But something similar is, I think, recognized by non-Kantians as well.

9. Thomas Scanlon, *What We Owe to Each Other* (Cambridge, MA: Harvard University Press, 1998), 25–30.

10. In *The Sources of Normativity* and in her lecture, "Realism and Constructivism in 20th Century Moral Philosophy," presented at the APA Eastern Division Meetings in 2001.

11. In "The Normativity of Instrumental Reason," 242.

12. I realize that some epistemologists would reject this claim, believing that claims about the justification of belief are not normative claims. With them, I must simply disagree. Others hold that the rationality of belief is a kind of instrumental rationality. Since claims of instrumental rationality are normative claims, I hope they can agree with what I say here.

13. "Normative and Recognitional Concepts," *Philosophy and Phenomenological Research* 64 (2002): 151–168.

14. Allan Gibbard, *Wise Choices, Apt Feelings. A Theory of Normative Judgment* (Cambridge, MA: Harvard University Press, 1990).

15. Something like this seems to be suggested by Simon Blackburn's remarks about the Apollonian character of realist claims about reasons and morality in Chapter 4 of *Ruling Passions*.

16. This fits with what Gibbard says in his discussion of "normative authority" in Chapter 6 of *Wise Choices, Apt Feelings*.

17. I leave aside here pragmatic, "state-given" reasons for *having* a belief that p. As in the parallel case of sate-given reasons for intention, I believe that what I say here can be modified to take account of such reasons, but I leave these complications aside for the present.

18. I call this view external because it is not the point of view of an agent making a decision but rather a reflection on that process. The quotation marks are a recognition of the fact than agents can take this view of themselves.

19. I discuss this issue more fully in "Structural Irrationality."

20. It might be said that I am here claiming that beliefs about reasons have two "directions of fit": as *beliefs* they must "fit the world" (that is to say be correct), but as beliefs about reasons they are rationally connected with action (and thus demand that the world should fit them). Michael Smith argues that it is "just plain incoherent" that a state should have both of these directions of fit. His argument is that a state cannot both, as a belief that p, be a state that a person must withdraw if she comes to see that p is false, and, as a desire that p, be a state that required the per-

son to try to bring the world into conformity with it. But this merely shows that p cannot be a correctness condition for a desire that p. It does not show, in general, that a state cannot both have correctness conditions and rationally require action. And indeed, as I observed in note 7 above, this is essentially the possibility that Smith himself goes on to argue for.

9

THE NATURALIST GAP IN ETHICS

Erin I. Kelly and Lionel K. McPherson

There is a truth in moral naturalism that moral cognitivists typically have shunned. Moral naturalists challenge the reason-giving authority of morality. To a certain extent, we are sympathetic to their challenge. We argue that moral reasons gain their "authority" only when they are accepted by moral agents and that persons qua rational agents do not have to accept moral reasons. Yet we agree with moral cognitivists that moral judgments are reason-sensitive and that this feature of moral judgments cannot be understood exhaustively or reducibly in the terms of psychology or biology (or, more specifically, evolution).

Moral naturalism, as we will understand it, shifts questions of justification heavily in the direction of explanation. For example, naturalists have stressed the primacy and functional role of various sentiments and reactive attitudes in accounting for the normativity of moral judgments. The idea is that psychological elements that characteristically explain why people are moved by certain considerations are the very basis of normative reasons. Yet an apparent gap lies between explanatory reasons and normative reasons: explaining why people reach the moral judgments they reach is not clearly the same as showing that such judgments are right or reasonable.

Probably the most prominent type of moral naturalist are philosophers in the tradition of Hume, who seek to describe the thought processes that lead people to come to identify and respond to certain considerations as moral reasons. Hume's view is that moral judgments are expressions of disinterested approval and disapproval to which a person is led through a complex psychological mechanism of sympathy with other people and a disposition to avoid conflict with them. A Humean moral psychology might seem both explanatory and normative: it locates the normativity of an agent's judgments in considerations that have motivating force from the agent's point of view. Simon Blackburn and Harry Frankfurt have developed contemporary versions of this idea. The problem is that the explanatory story remains more obvious than the normative story.

Blackburn argues that to affirm a moral standard is to value it or to be disposed to value it and that "to hold a value is typically to have a relatively stable disposition to conduct practical life and practical discussion in a particular way: it is to be disposed or *set* in that way, and notably to be set against change in this respect."[1] This is similar to Frankfurt's view that caring about something gives normative structure to a person's practical deliberation. Frankfurt argues that for you to care about something is to be disposed to support and sustain that which you care about, even when you would like to do something else.[2] For example, you might keep a vow of sexual fidelity to your spouse, despite having the opportunity to act (without complications) on sexual desires for a different person. Taking the vow is unlikely to preempt having outside sexual desires; otherwise, there would not be the familiar difficulty of keeping such a vow. Rather, assuming that you continue to care about the vow, you will try to limit your actions accordingly. Caring involves identifying with certain desires and thereby making them your own. When a person's identification with certain desires is "wholehearted," she establishes them as authoritative constraints that can be used to guide her decisions and considered preferences.[3] The freely faithful spouse can acknowledge having outside desires while not identifying herself, as an agent, with those desires: given that she continues to care about her vow of sexual fidelity, her outside desires do not provide her with grounds for action. By contrast, Frankfurt maintains, when a person has identified herself inconsistently or when she fails to identify with any of her desires, she fails to constitute herself as an agent and is to some degree passive with regard to the actions she performs.[4] Her desires would drive her to act, without regard to or even despite what she cares about. In this sense, such a person would be out of control and might even lack altogether a normative perspective for self-criticism.

These philosophers locate the normativity of an agent's judgments about what to do in the reflexive and hierarchical structure of an agent's desires—a structure that constitutes the agent's point of view. But the normative significance of the agent's point of view calls for further analysis. More specifically, when different people might reach different moral judgments, or when a conflict might arise in an agent's own thinking about what she has reason to do, this prompts the question whether there are independent criteria for resolving the inconsistencies. There is a range of views, for example, about the moral status of animals, particularly with regard to eating them, and a morally conscientious person may well be uncertain about how to balance the various considerations. In addition to the question about independent criteria for moral assessment, a related question arises when an agent reflects on whether she has adequate grounds for endorsing the values she holds. Why should an agent take the reasons that she favors to have normative significance? Moral naturalists appear to open a gap they cannot close: there appears to be no bridge between the conclusions that people actually reach, even after reflection, and conclusions about what is right or reasonable. Of course, moral naturalists might reject the idea that morality has cognitive content. Some hold the view that the normativity of morality has no source apart from the attitudes expressed by the moral commitments people have. This position, we argue, cannot be reconciled with the evaluative role that morality generally assumes vis-à-vis people's actual behavior and commitments.

Still, there is an important gap that moral naturalists can fill. This is the normative gap between, as we shall put it, "acknowledging" and "accepting" reasons.[5] You *acknowledge* a reason when you view certain considerations as counting in favor of something, for instance, that a person who plans to run a marathon has a reason to train. You *accept* a reason when you view it as bearing directly on your practical deliberation and decision making. Not only would you acknowledge that a marathon runner has a reason to train, but also you would accept that your plan to run a marathon gives you a reason to train. Acknowledging a reason and accepting a reason are rationally related: acknowledging reasons can be a rational basis for accepting them. But acknowledging a reason does not mean one must accept that it bears directly on one's own deliberation. You might acknowledge reasons that it would not be irrational or rationally defective for you to disregard. The reasons may support a course of action only in relation to a situation that you do not occupy, or they may be practically relevant only in view of sensibilities or values that you do not share. Your recognition of the sensibilities or values that make a course of action compelling need not entail that the reasons supporting this course of action are reasons for you. Opera aficionados have a reason to help sustain the art form by attending performances. But if opera does not move you or is not important to a culture with which you identify, you may have no such reason.

Morality does generate requirements of certain sorts. People who do not care about morality, however, do not accept the reasons supporting its requirements. While people who care about morality have these reasons, morally disinterested people will at most be willing to acknowledge moral reasons. We argue that such people are morally defective in a fairly obvious sense, though this defect does not necessarily point to a defect in their rational agency. The selfish and greedy who are willing to exploit others in order to promote their individual, family, or special interests are not necessarily irrational (assuming that the universality of moral reasons is not presupposed). Such people might take instrumentally rational means to accomplish their ends and might lead lives they reflectively endorse, without suffering crises of conscience. The morally disinterested do not directly have reason to do what morality requires. Moralists might hope that rational people must accept moral reasons. But the rational bindingness of moral reasons is obscure, apart from the relation of these reasons to some considerations that an agent accepts.

This means that psychological factors influence the reach of moral considerations as reasons for particular people. Specifically, the rational bindingness of moral imperatives depends on noncognitive elements. A moral agent not only will recognize that other people are sentient, concerned about their future, have meaningful relationships, and so on, but also will have a sensibility that moves her to care deeply about such facts. At the same time, we maintain that making sense of the content of morality requires a cognitivist analysis. There are truths about what morality requires or, at least, there are more and less reasonable judgments about the content of morality—judgments that roughly reflect the outer limits of defensible conduct. The Spanish Inquisition, chattel slavery, and genocide, for instance, are beyond the pale of any conception of decent treatment of people as such. Our point is that a cognitivist analysis of the content of morality cannot on its own provide an account of moral motivation nor, more controversially, of reasons that all rational persons have. A rational imperative to accept moral reasons depends on noncognitive elements at some level. Thus the directives of rational agency involve bootstrapping. This opens the door, in a limited way, to moral naturalism.

◎ ◎ ◎

A central aim of T. M. Scanlon's "Metaphysics and Morals" is to show that moral cognitivism can provide an account of the reason-giving and motivating quality of moral judgments—what he refers to as their "normative significance." According to Scanlon, sound moral judgments express beliefs and are supported by reasons. He addresses in separate steps the normative significance of the reasons that people address to one another and the intrapersonal normative significance of moral reasons for an agent's own decision making and action. We address each in turn.

The normative nature of the claims people make on one another, Scanlon argues, derives from the substance of morality. Normativity stems directly from the content of the reasons that support moral judgments. You claim that she should return the money she borrowed because she promised to pay it back. The normative implication of your claim is that there is reason for her to take it seriously. This implication is affirmed, Scanlon thinks, by typical responses to moral disagreements: people treat these disagreements not as reducing to differences of taste or feeling but, rather, as akin to disagreements about empirical fact, where each person claims that there are good reasons to support his or her view.[6] Because the subject matter of morality is practical, the reasons people affirm often direct somebody to act in a certain way. Still, the claims take a cognitive form. Your normative claim on her to return the money is supported by good arguments for her to believe that she is required morally to return it.

Behind the implication that others have reasons to accept moral judgments, Scanlon argues, is a claim about the nature of normative reasons in general: a reason for you would hold for other people as well, at least when they are similarly situated. Scanlon has referred to this as the "universality" of reason judgments. The apparent universality of reason judgments lends support to a cognitivist account of morality. Anyone who is cognitively sensitive to the sorts of considerations that are morally relevant (e.g., doing X would help someone in need, doing Y would cause harm) will be in a position to acknowledge that certain moral directives are reasonable. According to Scanlon, for instance, "When I defend my action by saying that revenge was a good reason for doing what I did, the normative claim I am making on you is that you *should accept* this view of reasons."[7] Here the agent would not merely be signaling that she feels strongly in favor of revenge under such circumstances. Rather, the agent believes that backing up her judgment are considerations that have merit independently of local feelings, dispositions, conventions, or traditions.

Scanlon helpfully stresses the reason-sensitive nature of moral judgments. Yet it is unclear how exactly to understand the demand that an agent accept reasonable moral judgments other people make. We need to know how to interpret the reason-giving force of moral judgments for particular agents. A person who acknowledges that certain considerations have moral significance might not accept that those considerations have direct bearing on her own practical deliberation. This raises a question about the factors that might mediate between considerations that a person identifies as supporting a moral judgment and considerations that she accepts should influence her conduct. A cognitivist account of moral motivation would need to show how an agent's belief that morality requires action of a certain sort rationally implies that the agent must herself intend to comply with morality's requirements.

Unlike many moral cognitivists, Scanlon admits that there are no moral reasons that all rational persons must accept on pain of irrationality. Nevertheless, he thinks that moral reasons have normative authority for all rational agents. This is because his account of the normativity of morality relies on claims about the relative bindingness of competing considerations for an agent—not from the perspective of the agent's existing dispositions and motives but, rather, from the perspective of what can be justified to other people. A moral reason applies to you not because you accept it or because, in view of your dispositions, motives, and values, you ought to accept it. According to Scanlon, the normative authority of a moral reason does not depend on whether the reason is supported by the agent's own judgments. Instead, this normative authority has its source in principles that no one could reasonably reject. Further, Scanlon's account affirms the value of standing in justifiable interpersonal relations as something that all persons have reason to accept.[8] Failure to act on a moral reason does not imply that an agent irrationally fails to act in accordance with her own judgments. The agent might not be, in this sense, irrational for failing to act on a reason that applies to her. In a broader sense, though, Scanlon believes that an agent is rationally defective in failing to accept the reasons there are.[9] He does not, in our view, give adequate content and support to this claim.

◎ ◎ ◎

In rejecting the idea that the content and normativity of morality derive from general principles that all rational persons as such must accept, Scanlon rejects what he refers to as "structural accounts." Structural accounts hold that the content and normative significance of moral reasons derive from unavoidable presuppositions of rational thought, communication, or agency. On these accounts, the substance of morality—its claims and directives—is unavoidable for beings who communicate, think, or act rationally. Structural accounts are attractive because they avoid metaphysical claims and are able to distinguish themselves from psychology. They avoid metaphysical claims by describing only what agents are committed to, not an independent realm of value; they distinguish themselves from psychology by focusing on the content of moral claims rather than their psychological role. Scanlon responds to Christine Korsgaard's ambitious account, and so will we.

Korsgaard argues that the normative authority of reasons is grounded in human nature, specifically, in a person's self-constitution as an agent. Agency *is* self-constitution, according to Korsgaard, and the normativity of reasons stems from the role they play in acts of self-constitution. She elaborates this idea as follows: Normative questions arise because, as reflective beings who act, people are faced with questions about what to do. Various inclinations provide candidate reasons for action, and an agent reflects on possible courses of action with the aim of choosing among them. In deliberating he presumes that what he does is up to him and not determined by some alien cause. This means that he in effect commits himself to choosing a course of action on the basis of a principle he gives to himself. Such a principle would be one with which he identifies. As Korsgaard puts it, "We endorse or reject our impulses by determining whether they are consistent with the ways in which we identify ourselves."[10] In this way, and only thus, does a person act autonomously. Since action is autonomous in the sense that in acting an agent thinks of himself as self-determining rather than determined, aligning himself with action-guiding principles is the only way in which he can truly act. Acting "forces us to have a *conception* of ourselves . . . the principle or law by which you determine your actions is one that you regard as being expressive of *yourself*."[11] Korsgaard refers to a person's self-conception as that person's "practical identity": it is, she says, "a description under which you value yourself, a description under which you find your life to be worth living and your actions to be worth undertaking."[12] An agent acts as a caring parent, a loyal friend, an aspiring writer, a patriotic citizen, a compassionate person, and so on. The normative authority of reasons stems from identities, such as these, that agents give themselves.

There is considerable affinity between Korsgaard and Frankfurt on these points, and the idea that the normative authority of reasons comes from an agent's commitment to them does have merit. Still, it is not obvious how best to analyze this idea. The features of a person's practical identity are contingent and varied. People value different pursuits and have different goals and attachments. This variation prompts the question of why practical identities should be thought to provide normative reasons. Why think, for instance, that loyalty or patriotism could license actions that otherwise would be unacceptable? The claims that a person makes about the importance of his pursuits, goals, and attachments as he thinks about what to do are open to normative challenge. Korsgaard's appeal to the fact that a person is characteristically moved in certain ways and identifies himself with particular aims, we are arguing, hardly seems sufficient to establish the normativity of his reasons for pursuing those aims.

Korsgaard's idea applies more narrowly: it seems true that at the moment of decision, an agent commits himself to the normative significance of the reasons he affirms. This confers on these reasons some rational force for him, as long as he maintains a commitment to them. If the agent's supposed commitment is not sustained over time, it cannot be understood to express or to have helped to constitute his agency or identity—and this serves to undermine the basis for maintaining that he actually has the reasons in question. In any case, the soundness of particular reasons may be questioned from the perspective of other people or later by the agent himself. This is especially true of reasons that are open to moral criticism, since the justification of moral reasons is often directed toward the interests of people who may be of no special concern to the agent.

What is left unclear in Korsgaard's argument is how recognizing that rational agency requires action-guiding principles commits an agent to any robust, interpersonally oriented morality. Even if it is granted that all people, as rational agents, require action-guiding principles, this does not imply that all people must accept morality's requirements.[13] So self-constitution can be understood as the source of the rational force or bindingness of reasons for the agent who accepts them—but in a limited way that seems inadequate to establish that all people must accept the full range of moral reasons. You can be said to have moral reasons only when you accept certain kinds of considerations as fundamental, such as promoting the best state of affairs or being able to justify one's conduct to others. The "authority" of moral reasons cannot be given a broader, structural derivation. What does seem convincing in Korsgaard's account, though, is the idea that an agent confers rational authority on his reasons by accepting them or by accepting other reasons that commit him to the reasons in question.

The failure of structural accounts to establish that moral reasons bind all rational agents opens the possibility that a rational person may understand how reasons can be rooted in certain values without sharing those values. The reasons would lack normative bearing on such a person. Nonetheless, certain values—for example, that suffering is bad and generally worth preventing[14]—might be so central to moral thought that they are constitutive of it. The cognitivist, to this extent, is right.

◎ ◎ ◎

Scanlon admits the relevance of structural claims about rationality when he analyzes the intrapersonal normative significance of moral judgments. The structural claim that he exploits to establish that moral judgments have intrapersonal normative significance is that there is a rational connection between judging that one has a conclusive reason and intending to act on the reason. The language of conclusive reasons implies that a person who understands all the relevant considerations would accept the reason in question as a decisive basis for action. The step from accepting a reason as a decisive basis for action to intending to act on it is short enough that the following seems true: a constitutive feature of one's judging a reason to be conclusive is one's intending to act accordingly. As Scanlon puts the point, intention is a judgment-sensitive notion, and so it follows that a rational agent must intend to act from a reason she judges to be conclusive.

The scope of Scanlon's structural claim about rationality is, however, narrower than he recognizes. An agent's acknowledgment of the cognitive substance of moral reasons, which carries with it some sort of universal claim, does not support a structural derivation of an intention to act. There is a gap between the claim that rational people would acknowledge that C counts in favor of A and the claim that a person should rationally accept A as bearing directly on her deliberation. Scanlon himself, recall, is skeptical that there are moral considerations that all rational agents must accept. Yet he takes his universality thesis to suggest that all reason-sensitive people would see the reasons there are as having some normative significance for them.[15] Scanlon thinks this means that these reasons rationally imply certain motives and intentions. We resist this suggestion. You might acknowledge the normative significance of certain reasons without accepting their relevance to you.

Let's look more closely at the kind of failure that failing to act on moral reasons might consist in, which helps to shed light on the source of those reasons. The normative significance of a reason is closely related to its motivating force, but the two notions should be distinguished. When you accept some consideration as a moral reason for you, this influences your deliberation. You believe that the reason ought to bear directly on your decision about what to do. Yet you might not be motivated to so act. If this happens, you seemingly can be faulted: your lack of moral motivation would reveal something like weakness of will. You would be going against your own judgment that moral considerations are relevant to your deliberation and hence acting against your own commitments. The inconsistency arises because when you accept a reason, you have committed to acting in a way that is consistent with the reason. There is a rational, if not an empirical, connection between accepting a reason and being motivated to act accordingly. The judgment that a person who fails to so act is irrational does not seem particularly rationalistic.

If you merely acknowledge a reason as counting in favor of a certain course of action, however, and do not accept the reason as bearing directly on your deliberation, there may be no inconsistency or tension in your will. It may be true that you lack moral virtue, but this would not represent a defect in your rational faculties. A potential disparity between the reasons a person merely acknowledges and those she accepts may be what leads Scanlon to focus on the reasons an agent acknowledges as being conclusive, in which case the agent's acceptance of the reasons seems built in. It is plausible to think that intending to act accordingly is constitutive of judging a reason to be conclusive. But it does not follow that one's judging consideration C to be a reason in favor of doing A in situation S entails that in situation S one would, if rational, feel some motivation to do A, other things equal. One might not care much about the type of considerations that favor doing A.

The point of moral theory is best understood as a matter of illuminating the normative significance of certain reasons for people with certain sensibilities and values. This leads to an important contrast Scanlon draws between a rational person and a reasonable person. We understand this contrast in terms of the distinction we have introduced between acknowledging a reason and accepting a reason. The cognitive character of reason judgments implies that a rational person could understand how considerations C support action A in situation S. This motivationally detached understanding is a matter of acknowledging reasons.

We have contrasted the possibility of thus acknowledging reasons with accepting reasons—where accepting a reason entails believing that one must, in a self-generated sense, be practically guided or motivated by it. With respect to morality, this represents the domain of the reasonable. Reasonable people rationally commit themselves to acting in accordance with the requirements of morality. Questions about the importance of moral reasons are answered by analyzing their content. Such an analysis would show what is involved in taking a commitment to morality seriously. This may enable people who care about the considerations underlying moral directives to better understand what they care about and why. The reasons morally concerned and motivated people accept are substantively justified insofar as the domain of the reasonable points to values that are plausibly thought to be constitutive of morality. But moral reasons, while justified on substantive grounds, are rationally binding only for reasonable persons—who bind themselves.

Moral naturalists would resist a structural account of rationality that aims to forge a rational connection between the belief that certain considerations are morally important and accepting the direct bearing of these considerations on one's own practical deliberation. This resistance is plausible. At some level, noncognitive elements are indeed needed to fill the gap between acknowledging a reason and a rational imperative to accept a reason oneself. As Frankfurt and Blackburn recognize, a person's cares and commitments are needed to account for the rational bindingness of moral reasons. Expressivists and other moral naturalists are to this extent correct about the contingency of morality as a source of action-guiding reasons. Scanlon should make this concession to the noncognitivist. Moral reasons have their "authority" only for people who have a certain kind of sensitivity to or regard for other people. But contingency in the reach of moral reason judgments is compatible with a cognitivist account of the content of morality. It would be an overstatement to think that moral judgments only express a person's sentiments or attitudes. Rather, having certain sentiments or attitudes is a condition for having moral concern at all. When this condition is met, moral judgments become reason-sensitive. For morally interested persons, sensibilities of a characteristically moral sort frame deliberative problems such that certain considerations become important. Expressivists and other moral naturalists conflate a condition of moral concern with the content of moral reasons.

NOTES

1. Simon Blackburn, *Ruling Passions: A Theory of Practical Reasoning* (Oxford: Oxford University Press, 1998), 67.
2. Harry Frankfurt, "On Caring," in *Necessity, Volition, and Love* (Cambridge: Cambridge University Press, 1999), 160.
3. See Harry Frankfurt, "Identification and Wholeheartedness," in *The Importance of What We Care About* (Cambridge: Cambridge University Press, 1988), 175–76.
4. Ibid., 163.
5. Here we follow and elaborate the distinction Lionel McPherson draws between there being a reason and an agent's having a reason; see "Normativity and the Rejection of Rationalism," *Journal of Philosophy* 104 (2007): 65. A similar distinction is made Sigrun Svavarsdottir in "How do Moral Judgments Motivate?" in *Con-

temporary Debates in Moral Theory, ed. James Dreier (Malden, Mass.: Blackwell, 2006), 171–72. See also her "Moral Cognitivism and Motivation," *Philosophical Review* 108 (1999): 192.

6. Scanlon, "Metaphysics and Morals," this volume, chap. 8.

7. Scanlon, "Metaphysics and Morals," 186.

8. Scanlon, *What We Owe to Each Other* (Cambridge, Mass.: Harvard University Press, 1998), 154.

9. Scanlon, *What We Owe to Each Other*, 370.

10. Christine Korsgaard, *The Sources of Normativity* (Cambridge: Cambridge University Press, 1996), 120.

11. Korsgaard, *Sources of Normativity*, 100.

12. Korsgaard, *Sources of Normativity*, 101.

13. Korsgaard (*Sources of Normativity*, 120) argues that because all people have reasons that stem from their common humanity, they must have moral obligations to one another. This conclusion, at least with regard to anything that plausibly would be recognized as the range of morality's requirements, does not seem to follow from the argument she gives.

14. Cf. Peter Singer, "Famine, Affluence, and Morality," *Philosophy and Public Affairs* 1 (1972): 231.

15. Scanlon, *What We Owe to Each Other*, 367.

PHENOMENOLOGY AND THE NORMATIVITY OF PRACTICAL REASON

Stephen L. White

THE HUMEAN CONCEPTION OF PRACTICAL REASON

Hume's account of the relation of reason to the passions implies that we can reason only about means, not ultimate ends. As Hume says, "Where a passion is neither founded on false suppositions, nor chuses means insufficient for the end, the understanding can neither justify nor condemn it. 'Tis not contrary to reason to prefer the destruction of the whole world to the scratching of my finger."[1] Should someone fail to care about the things we value, about the welfare of others, or even about his or her own welfare, there are, on Hume's account, no rational grounds for criticism.[2]

For contemporary Humeans, something like Hume's conclusion regarding reason in its practical applications is thought to follow from a Humean moral psychology—a moral psychology distinctly different from Hume's but assumed to be similar in its spirit and import. According to the *Humean moral psychology*, beliefs and desires are "distinct existences" defined by their different directions of fit. If the propositional content of a belief fails to fit the world, it is in the nature of belief that it is to change or be changed to match the world; if the propositional content of a desire fails to fit the world, it is in the nature of desire that the world is to be changed to match the desire.[3]

The implication is that there are no entailment relations or conceptual connections between a subject's holding a set of beliefs and his or her having, or being justified in having, any particular set of noninstrumental or underived desires. Regardless of how outrageous or alien a subject's noninstrumental desires may seem, there need be no way in which the subject's beliefs differ from ours and no way in which they fail to match the world. Indeed, there is nothing about the factual nature of the world that the subject must be missing. Thus we have the apparent impossibility of appealing to reason to change the subject's ends construed as the objects of underived desires. Moreover, we have the characteristically Humean conception of practical reason as concerned only with the means to antecedently established ends. In addition, it is an assumption of the Humean moral psychology that the explanation of action is in terms of beliefs and desires so understood. And in the context of this Humean framework, it has seemed natural to assume that the agent's reason for acting is what explains why the agent did what he or she did.[4]

THE SIDGWICKEAN OBJECTION (SIDGWICK, NAGEL, PARFIT)

A number of considerations have been advanced to counter or undermine the Humean conception of practical reason. Consider the following argument.

i. Extreme imprudence is irrational.

Suppose you know now, at age twenty, that at age seventy you will want very much to have an adequate income. And suppose you also know now that by making a very slight sacrifice—saving a very small amount on a regular basis—you can make a very large contribution to your income at seventy. Suppose, however, that because you now care nothing about the satisfaction of your desires at seventy, you refuse to save. Such extreme imprudence is, we normally assume, the very paradigm of irrational behavior.

ii. If a subject does not care *now* about his or her future desires, the Humean (strictly, the Humean *present-aim theorist*, who holds that rationality requires that one choose the means to the satisfaction of the strongest of one's actual *present* desires) is committed to holding that for that subject even extreme imprudence is rational, or at least not irrational.[5]

iii. Hence, the Humean account of irrationality (and so of rationality) is inadequate.

The intended implication, of course, is that we should adopt a theory of rationality and practical reason according to which we can, on rational grounds, justify or criticize at least some ultimate ends or noninstrumental (underived) desires.[6] There is a further intention associated with at least some of the appeals to Sidgwickean considerations: to call into question the Humean moral psychology that gives rise to the purely instrumental conception of rationality.[7]

But the argument that the Humean cannot account for the intuitive irrationality of extreme imprudence is hardly conclusive. The Humean can simply bite the bullet (as Hume himself seems to) and deny (i).

THE STRENGTHENED SIDGWICKEAN OBJECTION (PARFIT)

The anti-Humean, however, can point to forms of behavior that are, intuitively, even more obviously irrational than imprudence: the varieties of so-called "pathological indifference."[8] And these are, allegedly, kinds of behavior whose irrationality the Humean must deny. Examples include:

a. *Future-Tuesday Indifference.* One cares about one's future in the normal way except as regards future Tuesdays. Though one cares now about one's desires on future days other than Tuesday, one cares nothing about those that occur on a Tuesday. (Of course, when Tuesday comes, one will care about the satisfaction of one's [then] present desires in the usual way.) Thus one would choose now to experience the greatest pain on a future Tuesday in order to spare oneself the smallest discomfort on any other day and to forego the greatest pleasures on a future Tuesday in order to experience the mildest gratification at some other time.[9]

b. *SoHo Indifference.* This is the spatial analogue of future-Tuesday-indifference. One has the normal concern about the satisfaction of one's future desires except those that occur in SoHo, which one discounts completely.[10]

c. *Desire under the M's.* John loves Mary passionately, but only as long as she remains in a state whose name begins with *M*. When she crosses the border from Maine to New Hampshire or Massachusetts to Rhode Island, he cannot see her as any different from anyone else.

Since we do normally discount the future (and sometimes quite heavily), it is not completely implausible for the Humean present-aim theorist to claim that *imprudence* need not be irrational. It is far more difficult to make this claim about pathological indifference, even though such a Humean theorist is committed to doing so.

THE NORMATIVITY OBJECTION
(KORSGAARD, KENNETT)

Some recent critics of the Humean conception of rationality have argued that it cannot make sense of the normativity of practical reason and so cannot make sense of practical reason at all. According to Christine Korsgaard, this is because whatever one does will be a case of one's acting on one's strongest desire. Thus it will be a case of one's acting rationally in the only sense to which the Humean can appeal. In particular, according to Korsgaard, the Humean lacks the resources to distinguish between a weak-willed agent and one with an unusual desire set. That is, the Humean account has no room for the distinction between two kinds of subjects: one whose failure to take the means to one of his or her desired ends is irrational and one who acts rationally on an unusual set of desires—a set in which the desire not to take the means is simply stronger than the desire for the end. Indeed, the Humean cannot avoid assimilating the former case to the latter.[11]

We might think here of Ahab of the original film adaptation of *Moby Dick*. In a scene that does not occur in the book, Ahab, after losing his leg, must have the stump cauterized to save his life. When the time comes, Ahab resists, and the procedure is carried out against his will. The Humean, it seems, cannot interpret Ahab's resistance as anything other than his having a stronger desire to avoid the pain of the cauterization than to go on living—and his acting rationally on that basis. It seems, however, that although this is a possible interpretation of Ahab's resistance, there is another, and possibly more natural, one: that his behavior is weak willed and irrational.[12] And that the Humean cannot make sense of this latter interpretation seems to support the contention that the Humean conception of rationality is fundamentally flawed.

Jeanette Kennett pursues this point in her critique of the Humean conception of rationality and her argument for "the possibility of the moral or prudential considerations attended to in deliberation providing reasons for action, independently of what we antecedently desire."[13] Kennett follows James Dreier in arguing that even Humeans must accept the normativity of practical reason.[14] Kennet claims that "if someone does not accept the means-end principle, and so sees no reason to adopt the acknowledged means to their ends, this deficiency cannot be remedied by furnishing them with another desire, for in so far as this gives them another end it will be subject to the same deficiency."[15] And she goes on: "An agent is not motivated simply by the belief-desire pair cited in a means-end explanation of action, she is motivated by the reason it provides; that is, she must be capable of recognizing that the *fact* that a particular action is a means to her end gives her a *reason* to perform that action."[16] She concludes:

> I suggest that the conclusions of practical reflection are not inescapably the product of existing desires and ends and may conflict with these. If this is so, it is difficult to see a principled way of ruling out the possibility of moral or prudential considerations attended to in deliberations providing reasons for action, independently of what we antecedently desire. Certainly such considerations may fail to move us to action. But there can be no special problem for an account of practical reason here that is not also faced in the case where, though we desire our end, we fail to desire the means to that end.[17]

THE SOPHISTICATED HUMEAN REPLY
(FRANKFURT, LEWIS, WHITE)

The sophisticated Humean retains the fundamental Humean moral psychology of beliefs and desires, the standard belief-desire forms of explanation of action, and the assumption that (roughly) the agent's reasons for acting are what explain why the agent acted as he or she did. But the sophisticated Humean distinguishes between the motivational and evaluational strength of the agent's desires.

The distinction between motivational and evaluational strength might be made (as it is by Harry Frankfurt and David Lewis) as follows: Take as special those first-order desires for which there are second- (or higher-) order desires that those first-order desires should be the ones on which one acts.[18] The intuition would be that the first-order desires for which such second- (or higher-) order desires exist would be one's "real" desires, one's commitments, or one's values. This intuition is difficult to sustain in light of the fact that one's second- and higher-order desires may themselves be just as irrational as desires of the first order. One might, for example, have a second-order desire to act on all and only those desires not formed on Tuesdays.

This problem can be avoided if, instead of desires of higher order, we appeal to relations of support among first-order desires.[19] Consider the desire to be a first-rate mathematician and the desire to be a famous mathematician (or at least one whose work is well known and well appreciated in mathematical circles). These desires are independent, and because they support one another, one would normally desire both. But (again normally), neither is desired as a means to the other. Rather, one's desire to share one's work with a wide, intelligent, and appropriately appreciative audience gives meaning to the desire for excellence, and vice versa. The character of Will in *Good Will Hunting* shows how the lack of a desire for recognition can undermine and trivialize the desire for excellence. Because he has so little respect for the mathematical abilities of others, he has little motivation to strive for excellence in anything but a shallow way. Although he takes pride in his mathematical talent, he is unmotivated to work at a level that would demand more than he can "toss off" without significant effort or sacrifice.

Consider also the following examples.

Jet-Set Peace Corps Worker. Imagine a Peace Corps worker whose desire to alleviate suffering in impoverished villages is beyond question. He works tirelessly, sixteen to eighteen hours a day from Monday to Friday, and has done so for several years. When Friday evening comes, however, he takes advantage of his family's connections and jets off to the most fashionable capitals of the world where he parties until Monday morning. Asked if there isn't something contradictory in his behavior, he simply shrugs and says, "I work hard and I party hard."

In such a case, it seems, we may not be able to object to the strength or the sincerity of the desire to relieve suffering, but we can point to its shallowness. First, there is something shallow in this person's relation to the villagers with whom he works, since they may experience crises that do not end on Friday night. Thus, his desire to alleviate suffering is not supported by a network of desires constituting the usual sorts of personal connections with the actual individuals with whom he works. Second, because the desire to alleviate suffering has this abstract and isolated character, if it were lost, there would be nothing to motivate this person to try to reinstate it. Were he to lose the desire, he could simply walk away from his Peace Corps work without a second thought. By contrast, were one to lose one's desire to do philosophy, one would have a great many desires that would motivate one to try to get it back.

Person Who Doesn't Believe in Punishment. Imagine a person who doesn't believe in punishment because he takes a self-consciously and resolutely forward looking perspective. Given this perspective, he sees no point, even if someone has caused serious suffering to others, in adding to the amount of suffering by punishing the offender. Of course, this person believes in some

---

form of social control, but he has, he claims, no beliefs that would underwrite the idea of *punishment* as required by *justice*.

Now imagine that a friend, recently deceased, who had done seminal work in his field, has been falsely accused of plagiarism. The person, as he sees it, has three alternatives. If he decides to use his remaining productive years to making the greatest contribution to aggregate utility, he will ignore the unjust accusation. For his friend has no other remaining friends and no relatives, and the work will have the same social utility regardless of who gets credit. Thus the person's contribution to aggregate utility will be maximized if he concentrates on research with the most direct social applications. If, however, he decides to maximize his own utility, he will concentrate on research projects with the greatest intrinsic intellectual interest. Finally, he can use the time that remains to vindicate his friend.

Imagine that he opts for the third alternative. When asked why, he says simply that he wants to "set the record straight." When pressed, however, he points to a number of analogies that make the decision more intelligible. He cites as an example his admiration for the medieval crafts ideal that allowed sculptors to devote as much attention to the parts of statues that would never be seen as to those that would. And he compares this ideal to his desire to set the record straight in the face of nearly complete public indifference. He also cites his taste for historians of marginalized groups, such as Richard Cobb and Eric Hobsbawm, his instinctive tendency to side with the underdog, and his admiration for the fact that his friend's work was done for it own sake. And he points to the analogy between such tastes, dispositions, and ideals and his current nonconsequentialist choice. And when pressed as to whether he sees an analogy between his attitude toward his friend and the backward-looking perspective in which punishment as opposed to treatment makes sense, he is forced to admit this much: that although he is no supporter of contemporary institutions of punishment, he cannot maintain, as he once did, that they presuppose a perspective he finds wholly unintelligible.

In terms of this relation of support among nonderived and noninstrumental desires, we can draw a plausible distinction between our *mere* desires and our values or commitments. Thus we can distinguish between the motivational and the evaluational strengths of our desires. Evaluational strength is largely a matter of support, rather than motivational strength. Well-supported desires, as we have seen, are those we would be motivated to strengthen or replace were they to wane or disappear. And poorly supported desires are those we would be motivated to eliminate were they to conflict with those whose support was significantly better. For example, given a strong interest in Bach and Brahms, were one's desire to hear Beethoven to wane, one would very likely be motivated to try to reinvigorate it—by listening to his less familiar works, by listening to overly familiar work played on original instruments, by enrolling in a course, and so forth.

On this sophisticated Humean account, rationality is a matter of one's acting on one's strongest desires in the evaluational sense, not on one's strongest desires in the motivational sense. It is of course a truism that if one acts, one acts on the desire that is motivationally strongest. But in doing so one may, on this account, act in the face of one's deepest values and commitments and in the face of what one thinks one ought to do in a suitably internal sense. When one does so, one acts irrationally—for example, out of weakness of the will, compulsion, or the like. And in so doing, one acts no differently than one would in acting on a desire to smoke that one would rather not have and rather not act upon, given that one had it.

In the case of prudence, there are many desires that would support a desire that one's desires when one is seventy be satisfied. Thus there are many desires that would motivate one's trying to cultivate this desire were it absent or to reinstate it were it lost. One might be motivated, for example, by the connections between remaining self-sufficient and the continuation of important relationships with others, one's ability to continue to pursue meaningful projects, one's present and continued sense of self-respect, and so forth. And, as we saw in the mathematics example, these relations of support need not be construed as means-end. Self-sufficiency, self-respect, meaningful relations, and meaningful projects are mutually supporting, each one lending content, meaning, and depth to the others.

Moreover, on the sophisticated Humean account (in contrast to the Humean present-aim theory), there are several useful things we can say about the alleged irrationality of pathological indifference. We are likely to assume that the pathological lack of desire (e.g., not to avoid pain in SoHo), will be unsupported, since it is based on an apparently arbitrary distinction. If it is, the sophisticated Humean (unlike the present-aim theorist) can say that action on the basis of such a desire set *is* irrational. And this could be for either of two reasons. If it is badly supported and conflicts with desires that are well supported, it would be like the desire to smoke, and to act on it would typically be weak willed. Alternatively, if all the desires were equally arbitrary and badly supported, this would in itself undermine one's autonomy and, hence, one's rationality. If, however, the desires of the SoHo-indifferent subject were all extremely well supported, then it seems that either the distinction would not be arbitrary or the motivational makeup have some instrumentally rationale. (One might, for example, schedule all one's dental work in SoHo and otherwise avoid it.)

With regards to Korsgaard's problem of the normativity of practical reason, the sophisticated Humean can make precisely the distinction that Korsgaard thinks is required. In the case of a painful procedure, we can interpret resistance in terms of a fully rational belief that continuing to live is not worth the pain of the treatment. Terminally ill patients, for example, often make such rational decisions. Equally, however, we can, at least in some cases, make good sense of the idea that resistance is weak willed and irrational.

THE REAL PROBLEM OF WEAKNESS OF THE WILL

When we perform a weak-willed action, what makes it an *action*? For example, one takes a (fifth) martini, even though one judges that all things considered it would be best to refrain. One's all-things-considered judgment never wavers as one takes it, and one's judgment is in no way confused or contradictory. It seems clear that in this case taking the martini is an action. But what is it in virtue of which it is something one does, not merely something that happens to one?

The suggestion that it is an action because it is behavior caused by an appropriate belief and desire in not the basis of an adequate account. In Davidson's well-known counterexample,

A climber might want to rid himself of the weight and danger of holding another man on a rope, and he might know that by loosening his hold on the rope he could rid himself of the weight and danger. This belief and want might so unnerve him as to cause him to loosen his hold, and yet it might be the case that he never chose to loosen his hold, nor did he do it intentionally.[20]

Moreover, we cannot, as some have suggested, require that the causal chains be "of the right kind," if this is understood as a matter of neurophysiology. For the detailed neurophysiology is a matter of subpersonal facts to which we normally have no access. And yet we know, ordinarily, which pieces of our behavior are actions and which are not.

Nor can we require that the causal chain go through an appropriate piece of practical reasoning. For in the case in which one takes the martini, the appropriate piece of reasoning tells one to refrain. Indeed, as we have seen, one may be completely lucid in one's judgment about what one should do, and one's judgment may never waver, even as one's hand moves toward the glass. Thus, it seems, it cannot be anything at the level of judgment that makes the movement an action.

THE PHENOMENOLOGY OF AGENCY AND VALUE

It will help in seeing what kinds of noncausal and nonjudgmental conditions are relevant to a piece of behavior's being an action, if we consider the following example.[21]

The Passive Subject. Imagine that you have agreed to make a trip with a friend, and its success requires an early start. When you arrive at your friend's house at the appointed time, you are dismayed to find that he is still in bed. You suggest tactfully that the situation is not hopeless but that the success of the trip depends on his being up and ready *very* shortly. Your friend readily agrees and professes his continued strong desire to make the trip, but makes no effort to move.

In response to your evidently increasing consternation, your friend hastens to explain. Whereas he had been normal the night before, when he awoke he found that he no longer understood the concept of action. Though your friend agrees with you about all the objective facts that can be stated without the agential vocabulary—that the trip will only have a chance of success if you are underway very soon and so forth—the only response that he can make to your direct suggestion that he get up is that he hopes it *happens*. He finds himself forced, in other words, to translate your talk of *doings* into the only language he now understands—the language of *happenings*.

Interestingly, however, he has not lost the conceptual roles of the elements of the action vocabulary. He still knows all such "truths" such as that normally we hold people responsible only for what they do, that only creatures with mental states are capable of action, and so forth. What he professes to have lost are not the inferential relations between propositions containing agential concepts. Rather, it is the intelligibility of actually applying the agential vocabulary. In this respect he is like the anthropologist who knows the inferential roles of the witchcraft vocabulary employed by the people he studies but sees it as having no application to the real world.

What is it we have that the passive subject lacks? By hypothesis the passive subject knows all the facts we know that can be expressed without the agential vocabulary. Moreover, he still retains the inferential role of that vocabulary, even if he sees no possibility of applying it to the world. Thus it seems that all he could lack is an experiential grounding—one that could give the vocabulary genuine content that would distinguish it from a formal calculus.

But what could such an "experiential grounding" be *like*? What does something look like when it looks like an opportunity or a threat to which one has to react? If the challenge is to provide an account in terms of apparent shapes, colors, and relative sizes, then, as I shall argue below, the question has no answer. But there is a large empirical and experimental literature on what we might call "rich perception," stemming from the work of Michotte on the perception of causation and J. J. Gibson on the perception of "affordances."[22] These latter perceptual experiences include the perception of opportunities for action and the perception of things as given in their agentially relevant functional potentialities—things given, for example, as shelters, hiding places, or escape routes, as predators or prey. Recent work on such experience includes studies of the perception of animateness (i.e., being an original source of motion), intentionality, and agency.[23] This literature does not discuss the perception of values explicitly, but the agential and functional properties themselves, such as something's being a shelter, are clearly not value neutral.

There is, however, a deeper, nonempirical, reason why we must be given things as valuable. The Humean moral psychology explains action (and what it is to be an action) in terms of our having an appropriate belief and desire. But how are our *desires* given to us? Certainly our desires *can* be given to us from an objective point of view or, as I shall say, "as objects." A psychologist or neurophysiologist, for example, might tell you that you had a desire of which you had been completely unaware. But being the product of desires given in this way is not what being an action consists in, not what makes an action an action. For such desires have no appropriate internal connection to action. Their presence, and indeed our recognition of their presence (as such), provides no reason to act. Thus it cannot rationalize anything we do. With regard to desires given as objects, it makes perfect sense to ask why, by itself, the existence of such a desire of mine should be of any concern to me— or at least why it should be of any more concern to me than the existence of such a desire someone else's. Were a desire present and given only as an object, it would give one a reason to aim at its satisfaction only if one had some further desire to do so.

Imagine that a psychologist or neurophysiologist gives you the following information: You have a desire—as yet unrecognized, indeed as not yet manifest—for great wealth. Clearly this gives you no reason to leave philosophy and enroll in business school—none, at any rate, in the absence of some desire that this desire be satisfied. And where the desire for vast wealth is concerned you need have no such inclination. You might, in fact, pay a significant sum to have it removed.

This fact about desires given as objects would generate an infinite regress were there no other way in which desires could be given. And there is. Desires may be given as transparent. This is to be understood by analogy with the way in which tools and prosthetic devices, as well as perceptual experience, may be given. The blind person's stick, for example, is transparent when what he or she is given most directly is the object at the end of it. And our perception is transparent when what we are given most directly are real, as opposed to apparent, shapes, sizes, and colors—or indeed the real objects themselves. The analogy is this: when our desires are given transparently, they are given *implicitly* in the way the objects of those desires are given. And the objects of those desires are given as desir*able*.

But why shouldn't the objects of desires (when the desires are given to us transparently) be given as desir*ed* (by us)? But suppose that all I understand by "desire" is desire as an object. (Such a desire might be understood in turn, for example, as a functional state of an organism.) Then, again, the fact that something is given as desired by me in this sense gives me no reason whatsoever to pursue it. It is merely one more fact about the objective world, like the fact (which might also be given in perception) that a particle is exerting an attraction on the ones around it. And, as such, this fact leaves it completely open how I should, by my own lights, respond. To say that the objects of desires are given as desired by us, in other words, cannot prevent the infinite regress that we have been at pains to avoid.

Moreover, it seems clear that the perception of something as prey is the perception of something "to be pursued"; the perception of something as an escape route is the perception of something "to be taken." Thus, there is an internal connection of the right kind between the perception of functional properties and affordances of the Gibsonian kind and motivation. Therefore, the representation of things as desirable in this sense seems obviously more basic and more primitive than the representation of things as desired by me. The thought that something might be desired by me though not necessarily desirable is an extremely sophisticated one. It requires, among other things, some notion of intentional states in general and at least some particular kinds of intentional states, the distinction between my intentional states and those of others, and some self-conscious access to my own intentional states, particularly my desires. And the problem of my access to my own desires simply raises the prospect of an infinite regress once again. Objective access is no help, and, in any case, obviously unavailable to such unsophisticated subjects as animals and children.

The Humean has no account of how a desire is given (how it is given to the subject at the personal level) when it is given "subjectively" in the normal way. (Hume's own account in terms of a purely qualitative feeling attaching to an inert and neutral mental representation has, for good reason, not been seriously entertained by contemporary Humeans.) The only possibility, as we have seen, seems to be that it is given implicitly in its object's being given as desirable or to be pursued, and so forth. The conclusion is that the perception of things as desirable is more basic than the explicit representation of them as (merely) desired by us, and that the latter is made possible by the former.[24]

We can see what this implies for phenomenology if we consider contemporary sense-datum theories of visual perception. According to such theories, what we are given most directly are the alleged visual properties of objects—apparent shapes, colors, and relative sizes. (This is the contemporary analogue of the empiricist phenomenologies of Locke, Berkeley, and Hume.)[25] My claim is that relative to these theories, we have an *inflationary/ deflationary phenomenology*. What we are given most directly in visual perception (at the personal level) is in some respects much richer and in other respects much more impoverished than sense-datum theories allow.[26]

i. *On the Side of Deflation.* Asked to draw the apparent shape of their car windshield as seen from the driver's seat, most subjects are incapable of producing an even rough approximation. Similarly, subjects cannot accurately compare the apparent lengths of their hands (when held four inches from their eyes) with the apparent widths of their feet. Examples such as these suggest that normally the apparent visual properties of objects are accessible to us only indirectly, if at all. (The deflationary point is relevant to virtue theory discussed later.)

ii. *On the Side of Inflation.* Seeing things as desirable is, as we have seen, the way in which our desires are normally given to us. And our desires being given to us in this way (transparently) is the only way in which they have their normal motivational efficacy.

Value perception is, we might say, nonjudgmental, at least in the sense that it is not linguistic-descriptive. The perceptual system, at least with respect to what is given to the subject at the personal level, is not structured like a language, and its deliverances lack an inferential structure that could put them in direct conflict with our linguistically formulated judgments.

THE NORMATIVITY OF PRACTICAL
REASON RECONSIDERED

How, then, does the inflationary/deflationary phenomenology solve the problem of the normativity of practical reason? The weak-willed action is an action in virtue of its object's being given as valuable in perception. This perception provides what John McDowell evidently, and rightly, requires—the presentation of the object as desir*able*.[27] Such perception, however, need not be assumed to be veridical. There is an analogy with the Müller Lyer lines and other examples of visual illusion—the perceptual experience can persist in the face of disbelief. In the case of value perception, the paranoid may see two strangers who are merely talking as conspiring against him, even though he is fully aware of how unlikely this is.

Because the perceptual content is nonconceptual or nonjudgmental content, one does not contradict oneself or waver or change one's mind at the level of judgment. Indeed, one's judgment need be in no way clouded at all. This is necessary for a case of genuine weakness of the will. Desirability, as we have seen, is given in a more fundamental way than our own desires. Indeed, we have seen something stronger—it is only as given implicitly in the perception of the desirability of their objects that our recognition of our desires has any internal connection with motivation.

THE ANSWER TO "THE MORAL PROBLEM"

Michael Smith's discussion of what he terms "the moral problem" suggests that the following triad is inconsistent.[28]

a. *Cognitivism*: Evaluative statements express genuine beliefs (as opposed to expressing attitudes or emotions or having an imperatival analysis).

b. *Internalism*: The sincere belief in a genuine evaluative statement is internally (conceptually) connected with an appropriate form of motivation—e.g., motivation to pursue or support the thing which is positively evaluated.

c. *Humean Moral Psychology*: Beliefs and desires are separate existences defined by their different directions of fit. This means that what we believe implies nothing about what we desire and hence that there can be no internal connections between beliefs and desires.

A phenomenology that allows for the direct perception of things as valuable allows us to deny (c). Such perception can be, as we have seen, accurate or inaccurate. Hence the requirement of cognitivism is satisfied. But there is also an internal connection with motivation. That is, one will be motivated to pursue or support the things one perceives as valuable *to the extent that one is rational.* And notice that the italicized clause does not trivialize the criterion of internalism, since the irrationality in question will take the form of weakness of the will (or some related form), and the inflationary/deflationary phenomenology gives us a substantive account of such weakness consists in. In particular, we have an account that distinguishes the weak-willed subject from the rational subject who simply has an unusual set of desires.

THE IMPLICATIONS FOR VIRTUE THEORY

The standard examples of weakness of the will (e.g., taking the martini) lead us to assume that when value perception conflicts with judgment, rationality requires that we side with the latter. We assume, that is, that in such conflicts we must identify with our all-things-considered judgment about what it is best to do. But just as the deliverances of the perceptual system may be illusions of value (martini case), so the deliverances of the judgmental system may be shallow or self-deceived or for various other reasons amount to little more than rationalizations.

Huckleberry Finn. Huck Finn believes that because his friend and companion is a fugitive slave, it is his moral duty to turn him in to the authorities. However, because of the depth of his feelings for Jim and the particularities of their shared experiences and context, he cannot bring himself to do so—even though he is seriously disturbed by his failure to "do right." From our vantage point, however, it is clear that his decision not to turn Jim in represents his best and deepest self. And we see his judgment to the contrary as the product of a superficial commitment to the immoral ideology of his community.[29]

Moreover, as the example suggests, the perceptual system may deliver results that would be otherwise unavailable. The inflationary phenomenology is necessary to account for the personal-level psychology of expertise. As Hubert Dreyfus and others have argued, the expertise of the best chess players is largely the result of their perceiving chess positions immediately as highly structured contexts of strengths and weaknesses. They are given, then, in a form of perception with features importantly like Gibson's affordances. And the positions are perceived in the light of a network of analogies and disanalogies with previously encountered positions and games—perception that allows the experts to see, in a literal and immediate way, what is to be done.[30] This is in marked contrast to the style of play of nonexperts and of the most sophisticated chess playing programs. And it means that neither the deliverances of the most sophisticated rules internalized by those who lack expertise, nor the deliverances of the rules embodied in the most sophisticated programs can match the perceptual capacities of the best players.

This account of perception is important in helping us appreciate Aristotle's claim that the virtuous person is in some sense superior to the person who is merely continent—a claim that McDowell has been criticized for endorsing.[31] For suppose the advantage that the virtuous person has over the merely continent person is simply motivational in character—the virtuous person sees what is required and immediately acts on the perception without being tempted by other alternatives. In this case it is not at all clear why we should value the virtuous person more. Intuitively, the continent person who is tempted to run from the battle but stands his ground seems at least as admirable as the virtuous person who is never tempted at all. The puzzle is removed if we think of the advantage that the virtuous person has as importantly cognitive in nature. The virtuous person arrives at decisions about what to do simply by seeing what the context requires. And in the relevant contexts such decisions are superior to those of the continent person who weighs alternatives and balances competing claims.

We would, then, be well advised not to think of virtue theory exclusively or primarily in terms of virtues like courage in battle. In such cases there is usually little doubt as to what we ought to do and the problem is one of motivation—the necessity of overcoming fear for example. Rather we should think in terms of virtues such as kindness and in cases of some genuine complexity. In such cases there is often no lack of motivation, but it is frequently difficult to know what kindness in the truest sense requires.

Cries and Whispers. In Ingmar Bergman's film, a woman's relatives all try to give her solace and comfort as she is dying, but each one fails. Although the efforts of each of her relatives are sincere, only the family servant is capable of doing what kindness requires under the circumstances.

Though Dreyfus, who defends a version of virtue theory, has, following Merleau-Ponty, long seen the necessity of an inflationary phenomenology, he has no real counterpart of the deflationary phenomenology that lets us say that in some cases we are given external objects directly, without there being anything to which we have access independently of our access to those objects. This is not to deny that there is a sequence of causal connections in virtue of which we perceive the objects in the way we do. Rather it is to say that at the personal level there is nothing we are given more directly than the objects themselves. And this inability to make sense of our being given external objects directly leaves Dreyfus vulnerable to skeptical worries of exactly the sort raised by Hume's argument for skepticism about the external world.[32] As Dreyfus says in connection with *The Matrix*

> Given the conceivability of the brain-in-the-vat fantasy, the most that we can be sure of is that our coping experience reveals that we are directly up against some boundary conditions independent of our coping—boundary conditions with which we must get in sync in order to cope successfully. In this way, our coping experience is sensitive to the causal powers of these boundary conditions. Whether these independent causal conditions have the structure of an independent physical universe discovered by science, or whether the boundary conditions as well as the causal structures are both the effect of an unknowable thing in itself that is the ground of appearances as postulated by Kant, or even whether the cause of all appearances is a computer is something we could never know from inside our world.[33]

McDowell, who is far more attuned to the threat of skepticism, is unwilling to allow for the existence of a perceptual system that is sufficiently independent of the system of linguistically formulated judgments to provide an account of the possibility of weakness of the will.[34] What is required, then, is that we see the possibility of a phenomenology that, relative to that of the sense-datum theorist, is *both* inflationary and deflationary.

THE ADJUDICATION OF CONFLICTS BETWEEN VALUE PERCEPTIONS AND JUDGMENTS

The sophisticated Humean scheme for adjudicating between conflicting (nonderived) desires can be retained even though we abandon the Humean moral psychology. Relations of support in the form of meaning-giving analogies are now thought of as holding between values. And someone who (say) does not understand the backward-looking perspective integral to punishment can come to appreciate it in ways that it is perfectly reasonable to call rational. This essentially coherentist approach works between value perceptions, between value judgments, and between value perceptions and value judgments.

As the chess analogy suggests, value perception is highly theory (and practice) laden, so the perceptions and judgments are to some extent made for each other. Moreover, we can, if circumstances require, bring the contents of particular items in our perceptual system into the explicit linguistic form of our system of judgments. We do so through a process by which we articulate to ourselves, and thereby make discursively explicit, what had been implicit in various relevant perceptual experiences—a process like that by which the person who does not believe in punishment comes, nonetheless, to recognize the significance of the backward-looking perspective in the domain of his value commitments. Such explicit forms of recognition, however, are necessarily given against the background of the implicit deliverances of the perceptual system.

Thus the perspective that takes virtue as basic is fully compatible with the full critical use of our judgmental capacities to correct our tendencies toward value illusions such as those of the paranoid. But short of this coherentist approach, there is no quick route to the adjudication between what purport to be value perceptions and our evaluative judgments—even philosophers are capable of rationalization, and even paranoids have enemies.

NOTES

1. David Hume, *A Treatise of Human Nature*, ed. L. A. Selby-Bigge (Oxford: Clarendon Press, 1896), 416.

2. Hume, *A Treatise of Human Nature*, 413–18.

3. On Humean moral psychology, see Michael Smith, *The Moral Problem* (Oxford: Blackwell, 1994), 8–13. On the complex relations between the Humean theory of moral psychology and Hume's actual theories, official and otherwise, see John Bricke, *Mind and Morality: An Examination of Hume's Moral Psychology* (Oxford: Clarendon Press, 1996), chaps. 1 and 2.

4. See Bernard Williams, "Internal and External Reasons," in *Rational Action*, ed. Ross Harrison (Cambridge: Cambridge University Press, 1980), reprinted in Bernard Williams, *Moral Luck: Philosophical Papers, 1973–1980* (Cambridge: Cambridge University Press, 1981), 102.

5. On the present-aim theory, see Derek Parfit, *Reasons and Persons* (Oxford: Clarendon Press, 1984), 92–95.

6. This point seems implicit in Sidgwick's remarks on egoism when he says, "I do not see why the axiom of Prudence should not be questioned, when it conflicts with present inclination on a ground similar to that on which Egoists refuse to admit the axiom of Rational Benevolence. If the Utilitarian has to answer the question, 'Why should I sacrifice my own happiness for the greater happiness of another?'

it must surely be admissible to ask the Egoist 'Why should I sacrifice a present pleasure for a greater one in the future. Why should I concern myself about my own future feelings any more than about the feelings of other persons?'" (Henry Sidgwick, *The Methods of Ethics*, 7th ed. [London: Macmillan, 1907; reprint, New York: Dover, 1966], 418). Also see Thomas Nagel, *The Possibility of Altruism* (Oxford: Clarendon Press, 1970), 15–17; Parfit, *Reasons and Persons*, 137–49.

7. Nagel, *The Possibility of Altruism*.

8. See my "Rationality, Responsibility, and Pathological Indifference," in *Identity, Character, and Morality*, ed. Owen Flanagan and Amelie Rorty (Cambridge, Mass.: MIT Press, 1990) 401–26, reprinted, with revisions in my *The Unity of the Self* (Cambridge, Mass.: MIT Press, 1991) 259–84.

9. Parfit, *Reasons and Persons*, 124

10. The example of SoHo indifference was suggested to me by Jerrold Katz.

11. Christine Korsgaard, "The Normativity of Instrumental Reason," in *Ethics and Practical Reason*, ed. Garrett Cullity and Berys Gaut (Oxford: Clarendon Press, 1997), 229–31.

12. For a discussion of the difficulties in choosing between these interpretations and considerations suggesting that examples of extreme pain do not best illustrate the distinction between them, see Thomas Schelling, "Self-Command in Practice, in Policy, and in a Theory of Rational Choice," *AEA Papers and Proceedings* 74 (1984): 1–11, reprinted in his *Strategies of Commitment and Other Essays* (Cambridge, Mass.: Harvard University Press, 2006), 76. In this connection it is interesting to consider the John Dunbar character in *Dancing with Wolves*, who resolves to commit suicide rather than allow his leg to be amputated without anesthesia.

13. Jeanette Kennett, *Agency and Responsibility: A Common-sense Moral Psychology* (Oxford: Clarendon Press, 2001), 96.

14. James Dreier, "Humean Doubts About the Practical Justification of Morality," in *Ethics and Practical Reason*, ed. Garrett Cullity and Berys Gaut (Oxford: Clarendon Press, 1997), 93–96.

15. Kennett, *Agency and Responsibility*, 91.

16. Ibid., 91–92.

17. Ibid., 96.

18. Harry Frankfurt, "Freedom of the Will and the Concept of a Person," *Journal of Philosophy* 68 (1971): 5–20; David Lewis, "Dispositional Theories of Value," *Proceedings of the Aristotelian Society*, Supplementary Volume 63 (1989): 113–37, reprinted in his *Papers in Ethics and Social Philosophy* (Cambridge: Cambridge University Press, 2000), 71.

19. See my "Rationality, Responsibility, and Pathological Indifference." As will become clear, this *über*-Humean position is not the position that I currently hold and defend below. It is merely the position of an increasingly distant former self.

20. Donald Davidson, "Freedom to Act," reprinted in Davidson, *Essays on Actions and Events* (New York: Oxford University Press, 1980), 79.

21. See also my "Subjectivity and the Agential Perspective," in *Naturalism in Question*, ed. Mario De Caro and David Macarthur (Cambridge, Mass.: Harvard University Press, 2004), 201–30.

22. Albert Michotte, *The Perception of Causality* (New York: Basic Books, 1963); J. J. Gibson, *The Ecological Approach to Visual Perception* (Hillsdale, N.J.: Laurence Erlbaum, 1986), 127–28, 134.

23. See, for example, Dan Sperber, David Premack, and A. J. Premack, eds., *Causal Cognition* (New York: Oxford University Press, 1995).

24. The priority of the perceptual experience of things as valuable over explicit, linguistically formulated judgments that they are desired is defended at greater length in the context of a discussion of transcendental arguments and skepticism in my "The Transcendental Significance of Phenomenology," *Psyche* 13, no. 1 (April 2007), available at http://journalpsyche.org/ojs-2.2/index.php/psyche/article/view File/2804/2669.

25. See Bertrand Russell, *Problems of Philosophy* (New York: Oxford University Press, 1959); A. J. Ayer, *The Foundations of Empirical Knowledge* (London, Macmillan, 1962); R. J. Swartz, ed., *Perceiving, Sensing, and Knowing* (New York: Doubleday, 1965), sec. 2.

26. On the notions of an inflation and deflation, see my "Skepticism, Deflation, and the Rediscovery of the Self," *The Monist* 87 (2004): 275–98; "Subjectivity and the Agential Perspective"; "Empirical Psychology, Transcendental Phenomenology, and the Self," in *Cartographies of the Mind: Philosophy and Psychology in Intersection*, ed. Massimo Marraffa, Mario De Caro, and Francesco Ferretti (Dordrecht: Sprinter, 2007); and "The Transcendental Significance of Phenomenology."

27. McDowell says, for example, "The idea of value experience involves taking admiration, say, to represent an object [for example, virtue] as having a property that… is conceived to be not merely such as to elicit [admiration]… but rather such as to *merit* it" ("Value and Secondary Qualities," in *Morality and Objectivity*, ed. Ted Honderich [London: Routledge and Kegan Paul, 1985], reprinted in John McDowell, *Mind, Value, and Reality* [Cambridge, Mass.: Harvard University Press, 1998], 143).

28. Michael Smith, *The Moral Problem* (Oxford: Blackwell, 1994), 11–13. Smith's triad involves moral judgments rather than evaluative statements, but the latter seem to raise the same problems. Arguably they do so in an even more pressing form. And in focusing on evaluative statements rather than, say, "ought" statements, I do not intend to beg any questions on the issues that divide consequentialists and deontologists. The value perception with which I'm primarily concerned is equally well described as our seeing things as valuable (in specific ways), as our seeing things as "to be pursued," as our experiencing the world as making certain demands on us, and so forth.

29. Mark Twain, *The Annotated Huckleberry Finn*, Michael Patrick Hearn, ed. (New York: Norton, 1981), 153–57.

30. Hubert Dreyfus and Stuart Dreyfus, "What Is Morality? A Phenomenological Account of the Development of Ethical Expertise," in *Universalism and Communitarianism*, ed. David Rasmussen (Cambridge, Mass.: MIT Press, 1990), esp. 242–49.

31. See John McDowell, "Are Moral Requirements Hypothetical Imperatives," *Proceedings of the Aristotelian Society*, Supplementary Volume 52 (1978): 13–29, reprinted in *Mind, Value, and Reality* (Cambridge, Mass.: Harvard University Press, 1998), 91–93. Kennett, *Agency and Responsibility*, 35–37.

32. Hume, *A Treatise of Human Nature*, 212. Cf. Barry Stroud, "Skepticism and the Possibility of Knowledge," *Journal of Philosophy* 81 (1984): 545–51, reprinted in *Epistemology: The Big Questions*, ed. Linda Martin Alcoff (Oxford: Blackwell, 1998), 363–64.

33. Hubert Dreyfus and Stuart Dreyfus, "Existential Phenomenology and the Brave New World of *The Matrix*," in *Philosophers Explore* The Matrix, ed. Christopher Grau (New York: Oxford University Press, 2005), 79.

34. For this criticism of McDowell, see Kennett, *Agency and Responsibility*, 28–30.

PART V

EPISTEMOLOGY AND NORMATIVITY

11

TRUTH AS CONVENIENT FRICTION

Huw Price

1. INTRODUCTION

In a recent paper, Richard Rorty begins by telling us why pragmatists such as himself are inclined to identify truth with justification:

> Pragmatists think that if something makes no difference to practice, it should make no difference to philosophy. This conviction makes them suspicious of the distinction between justification and truth, for that distinction makes no difference to my decisions about what to do.[1]

Rorty goes on to discuss the claim, defended by Crispin Wright, that truth is a normative constraint on assertion. He argues that this claim runs foul of this principle of no difference without a practical difference:

> The need to justify our beliefs to ourselves and our fellow agents subjects us to norms, and obedience to these norms produces a behavioral pattern that we must detect in others before confidently attributing beliefs to them. But there seems to be no occasion to look for obedience to an *additional* norm—the commandment to seek the truth. For—to return to the pragmatist doubt with which I began—obedience to that commandment will produce no behavior not produced by the need to offer justification.[2]

Again, then, Rorty appeals to the claim that a commitment to a norm of truth rather than a norm of justification makes no behavioral difference.

This is an empirical claim, testable in principle by comparing the behavior of a community of realists (in Rorty's sense) to that of a community of pragmatists. In my view, the experiment would show that the claim is unjustified, indeed false. I think that there is an important and widespread behavioral pattern that depends on the fact that speakers do take themselves to be subject to such an additional norm. Moreover, it is a behavioral pattern so central to what we presently regard as a worthwhile human life that no reasonable person would knowingly condone the experiment. Ironically, it is also a pattern that Rorty of all people cannot afford to dismiss as a pathological and dispensable by-product of bad philosophy. For it is conversation itself, or at any rate a central and indispensable part of conversation as we know it—roughly, interpersonal dialogue about "factual" matters.[3]

In other words, I want to maintain that in order to account for a core part of ordinary conversational practice, we must allow that speakers take themselves and their fellows to be governed by a norm stronger than that of justification. Not only is this a norm which speakers acknowledge they may fail to meet, even if their claims are well-justified—this much is true of what Rorty calls the cautionary use of truth[4]—but also, more significantly, it is a norm which speakers immediately assume to be breached by someone with whom they disagree, *independently of any diagnosis of the source of the disagreement*. Indeed, this is the very essence of the norm of truth, in my view. It gives disagreement its immediate normative character, a character on which dialogue depends, and a character which no lesser norm could provide.

This fact about truth has been overlooked, I think, because the norm in question is *so* familiar, so much a given of ordinary linguistic practice, that it is very hard to see. Ordinarily we look through it, rather than at it. In order to make it visible, we need a sense of how things would be different without it. Hence, in part, my reason for beginning with Rorty. Though I disagree with Rorty about the behavioral consequences of a commitment to "a distinction between justification and truth," I think that the issue of the behavioral consequences of such a commitment embodies precisely the perspective we need, in order to bring into focus this fundamental aspect of the normative structure of dialogue.

In sharing Rorty's concern with the role of truth in linguistic practice, I share one key element of his pragmatism. But my kind of pragmatism about truth is not well marked on contemporary maps, and hence my second reason for beginning with Rorty. Rorty has explored the landscape of pragmatist approaches to truth more extensively than most pragmatist writers, past or present, and at different times has been inclined to settle in different parts of it. By locating my own kind of pragmatism with respect to views that Rorty has visited or canvassed, I hope to show that there is a promising position that he and others pragmatists have overlooked.

As noted, my view rests on the claim that a norm of truth plays an essential and little-recognized role in assertoric dialogue. In pursuit of this conclusion, it will turn out to be helpful to distinguish three norms, in order of increasing strength: roughly, sincerity, justification and truth itself. Though somewhat crudely drawn, these distinctions will suffice to throw into relief the crucial role of the third norm in linguistic practice. My strategy will be to contrast assertion as we know it with some non-assertoric uses of language. In these latter cases, I'll argue, the two weaker norms still apply. Moreover, it turns out that some of the basic functions of assertoric discourse could be fulfilled in an analogous way, by a practice which lacked the third norm. But it will be clear, I hope, that that practice would not support dialogue as we know it. What is missing—what the third norm provides—is the automatic and quite unconscious sense of engagement in common purpose that distinguishes assertoric dialogue from a mere roll call of individual opinion. Truth is the grit that makes our individual opinions engage with one another. Truth puts the cogs in cognition, at least in its public manifestations.[5]

To use a Rylean metaphor, my view is thus that truth supplies factual dialogue with its essential esprit de corps. As the metaphor is meant to suggest, what matters is that speakers think that there is such a norm—that they take themselves to be governed by it—not that their view be somehow confirmed by science or metaphysics. Science has already done its work, in pointing out the function of the thought in the lives of creatures like us. This may suggest that a commitment to truth is like a commitment to theism, an analogy which Rorty himself draws, against Wright, in the paper with which I began. In effect, Rorty's point is that it is one thing to establish that we do employ a realist notion of truth, a normative notion stronger than justification; quite another to establish that we ought to do so. As in the case of theism, we might do better to wean ourselves of bad realist habits.

However, there are several important differences between the two cases. First, the behavioral consequences of giving up theism are significant but hardly devastating.[6] But if I am right about the behavioral role of truth, the consequences of giving up truth would be very serious indeed, reducing the conversation of mankind to a chatter of disengaged monologues.[7]

Second, it is doubtful whether giving up truth is really an option open to us. I suspect that people who think it is an option haven't realized how deeply embedded the idea of truth is in linguistic practice, and therefore underestimate the extent of the required change, in two ways. They fail to see how radically different from current practice a linguistic practice without truth would have to be, and they overestimate our capacity to change our practices in general to move from here to there (underestimating the practical inflexibility of admittedly contingent practices[8]).

Third, and most interestingly of all, the issue of the status of truth is enmeshed with the terms of the problem, in a way which is quite uncharacteristic of the theism case. Metaphysical conclusions tend to be cast in semantic vocabulary. Theism is said to be in error in virtue of the fact that its claims are not *true,* that its terms fail to *refer.* For this reason, it is uniquely difficult to formulate a meaningful antirealism about the semantic terms themselves. In my view the right response to this is not to think (with Paul Boghossian[9]) that we thereby have a transcendental argument for semantic realism. Without an intelligible denial, realism is no more intelligible than antirealism. The right response—as Rorty himself in any case urges—is to be suspicious of the realist-antirealist debate itself.[10] However, Rorty ties rejection of the realist-antirealist debate to rejection of a notion of truth distinct from justification, and of the idea of representation. I think this is the wrong path to the right conclusion. We should reject the metaphysical stance not by rejecting truth and representation, but by recognizing that in virtue of the most plausible story about the function and origins of these notions, they simply don't sustain that sort of metaphysical weight.

Concerning his own view of truth, Rorty describes himself as oscillating between Jamesian pragmatism, on the one hand, and deflationism, on the other: "swing[ing] back and forth between trying to reduce truth to justification and propounding some sort of minimalism about truth."[11] My own view is neither of these alternatives, but has something in common with each. On the one hand, it is certainly some sort of minimalism about truth, but not the familiar sort that Rorty has in mind—not "Tarski's breezy disquotationalism," as he calls it.[12] I agree with familiar disquotationalist minimalists such as Quine[13] and Paul Horwich[14] that truth is not a substantial property, about the nature of which there is an interesting philosophical issue. Like them, I think that the right approach to truth is to investigate its function in human discourse—to ask what difference it makes to us to have such a concept. Unlike such minimalists, however, I don't think the right answer to this question is that truth is merely a grammatical device for disquotation. I think that it has a far more important function, which requires that it be the expression of a norm. But like other minimalists, again, I think that there is no further question of interest to philosophy, once the question about function has been answered.

On the other hand, my view of truth is also pragmatist, for it explicates truth in terms of its role in practice. (This is also true of standard disquotational views, of course, although they ascribe the truth predicate a different role in practice.) In another sense, it conflicts with pragmatism, for it opposes the proposal that we identify truth with justification. This contrast reflects a deep tension within pragmatism. From Peirce and James on, pragmatists have often been unable to resist the urge to join their opponents in asking "What is truth?" (Indeed, the pragmatist position as a whole is often characterized in terms of its answer to this question.) Pragmatism thus turns its back on alternative paths to philosophical illumination about truth, even though these alternative paths—explanatory and genealogical approaches—are at least compatible with, if not mandated by, the pragmatist doctrine that we understand problematic notions in terms of their practical significance.

Rorty himself is well aware of this tension within pragmatism. In "Pragmatism, Davidson and Truth," for example, he notes that James is less prone than Peirce to try to answer the "ontological" or reductive question about truth, and suggests that Davidson may be thought of as a pragmatist in the preferable non-reductive sense.[15] As he swings between pragmatism and deflationism, then, Rorty himself is at worst only intermittently subject to this craving for an *analysis* of truth. All the same, it seems to me that he is never properly aware of the range of possibilities for non-reductive pragmatism about truth. In particular, he is not properly aware of the possibility that such a pragmatism might find itself explaining the fact that the notion of truth in ordinary use is (and perhaps ought to be, in whatever sense we might make of this) one that conflicts with the identification of truth with justification: a normative goal of inquiry, stronger than any norm of justification, of the very kind that realists about truth—opponents both of pragmatism and of minimalism—mistakenly sought to analyze. In other words, Rorty seems to miss the possibility that the right thing for the explanatory pragmatist to say might be that truth is a goal of inquiry distinct from norms of justification, and that the realist's mistake is to try to *analyze* this normative notion, rather than simply to investigate its function and genealogy. It is this latter possibility that I want to defend.

2. FALSITY AND LESSER EVILS.

As I have said, I want to argue that truth plays a crucial role as a norm of assertoric discourse. It is not the only such norm, however, and a good way to highlight the distinctive role of truth is to distinguish certain weaker norms, and to imagine a linguistic practice which had those norms but not truth.[16] By seeing what such a practice lacks, we see what truth adds.

There are at least two weaker norms of assertion, in addition to any distinctive norm of truth.[17] The weakest relevant norm seems to be that embodied in the principle that it is prima facie appropriate to assert that p only when one believes that p—prima facie, because of course many other factors may come into play, in determining the appropriateness of a particular assertion in a particular context. Let's call this the norm of *subjective assertibility*.[18] The norm is perhaps best characterized in negative form—that is, in terms of the conditions under which a speaker may be censured for failing to meet it:

(Subjective assertibility)
A speaker is incorrect to assert that p if she does not believe that p; to assert that p in these circumstances provides prima facie grounds for censure, or disapprobation.

The easiest way to see that this norm has very little to do with truth is to note that it is analogous to norms which operate with respect to utterances which we don't take to be truth-apt. Prima facie, it is inappropriate to request a cup of coffee when one doesn't want a cup of coffee, but this doesn't show that requests or expressions of desires are subject to a norm of truth. In effect, this norm is simply that of sincerity, and some such norm seems to govern much conventional behavior. Conventions often depend on the fact that communities censure those who break them in this specific sense, by acting in bad faith.

The second norm is that of (personal) *warranted assertibility*. Roughly, 'p' is warrantedly assertible by a speaker who not only believes that p, but is *justified* in doing so. The qualification 'personal' recognizes the fact that there are different kinds and degrees of warrant or justification, some of them more subjective than others. For example, is justification to be assessed with reference to a speaker's actual evidence as she (presently?) sees it, or by some less subjective lights? For the moment, for a degree of definiteness, let us think of it in terms of subjective coherence—a belief is justified if supported by a speaker's other current beliefs. This is what I shall mean by *personal* warranted assertibility.

Again, this second norm is usefully characterized in negative or censure form:

(Personal warranted assertibility)
A speaker is incorrect to assert that p if she does not have adequate (personal) grounds for believing that p; to assert that p in these circumstances provides prima facie grounds for censure.

A person who meets both the norms just identified may be said to have done as much as possible, *by her own current lights*, to ensure that her assertion that p is in order. Obviously, realists will say that her assertion may nevertheless be incorrect. Subjective assertibility and (personal) warranted assertibility do not guarantee truth. To an extent, moreover, most pragmatists are likely to agree. Few people who advocate reducing truth to (or replacing truth by) a notion of warranted assertibility have personal warranted assertibility in mind. Rather, they imagine some more objective, community-based variant, according to which a belief is justified if it coheres appropriately with the other beliefs of one's community. If we call this *communal* warranted assertibility, then the point is that we can make sense of a gap between the personal and communal notions. A belief may be justified in one sense but not the other.

Pragmatists and realists may thus agree that there is a normative dimension distinct from subjective assertibility and personal warranted assertibility—an assertion may be *wrong*, despite meeting these norms. This does not yet establish that the normative standard in question need be marked in ordinary discourse. In principle, it might be a privileged or theoretical notion, useful in expert second-order reflection on linguistic practice but unnecessary in folk talk about other matters. In practice, however, there seems a very good reason why it should not remain restricted in this way. Unless individual speakers recognize such a norm, the idea that they might *improve* their views by consultation with the wider community is simply incoherent to them. (It would be as if we gave a student full marks in an exam, and then told him that he would have done better if his answers had agreed with those of other students.)

It may seem that as yet, this argument doesn't favor realism over pragmatism. If the normative standard an individual speaker needs to acknowledge is that of the community as a whole, there is as yet no pressure to a notion of truth beyond community-wide warranted assertibility. But what constitutes the relevant community? At any given stage, isn't the relation of a given community to its possible present and future extensions just like that of the individual to her community? If so, then the same argument applies at this level. At each stage, the actual community needs to recognize that it may be wrong by the standards of some broader community.[19]

The pragmatist might now seem obliged to follow Peirce, in identifying truth with warranted assertibility in the ideal limit of inquiry. The useful thing about this limit, in this context, is that it transcends any actual community. But in my view, as I'll explain below (and as Rorty in some moods already agrees), a better move for a pragmatist is to resist the pressure to *identify* truth with anything—in other words, simply to reject the assumption that an adequate philosophical account of truth needs to answer the question "What is truth?" Better questions for a pragmatist to ask are the explanatory ones: Why do we have such a notion? What job does it do in language? What features does it need to have to play this role? And how would things be different if we didn't have it?

For the moment, we have the beginnings of an answer to the last question. If we didn't have a normative notion in addition to the norms of subjective assertibility and personal warranted assertibility, the idea that we might improve our commitments by seeking to align them with those of our community would be simply incoherent. I'll call this the passive account of the role of the third norm—passive, because it doesn't yet provide an active or causal role for a commitment to truth. Later, I'll argue that the third norm not only creates the conceptual space for argument, in this passive sense, but actively encourages speakers to participate.[20]

3. THE THIRD NORM IN FOCUS.

The best way to bring the third norm into focus is again to consider its negative or censure form:

(Truth)
If not-p, then it is *incorrect* to assert that p; if not-p, there are prima facie grounds for censure of an assertion that p.

The important point is that this provides a norm of assertion which we take it that a speaker may fail to meet, even if she does meet the norms of subjective assertibility and (personal) warranted assertibility. We are prepared to make the judgment that a speaker is *incorrect*, or *mistaken*, in this sense, simply on the basis that we are prepared to make a contrary assertion; in advance, in other words, of any judgment that she fails to meet one or other of the two weaker norms.[21]

One of the reasons why this third norm is hard to distinguish from the two weaker norms of assertibility is that when we apply it in judging a fellow speaker right or wrong, the basis for our judgment lies in our own beliefs and evidence. It is not as though we are in a position to make the judgment from the stance of reality itself, as it were. I think this can make it seem as if application of this norm involves nothing more than re-assertion of the original claim (in the case in which we judge it correct), or assertion of the negation of the original claim (in the case in which we judge it incorrect). Construed in these terms, our response contains nothing problematic for orthodox disquotational versions of the deflationary view, of course. Re-assertion of this sort is precisely one of the linguistic activities which disquotational truth facilitates. Construed in these terms, then, there is no need for truth to be a distinct *norm*.

However, our response is not merely re-assertion, or assertion of the negation of the original claim. If it were, it would involve no commendation or criticism of the original utterance. This non-normative alternative is hard to see, I think, because the norm in question is so familiar and so basic. As a result, it is difficult to see the immense difference the norm makes to the character of disagreements. But it comes into focus, I think, if we allow ourselves to imagine a linguistic practice which allowed re-assertion and contrary assertion, but without this third normative dimension. What we need to imagine, in other words, is a linguistic community who use sentences to express their beliefs, and have a purely disquotational truth predicate, but for whom disagreements have no normative significance, except in so far as it is related to the weaker norms of assertibility.

This imaginative project is not straightforward, of course. Indeed, it isn't clear that it is entirely coherent. If there is a third norm of the kind in question, isn't it likely to be constitutive of the very notions of assertion and belief? If so, what sense is there in trying to imagine an assertoric practice which lacked this norm?

Well, let's see. What we need is the idea of a community who take an assertion—or rather the closest thing they have to what we call an assertion—to be *merely* an expression of the speaker's opinion. The relevant idea is familiar in the case of expressions of desires and preferences. It is easy to imagine a community—we are at least close to it ourselves—who have a language in which they give voice to psychological states of these kinds, not by *reporting* that they hold them (which would depend on assertion), but directly, in conventional linguistic forms tailored specifically for this purpose.

Think of a community who use language primarily for expressing preferences in restaurants, for example. (Perhaps the development of such a restricted language from scratch is incoherent, but surely we might approach it from the other direction. Imagine a community of dedicated lunchers, whose language atrophies to the bare essentials.) In this community we would expect a norm analogous to subjective assertibility: essentially, a normative requirement that speakers use these conventional expressions sincerely. Less obviously, such a practice might also involve a norm analogous to personal warranted assertibility. In other words, expressed preferences might be censured on the grounds that they were not well-founded, by the speaker's own lights (for example, on the grounds they did not cohere with the speaker's other preferences and desires). However, in this practice there need be no place for a norm analogous to truth—no idea of an objective standard, over and above personal warranted assertibility, which preferences properly aim to meet.

At least to a first approximation, we can imagine a community who treat expressions of beliefs in the same way. They express their beliefs—that is, let us say, the kind of behavioral dispositions which we would characterize as beliefs—by means of a speech act we might call the *merely-opinionated assertion* (*MOA*, for short). These speakers—"Mo'ans," as I called them in another paper[22]—criticize each other for insincerity and for lack of coherence, or personal warranted assertibility. But they go no further than this. In particular, they do not treat a disagreement between two speakers as an indication that, necessarily, one speaker or other is mistaken—in violation of some norm. On the contrary, they allow that in such a case it may turn out that both speakers have spoken correctly, by the only two standards the community takes to be operable. Both may be sincere, and both, in their own terms, may have good grounds for their assertion.[23]

A speech community of this imagined kind could make use of a disquotational truth predicate, as a device to facilitate agreement with an expression of opinion made by another speaker. 'That's true' would function much like 'Same again', or 'Ditto', used in a bar or restaurant. Just as 'Same again' serves to indicate that one has the same preference as a previous speaker, 'That's true' would serve to indicate that one holds the same opinion as the previous speaker. The crucial point is that if the only norms in play are subjective assertibility and personal warranted assertibility, introducing disquotational truth leaves everything as it is. It doesn't import a third norm.

The difficulty we have in holding on to the idea of such a community stems from our almost irresistible urge to see the situation in terms of our own normative standards. There really is a third norm, we are inclined to think, even if these simple creatures don't know it. If two of them make incompatible assertions then one of them must be objectively incorrect, even if by their lights they both meet the only norms they themselves recognize. (I think even pragmatists will be inclined to say this, even though they want to equate the relevant kind of incorrectness not with falsity but with lack of some kind of justification more objective than that of personal warrant.) But the point of the story is precisely to bring this third norm into sharp relief, and hence I am quite happy to allow challenges to the story on these grounds, which rely on the very conclusion I want to draw. *For us*, there is a third norm. But why is that so? Where does the third norm come from? What job does it do—what difference does it make to our lives? And what features must it have in order to do this job?

4. WHAT DIFFERENCE DOES THE THIRD NORM MAKE?

Let's return to the Mo'ans, and their merely-opinionated assertions. Recall that Mo'ans use linguistic utterances to express their "beliefs," as well as other psychological states, such as preferences and desires. Where they differ from us is in the fact that they do not take a disagreement between two speakers in this belief-expressing linguistic dimension to indicate that one or other speaker must be at fault. They recognize the possibility of fault consisting in failure to observe one of the two norms of subjective assertibility or personal warranted assertibility, but lack the idea of the third norm, that of truth itself. This shows up in the fact that by default, disagreements are of a no-fault kind, in the way that expression of different preferences often are for us.

What does it take to add the third norm to such a practice? Do the Mo'ans need to come to believe that there is a substantial property that the attitudes they use MOAs to express may have or lack—perhaps the property of corresponding to how things are in the world, perhaps that of being what their opinions are fated to converge on in the long run? Does adoption of the third norm depend on a piece of folk metaphysics of this kind? Not at all, in my view. The practice the Mo'ans need to adopt is simply that whenever they are prepared to assert (in the old MOA sense) that p, they also be prepared to ascribe fault to anyone who asserts not-p, independently of any grounds for thinking that that person fails one of the first two norms of assertibility. Perhaps they also need to be prepared to commend anyone who asserts that p, or perhaps failure-to-find-fault is motivation enough in this case. At any rate, what matters is that disagreement itself be treated as grounds for disapproval, as grounds for thinking that one's interlocutor has fallen short of some normative standard.

At this point it is worth noting what may seem a serious difficulty. If the Mo'ans don't *already* care about disagreements, why should they care about disagreements about normative matters? Suppose that we two are Mo'ans, that you assert that p, and that I assert that not-p. If this initial disagreement doesn't bother me, why should it bother me when—trying to implement the third norm—you go on to assert that I am "at fault," or "incorrect"? Again, I simply disagree; and if the former disagreement doesn't bite then nor will the latter. And if what was needed to motivate me to resolve our disagreement was *my* acceptance that I am "at fault," then motivation would always come too late. If I accept this at all, it is only after the fact—after the disagreement has been resolved in your favor.

To get the sequence right, then, I must be motivated by your disapproval itself. This is an important point. It shows that if there could be an assertoric practice which lacked the third norm, we couldn't add that norm simply by adding a normative predicate. In so far—so very far, in my view—as terms such as 'true' and 'false' carry this normative force in natural languages, they must be giving voice to something more basic: a fundamental practice of expressions of attitudes of approval and disapproval, in response to perceptions of agreement and disagreement between expressed commitments. I'll return to this point, for it is the basis of an important objection to certain other accounts of truth.

Imagine for the moment that the Mo'ans could add the third norm by adding a normative predicate, or pair of predicates ('correct' and 'incorrect', say). What would be the usage rule for these predicates? Simply that one be prepared to assert that p is correct if and only if one is prepared to assert that p; and to assert that p is incorrect if and only if one is prepared to assert that not-p. In other words, the usage rule is something very close to the disquotational schema ('p' is true if and only if p). As a result, the present proposal, that the truth predicate is an explicit expression of the third norm, already seems well on the way to an explanation of the disquotational functions of truth. We have already noted that the converse argument does not go through. A practice which lacked the third norm could still make use of a disquotational truth predicate.[24]

For the moment, we are interested in the function of the third norm. Why might the invention of such a norm be useful? What distinctive job does it do? We already have one answer to the latter question, and hence possibly to the former, in the passive account. Without a norm stronger than that of warranted assertibility *for me*, or *for us*, the idea of improving *my*, or *our*, current commitments would be incoherent. The third norm functions to create the conceptual space for the idea of further improvement. To do this job, we need a norm stronger than that of warranted assertibility for any *actual* community. (Of course, this doesn't yet show that we need something more than Peircean ideal assertibility, but one thing at a time.)

However, we can do better than the passive account. The third norm doesn't just hold open the conceptual space for the idea of improvement. It positively encourages such improvement, by motivating speakers who disagree to try to resolve their disagreement. Without the third norm, differences of opinion would simply slide past one another. Differences of opinion would seem as inconsequential as differences of preference. With the third norm, however, disagreement automatically becomes normatively loaded. The third norm makes what would otherwise be no-fault disagreements into unstable social situations, whose instability is only resolved by argument and consequent agreement—and it provides an immediate incentive for argument, in that it holds out to the successful arguer the reward consisting in her community's positive evaluation of her dialectical position. If reasoned argument is generally beneficial—beneficial in some long-run sense—then a community of Mo'ans who adopt this practice will tend to prosper, compared to a community who do not.

I'll call this the active account of the role of the third norm. In effect, it contends that the fact that speakers take their belief-expressing utterances to be subject to the third norm plays a causal, carrot-and-stick role in encouraging them to settle their differences, in cases in which initially they disagree. The force of these carrots and sticks should not be over-stated, however. In any given case, we are free not to give voice to our third-norm-grounded disapproval. If we do express it, the speakers with whom we disagree are free not to rise to the bait. Many factors may determine what happens in any particular case. My claim is simply that the third norm adds something new to the preferential mix. In particular, it gives rise to a new preferential pressure towards resolution of the disagreement in question—a pressure which would not exist in its absence, which does not exist for the Mo'ans, and which could not exist for us, if we did not care in general about the approval and disapproval of our fellows. The third norm depends on the fact that (to varying extents in varying circumstances) we do care about these things. It exploits this fact about us to make disagreements matter, in a way in which they would not otherwise matter. But the third norm does not come for free, with a general disposition to seek the approval of our fellows. What we have but the Mo'ans lack is an additional, special purpose, disposition: the disposition to disapprove of speakers with whom we disagree. This disposition is the mark of the third norm.

As in the case of the passive account of the role of the third norm, we need to be careful that this active account does not viciously presuppose the very notions for which it seeks to account. The notion of disagreement requires particular care. For one thing, recognition that one differs from a previous speaker must take some form more basic than the belief that he or she has said something "false," for otherwise there could not be a convention of applying this normative predicate when one perceives that one differs. For another, there is an important sense in which on the proposed account, it is the practice of applying the third norm which creates the disagreement, where initially there was mere difference. Properly developed, the view seems likely to be something like this. There is a primitive incompatibility between certain behavioral commitments[25] of a single individual, which turns on the impossibility of both doing and not doing any given action A—both having and not having a cup of coffee, for example. All else—both the public perceived incompatibility of "conflicting" assertions by different speakers, and the private perceived incompatibility essential to reasoning—is by convention, and depends on the third norm.

Obviously, much more needs to be said about this. At another level, much also needs to be said about possible advantages of such a mechanism for resolving differences—about its long-run advantages, for example, both compared to the case in which there is no such mechanism and compared to the case in which there is some different mechanism, such as deference to social rank. For immediate purposes, however, my claim does not depend on this latter work. For the present, my claim is simply that truth does play the role of this third norm, in providing the friction characteristic of factual dialogue as we know it. (I also claim, roughly, that this is perhaps the most interesting fact about truth, from a philosophical perspective.) In principle, this claim could be true, even though the practice in question was not advantageous. In principle, truth, and with it dialogue, could turn out to be a bad thing for the species, biologically considered.[26] No matter. It would still be true that we wouldn't have understood truth until we understood its role in this debilitating practice.

Is talk of dialogue really essential here? Couldn't we say simply that the third norm is what distinguishes a genuinely assertoric linguistic practice from the "merely opinionated" assertoric practice of the Mo'ans? The distinguishing mark of genuine assertion is thus that by default, difference is taken as a sign of *fault*, of breach of a normative standard.

It would not strictly be incorrect to say this, in my view, but it ought to seem unsatisfying, by pragmatist lights. A pragmatist is interested in the practical significance of the notions of truth and falsity, in the issue of what difference possession of these notions makes to our lives. According to the view just suggested, the answer will be something like this. The difference that truth and falsity make is that they make our linguistic practice genuinely assertoric, rather than Mo'an. "I see that," the pragmatist will then say, "But what practical difference does *that* difference make, over and above the obvious difference—that is, over and above the fact that we approve and disapprove of some of our fellow speakers on occasions on which we wouldn't otherwise do so?"

My own answer to the new question is that these habits of approval and disapproval tend to encourage dialogue, by providing speakers with an incentive to resolve disagreements. It is true that at this point the pragmatist's question—"What difference does *that* make?"—can be (indeed, should be) asked all over again. The importance I have here attached to dialogue rests in part on the gamble that this question will turn out to have an interesting answer, in terms of the long-run advantages of pooled cognitive resources, agreement on shared projects, and so on. But not entirely. Dialogue seems such a central part of our linguistic and social lives, that the difference between a world without dialogue and our world is much greater than *merely* the difference between MOAs and genuine assertions. So even if were to turn out that the development of dialogue had been an historical accident, of no great value to the species biologically considered, it would still be true that the most interesting behavioral consequence of the third norm would be dialogue, and not merely the more-than-merely-Mo'an assertion which makes dialogue possible.[27]

Recall that I began by challenging Rorty's claim that no behavioral consequences flow from a distinction between justification and truth. In one sense, my challenge does indeed amount to pointing out that the third norm—a notion of truth stronger than justification—brings with it the following behavioral difference: a disposition to criticize, or at least disapprove of, those with whom one disagrees. But if this were all the challenge amounted to, Rorty would be entitled to reply that of course there is this difference, but that this difference makes no interesting further difference. Hence the importance of dialogue, in my view, which turns a small difference in normative practice into a big difference in the way in which speakers engage with one another (and thereby ensures that Rorty's claim fails in an interesting rather than an insignificant way).

5. PEIRCE REGAINED?

Now to the question deferred above. Does the third norm need to be other than a more-than-merely-personal notion of justification? In particular, couldn't it be a Peircean flavor of ideal warranted assertibility? I have several responses to this suggestion.

First, I think that the proposal is mistakenly motivated. As I said in the introduction, I think it stems from the tendency, still too strong in Peirce, to ask the wrong question about truth. If we think that the philosophical issue is "What is truth?," then naturally we'll want to find an answer—something with which we may identify truth. Then, given standard objections to meta-physical answers, it is understandable that Peirce's alternative should seem attractive. But the attraction is that of methadone compared to heroin. Far better, surely, from a pragmatist's point of view, to rid ourselves of the craving for analysis altogether. To do this, we need to see that the basic philosophical needs that analysis seemed to serve can be met in another mode altogether: by explanation of the practices, rather than reduction of their objects. (More-over, the explanatory project has the potential to allow us realist truth with-out the metaphysical disadvantages. The apparent disadvantages of realist truth emerge in the light of the reductive project, for it is from this perspec-tive that it seems mysterious what truth could be. If we no longer feel obliged to ask the question, we won't be troubled by the fact that it is so hard to answer. We lose the motivation for seeking something else—something less "mysterious" than correspondence—with which to identify truth.)

"I accept all that," the pragmatist might say. "Nevertheless, perhaps it is true of the notion of truth (as we find it in practice), that it is identical (in some interesting sense) to ideal warranted assertibility. Shouldn't you there-fore allow, at least, the possibility that a Peircean account is the correct one?"

Two points in response to this. The first, an old objection, is that it is very unclear what the notion of the ideal limit might amount to, or even that it is coherent. For example, couldn't actual practice be improved or idealized in several dimensions, not necessarily consistent with one another. In this sense, then, the Peircean pragmatist seems a long way from offering us a concrete proposal.[28]

The second point—also an old point, for as Putnam observes, it is essen-tially the naturalistic fallacy[29]—concerns the nature of the proposed identifi-cation of truth with ideal warranted assertibility. Truth is essentially a nor-mative notion. Its role in making disagreements matter depends on its immediate motivational character. Why should ideal warranted assertibility have this character? If someone tells me that my beliefs are not those of our infinitely refined future inquirers, why should that bother me? My manners are not those of the palace, but so what? In other words, it is hard to see how such an identification could generate the immediate normativity of truth.[30] (It seems more plausible that we begin with truth and define the notion of the ideal limit in terms of it: what makes the limit ideal is that is reaches truth. This doesn't tell us how and why we get into this particular normative circle in the first place.)

I haven't yet mentioned what seems to me to be the most telling argument against the pragmatist identification of truth with warranted assertibility (in

Peircean form or otherwise). It often seems to be suggested (by Rorty himself, among others—see the quotes with which I began), that instead of arguing about truth, we could argue about warranted assertibility. This seems to me to miss a crucial point. Without truth, the wheels of argument do not engage; disagreements slide past one another. This is true of disagreements about any matter whatsoever. In particular, it is true of disagreements about warranted assertibility. If we didn't already have truth, in other words, we simply couldn't argue about warranted assertibility. For we could be aware that we have different opinions about what is warrantedly assertible, without that difference of opinion seeming to matter. What makes it matter is the fact that we subscribe to a practice according to which disagreement is an indication of culpable error, on one side or other; in another words, that we already take ourselves to be subject to the norms of truth and falsity.[31]

The crucial point is thus that assertoric dialogue requires an intolerance of disagreement. This needs to be present already in the background, a pragmatic presupposition of judgment itself. I am not a maker of assertions, a judger, at all, unless I am already playing the game to win, in the sense defined by the third norm. Since winning is already characterized in terms of truth, the idea of a conversational game with some alternative point is incoherent. It is like the idea of a game in which the primary aim is to compete—this idea is incoherent, because the notion of competition already presupposes a different goal.[32]

There is a connection here with an old objection to relativism, which tries to corner the relativist by asking her whether she takes her own relativistic doctrine to be true, and if so in what sense. The best option for the relativist is to say that she takes the doctrine to be true in the only sense she allows, namely, the relativistic one. When her opponent replies, "Well, in that case you shouldn't be troubled by the fact that I disagree, because you recognize that what is true for me need not be true for you, and vice versa," the relativist has a reply. She can argue that truth is relative to communities, not to individual speakers, and hence that disagreements don't necessarily dissolve in this way.

My pragmatist opponents fare less well against an analogous argument, I think. The basic objection to their position is that in engaging with me in argument about the nature of truth (as about anything else), they reveal that they take themselves to be subject to the norm whose existence they are denying. If they didn't take themselves to be subject to it, they would be in the same boat as the Mo'ans, with no reason to treat the disagreement between us as a cause for concern. They affirm p, I affirm not-p; but by their lights, this should be like the case in which they say 'Yes' and I say 'No', in answer to the question 'Would you like coffee?' (This is what it should be like even if p is of the form 'q is warrantedly assertible'.) The disagreement simply wouldn't bite.

6. TRUTH AS CONVENIENT FICTION?

The third norm thus requires a notion of truth that differs from justification, even of a Peircean ideal variety. In this sense, then, the present account is realist rather than pragmatist about truth. In another sense, however, the view surely seems antirealist. After all, I have argued that what matters is that speakers take there to be a norm of truth, not that there actually be such a norm, in some speaker-independent sense. Isn't this antirealism, or more precisely, in the current jargon, a form of *fictionalism* about truth?

If so, could this be a satisfactory outcome? If truth does play the role I have claimed for it in dialogue, wouldn't the realization that it is a fiction undermine that linguistic practice, by making it the case that we could no longer consistently feel bound by the relevant norms?

Let's call this objection the threat of dialogical nihilism. In my view, it isn't a practical threat. I think that in practice we find it impossible to stop caring about truth. This isn't an argument for realism, of course. The discovery that our biological appetites are not driven by perception of pre-existing properties—the properties of being tasty, sexually attractive, or whatever—does not lessen the force of those appetites, but no one thinks that this requires a realist view the properties concerned. Even if nihilism were a practical threat, this wouldn't be reason for thinking that the claim that truth is a fiction is *false*, by the lights of the game as currently played. It might be a pragmatic reason for keeping the conclusion quiet, but that is a different matter altogether (especially according to my realist opponents).

So even if the present view is correctly characterized as a form of fictionalism about truth, the nihilism objection is far from conclusive. But are the labels 'fictionalism' or 'antirealism' really appropriate? The need for caution stems from the fact that this approach to truth threatens to deprive both sides of the realism–antirealism debate of conceptual resources on which the debate seems to depend. As I noted earlier, the relevant metaphysical issues tend themselves to be framed in terms of truth, and related notions. A theory is said to be in error if its claims are not *true*, or if its terms fail to *refer*, for example. So the issue of the status of truth is here enmeshed with the terms of the problem, in a way which is quite uncharacteristic of metaphysical issues about other notions. As a result, it may be impossible to formulate a meaningful antirealism or fictionalism about the semantic terms themselves. This doesn't mean that we have to be realists about semantic notions, but only that if we are not realists we should be cautious about calling ourselves antirealists (or fictionalists), if these categories presuppose the very notions we want to avoid being realist about.

This may sound like an impossible trick, but in fact the kind of distinction we need is familiar elsewhere. It is the distinction between someone who "talks god talk" and espouses atheism, and someone who rejects the theological language game altogether (on Carnapian pragmatic grounds, say). These are two very different ways of rejecting theism. In the present case, the point is that we may consistently reject semantically-grounded realism about the semantic notions themselves, so long as we do so by avoiding theoretical use of semantic notions altogether, rather than by relying on those notions to characterize our departure from realism. (Why "theoretical use"? Because there is nothing to stop us continuing to use these semantic notions in a deflationary or disquotational sense.)

It might be suggested that we can sidestep this difficulty altogether by casting the relevant metaphysical issues in ontological rather than semantic terms. On this view, the relevant issue is whether truth exists, not whether (some) truth-ascriptions are true. Against this suggestion, however, it is arguable that the relevant metaphysical issues arise initially from data concerning human linguistic usage, and only become metaphysical in the light of substantial semantic assumptions about the functions of the language concerned—for example, that it is truth-conditional, or referential, in function. If so, then truth is once again enmeshed with the terms of the problem. And even if we concede the possibility of the ontological shift, the authority of Quine, Carnap and others may perhaps be invoked in support of a deflationary attitude to ontology, with the result that the realist-antirealist issue still dissolves.[33]

These issues are complex, and deserve a much more detailed examination than I can give them here. For present purposes, I simply flag the following as a possible outcome (of considerable plausibility, in my view). In common with other deflationary approaches to truth, the present account not only rejects the idea that there is a substantial metaphysical issue about truth (a substantial issue about the truthmakers of claims about truth, for example). Because it is about truth, it also positively prevents "reinflation." In other words, it seems to support a general deflationary attitude to issues of realism and antirealism. If so, then deflationism about truth is not only not to be equated with fictionalism, but tends to undermine the fictional-nonfictional distinction, as applied in the metaphysical realm.[34]

As I noted at the beginning, the present account of truth is hard to find on contemporary maps. In part, as should now be clear, this is because it combines elements not normally thought to be compatible. In one sense it is impeccably pragmatist, for example, for it appeals to nothing more than the role of truth in linguistic practice. Yet it rejects the pragmatist's *ur*-urge, to try to identify truth with justification. Again, it defends a kind of truth commonly seen as realist, but does so from a pragmatist starting point, without the metaphysics that typically accompanies such a realist view of truth. So in thinking about how to characterize this account of truth, we should be sensitive to the possibility that our existing categories—fictionalism, realism, and perhaps pragmatism itself—may need to be reconfigured. If so, then putting the position on the map is not like noticing a small country (Lichtenstein, perhaps) that previously we'd overlooked. It is more like discovering a geographical analog of the platypus, a region which our pre-existing cartographical conventions seemed a priori to disallow.

I began with Rorty's claim that the distinction between justification and truth makes no difference in practical life, no difference to our "decisions about what to do." Rorty regards a commitment to a notion of truth stronger than justification as a relic of a kind of religious deference to external authority. He recommends that just as we have begun to rid ourselves of theism, we should rid ourselves of the "representationalist" dogma that our beliefs are answerable to standards beyond ourselves. For Rorty, then, realist truth is a quasi-religious myth, which we'd do better without.

Despite my reservations about the fictionalist label, I have agreed that truth is in some sense a myth, or at least a human creation.[35] But I have denied that in providing a norm stronger than justification, a commitment to truth makes no behavioral difference. On the contrary, I've argued, it plays an essential role in a linguistic practice of great importance to us, *as we currently are*. It is not clear whether we could coherently be otherwise, whether we could get by without the third norm. If so, however, then the result would be a very different language game. My main claim is that we haven't understood truth until we understand its role in the game we currently play.

NOTES

The first version of this paper was written for a conference in honour of Richard Rorty at ANU in 1999. I am grateful to participants at that conference and to many subsequent audiences for much insightful discussion of these ideas; and also to an anonymous referee for helpful comments on an earlier version.

1. "Is Truth a Goal of Enquiry? Donald Davidson versus Crispin Wright," in Richard Rorty, *Truth and Progress: Philosophical Papers*, vol. 3 (New York: Cambridge University Press, 1998), 19.

2. Rorty, "Is Truth a Goal of Enquiry?," 26.

3. Irony aside, nothing here turns on whether by 'conversation' I mean the same as Rorty. For me, what matters is the role of truth in the kind of interpersonal linguistic interaction I'll call factual or assertoric dialogue, or simply dialogue. I don't claim that dialogue exhausts conversation, in Rorty's sense or any other. I used scare quotes on 'factual' above in anticipation of the suggestion that the notion of factuality in play might depend on that of truth, in a way which would create problems for my own account of the role of truth in dialogue. There is no such difficulty, in my view. On the contrary, I take the perceived "factuality" or "truth-aptness" of the utterances in question to be part of the explanandum of the kind of account here proposed; cf. n. 25.

4. "Pragmatism, Davidson and Truth," in *Objectivity, Relativism and Truth: Philosophical Papers*, vol. 1 (New York: Cambridge University Press, 1991), 128.

5. If private cognition depends on the norms of public dialogue then truth plays the same role, at second hand, in the private sphere. This is a plausible extension of the present claim, in my view, but I won't try to defend it here.

6. At least compared to the alternative.

7. "Global *Waiting for Godot*," as a member of an audience in Dundee suggested I put it. Even more seriously, as noted above, giving up truth might silence our own "internal" rational dialogues.

8. Jonathan Rée makes a point of this kind against Rorty: "Contingencies can last a very long time. Our preoccupations with love and death may not be absolute necessities, but they are not a passing fad either, and it is a safe bet that they will last as long as we do." "Strenuous Unbelief," *London Review of Books* 20, No. 20, 15 October 1998.

9. Paul Boghossian, "The Status of Content," *Philosophical Review* 99, no. 2 (1990): 157–184.

10. A realist could object that a commitment to the third norm might be useful and yet in error, but Rorty can't. It is fair for him to object against Wright that this commitment might be like theism, because Wright takes metaphysics seriously. By Wright's professed standards, then, the theism objection poses a real threat.

11. Rorty, "Is Truth a Goal of Enquiry?," 21.

12. Rorty, "Is Truth a Goal of Enquiry?," 21.

13. W. V. O. Quine, *Philosophy of Logic* (Englewood Cliffs: Prentice-Hall, 1970).

14. Paul Horwich, *Truth* (New York: Basil Blackwell, 1990).

15. Robert Brandom makes a similar point in "Pragmatism, Phenomenalism, and Truth Talk," *Midwest Studies in Philosophy*, 23 (1988): 75–93.

16. For present purposes I can remain open-minded on the question as to whether such a practice is really possible. Perhaps a truth-like norm is essential to any practice which deserves to be called linguistic. At any rate, my use of the following linguistic thought experiment does not depend on denying this possibility.

17. In what sense "weaker"? In the sense, at least, that they apply to a wider range of linguistic behaviors. 'Less specialized' might be a better term.

18. This corresponds to a common use of the term 'assertibility condition', as for example when it is said that the subjective assertibility condition for the indicative conditional 'If p then q' is a high conditional credence in q given p.

19. Cf. Rorty's remark that "for any audience, one can imagine a better-informed audience." ("Is Truth a Goal of Enquiry?," 22.)

20. This account has prescriptive and non-prescriptive readings. The former uses the notion of improvement full-voice, saying that if speakers are to improve their commitments, they need the idea of the third norm. But as N. J. J. Smith pointed out to me, it could well be objected that the relevant notion of improvement simply presupposes the third norm, and therefore can't provide any independent rationale for adopting it. However, no such circularity undermines the non-prescriptive reading, whose point is that because our existing conversational practice does take for granted such a notion of improvement, it thereby reveals its commitment to a third norm.

21. Note the contrast with Rorty's cautionary use of true. In that use we say of a claim that we take to be well-justified that it might not be true. In the present use we say of a claim that we might even allow to be well-justified by its speaker's own lights that it is not true. It is the difference between mere caution and actual censure.

22. Huw Price, "Three Norms of Assertibility, or How the MOA Became Extinct," *Philosophical Perspectives* 12 (1988): 241–54. The present section and the next draw significantly on that paper.

23. As I noted earlier, my use of this example does not depend on the claim that such a linguistic practice be possible. It is doubtful whether notions such as belief, assertion and opinion are really load-bearing, in the imagined context. However, much of the effect of the example could be achieved in another way, by imposing suitable restrictions on real linguistic practices—by imagining self-imposed restrictions on what we are allowed to say. One way to approach the Mo'an predicament from our own current practice would be to adopt the convention that whenever we would ordinarily assert 'p', we express ourselves instead by saying 'My own opinion is that p'.

24. A defender of the disquotational view might argue that although there is a third norm, it is not the function of the truth predicate to express it. This will be a difficult position to defend, however. If any predicate—'correct', for example—expresses the third norm, then that predicate will function as a disquotational predicate, for the reason just mentioned. Hence it will have been pointless to maintain that true itself is not normative. So the disquotationalist needs to claim that the third norm is not expressed at all in this predicative form, and that seems implausible.

25. This is another place where circularity threatens. We need to be sure that the psychological states mentioned at this point are not thought of as already "factual" or "representational" in character, in a way which presupposes truth. In so far as it is truth-involving, the "factual" character of the domain in question needs to be part of the explanandum—something that emerges from, rather than being presup-

posed by, the pragmatic account of the origins and consequences of "truth talk." In my view, one of the attractive features of this approach is that it offers the prospect that the uniformity of "factual," truth-involving talk might be compatible with considerable plurality in the nature and functions of the underlying psychological states. It thus offers an attractive new form for expressivist intuitions. Cf. my *Facts and the Function of Truth* (New York: Basil Blackwell, 1988), Ch. 8; "Metaphysical Pluralism," *Journal of Philosophy* 89 (1992): §IV; and "Immodesty Without Mirrors—Making Sense of Wittgenstein's Linguistic Pluralism," in Max Kölbel and Bernhard Weiss (eds.), *Wittgenstein's Lasting Significance* (Boston: Routledge, 2004), 179–205.

26. Even if not dangerous on its own, the third norm might become so in combination with some particularly deadly source of intractable disagreements, such as religion commitment. More generally, the thought that argument is sometimes dangerous suggests a link between the concerns of this paper and the motivations of the Pyrrhonian skeptics. On the present view of truth, the question whether we could get by without truth seems closely related to that as to whether we could live as thoroughgoing Pyrrhonian skeptics.

27. This point would acquire new and even stronger force, if it were to be established that private cognition rests on the norms of public dialogue, in the way suggested in note 5.

28. As Rorty notes (in "Pragmatism, Davidson and Truth," 130), Michael Williams makes a point of this kind: "we have no idea what it would be for a theory to be ideally complete and comprehensive . . . or of what it would be for inquiry to have an end" (M. Williams, "Coherence, Justification and Truth," *Review of Metaphysics* 34 [1980]: 269).

29. Hilary Putnam, *Meaning and the Moral Sciences* (Boston: Routledge & Kegan Paul, 1978), 108.

30. It may seem that this argument begs the question against the pragmatist, by assuming that there is an epistemologically relevant gap between ideal warranted assertibility and truth. (I am grateful to a referee at this point.) But the issue is not whether we need some norm in addition to ideal warranted assertibility, but whether ideal warranted assertibility itself could be immediately normative, in the way in which truth is. No one disputes that the manners of the palace are normative for those who live there—that's what it is to be manners—but it is an open question whether they are or should be normative for the rest of us. Similarly for ideal assertibility, except that in this case no one lives at the limit, so that there is no one for whom the question is not open.

31. I noted above that the same point applies to the normative predicates themselves. If we weren't already disposed to take disagreement to matter, we couldn't do so simply by adding normative predicates, for disagreement about the application of those predicates would be as frictionless as disagreement about anything else. My claim is thus that the notions of truth and falsity give voice to more primitive implicit norms, which themselves underpin the very possibility of "giving voice" at

all. In effect, the above argument rests on the observation that this genealogy cannot be reversed: if we start with a predicate—warrantedly assertible or any other—then we have started too late. (I suspect that by "giving voice," I mean something close to "making explicit," in Brandom's sense.)

32. Here, incidentally, we see the essential flaw in the pious sentiments of Grantland Rice (1880–1954):.

For when the One Great Scorer comes,
To write against your name,
He marks—not that you won or lost—
But how you played the game.

The One Great Scorer might assign marks on this basis, for divine purposes. *Pace* Rice, however, we couldn't play the game in question with such marks as our primary goal, for then it would be a different game altogether.

33. I defend the first of these two options in "Naturalism Without Representationalism," in Mario De Caro and David Macarthur (eds.), *Naturalism in Question* (Cambridge, MA: Harvard University Press, 2004); and the second in "Metaphysical Pluralism," *Journal of Philosophy* 89 (1992): 387–409 and "Naturalism and the Fate of the M-worlds," *Proceedings of the Aristotelian Society*, Supp. vol. 71 (1997): 247–267.

34. Rorty often says that he wants to walk away from realist-antirealist disputes. In other words, he doesn't think that there is an interesting philosophical question as to whether our commitments "mirror" reality. The above argument suggests that like other deflationists about truth, I have reason to follow Rorty in walking away from these issues. (In particular, my defence of truth over justification does not force me to stay.)

35. In the light of the argument above, this is a point more about the genealogy than about the reality of truth.

12

EXCHANGE ON "TRUTH AS CONVENIENT FRICTION"

Richard Rorty and Huw Price

I. RORTY: INITIAL RESPONSE (FEBRUARY 2003)

My off-the-cuff reaction is: why wouldn't the need for cooperative action take the place of your third norm? We don't automatically disapprove when we encounter disagreement in belief. I think chocolate disgusting; you think it delicious. I affirm and you reject the filioque clause in the Creed. But, as genial, tolerant, easygoing types we wouldn't dream of disapproving of each other for that reason. But if we are building a house together and I think we need fifteen rafters to make sure the roof holds up and you think ten will do, we have a problem, and we start disapproving of each other. The line between the disagreements that we think worth resolving and those we don't is the same, I should think, as the line between the ones where we need to cooperate on some project and the ones where we don't. (The line shifts, of course. If the Ministry of Plenty declares that from now on only one flavor of ice cream will be produced, the issue about chocolate becomes what James called a "live, momentous and forced option" in the way that the filioque clause used to be but is, thank Heaven, no longer.)

In short, I think that the necessary friction is provided by the need to get together on what is to be done. The tradition says: some disagreements are over what is objectively the case, others, over mere matters of taste or value or something like that. I think it would be better to say: some disagreements are over what is to be done together, others, over what is to be done in independence of one another. It seems to me that these two lines fall in roughly the same place but that the former is drawn in rather mystifying terms and the latter is not.

II. PRICE: INITIAL REPLY (FEBRUARY 2003)

I don't think your principle divides the cases anywhere near where we in fact draw the line. For example, the present disagreement between us seems to matter in a sense in which our differing preferences concerning chocolate does not, even though (presumably) no cooperative action turns on the issue. Perhaps you might say that "disapproval" is too strong for our view of each other, but I think we do think each other *mistaken*, in a sense not true of the chocolate case. So there's a norm there, whatever we call it.

Or perhaps you might say that in drawing the line in the wrong place, we are still in the grip of the old realist mistake. In that case I'd say, first, that a major part of my claim is that conversation is like this, and that this is independent of the question as to whether it should be (in whatever sense we make of "should" here). But second, I'd be skeptical whether your alternative is really workable. What's supposed to happen when we simply don't know whether cooperative action turns on the issue in question? Or when we have different views about whether it does so? Does the meta-issue get in or out, and by whose lights?

I suspect that the only way it can work is the way it does work, namely for the application of the norm to be "positive presumptive"—we take disagreements to matter, and then cancel it where it seems appropriate, to avoid some unproductive conflicts (as in the chocolate case).

Of course, the more basic point is that if it's true that conversation needs to work like this, there's nothing here to alarm a pragmatist—truth is still explained in terms of its role in practice, not via metaphysics, even though the practice ends up looking more realist than pragmatists have usually recommended.

III. RORTY: FURTHER REMARKS (FEBRUARY 2005)

Huw Price argues that we need to distinguish "three norms, in order of increasing strength: roughly, sincerity, justification and truth" (231). The second of these is what he calls "personal warranted assertibility." A person has obeyed this norm if she has "done as much as possible, *by her own current lights*, to ensure that her assertion that p is in order" (235). But if she tries to go beyond those current lights, she must do more—she must try to justify her assertion that *p* to other people. When she takes that further step, Price says, she is obeying the third norm, the "norm of truth."

Price asks us to imagine a community in which there are no attempts at intersubjective justification, but in which its members nevertheless express "the kind of behavioral dispositions which we would characterize as beliefs . . . by means of a speech act we might call the *merely-opinionated assertion* (*MOA*, for short)" (238). He admits that one might doubt the possibility of such a community: perhaps, he says, "a truth-like norm is essential to any practice which deserves to be called linguistic" (249n. 16). But he thinks this possibility irrelevant to his thought experiment.

It seems relevant to me. I doubt that we can tell a plausible story about a Mo'an community. In particular, I do not see why a radical interpreter would construe as *assertions* the noises made by organisms that never attempt to correct one another's behavioral dispositions—never try to get others to make the same noise they do. I would advance arguments familiar from Wittgenstein, Davidson, and Brandom to urge that there must be social cooperation on projects of shared interest before language can get very far off the ground. One cannot justify by one's own lights if one does not know what it is to justify by the lights of others. Price's "chatter of disengaged mono-logues" (231) is possible only as an enclave within a culture in which there is lots of engaged dialogue.

But suppose we set the question of the possibility of a Mo'an community to one side. Then the disagreement between Price and myself boils down to whether the practice of intersubjective justification is evidence of obedience to a big wholesale norm ("the norm of truth") or just to a recognition of the many concrete benefits resulting from social cooperation. Price seems to think that we started off by adopting his "norm of truth" and then, as a result, began to justify our assertions to one another. He says that "unless individual speakers recognize such a norm, the idea that they might *improve* their views by consultation with the wider community is simply incoherent to them" (235). I think that we acquired the latter idea in the same way we acquired the idea that we might improve our chances of hunting down a woolly mammoth if we first consulted on tactics.

To state Price's big wholesale norm, one has to nominalize the adjective "true" and treat the result as the name of a goal (just as Plato nominalized "good" into "the good," thus luring his readers down the garden path of meta-physics). But we humans might have carried out our cooperative enterprises equally well if nominalization had never occurred to us and if no philosopher had ever suggested that we had wholesale goals in addition to retail ones. The use of "true" in such contexts as "What you have said is not true" would have sufficed to provoke conversational exchange. For that use is enough to put an interlocutor on notice that cooperation will remain difficult until a disagreement in belief (about which way the mammoth went, for example) has been resolved.

Price and I agree that "truth is not a substantial property, about the nature of which there is an interesting philosophical issue" (232). But since it is not, I cannot see what "Seek the truth!" could add to something like "Listen to other people's noises, figure out why they make different noises than you do, and try to find nonviolent ways of getting everybody to make roughly the same noises on the same occasions!"

Our ancestors' attempts to find such ways led to the adoption of lots of little retail norms. Examples are "If it looks like a cloudless sky, call it blue!"; "Don't all talk at once!"; "If you have asserted p and 'if p, then q,' do not deny q!"; "Hear the other side!"; and "Describe the result of putting two pairs of things together as four things!" When enough such norms are in place, social cooperation of a sort unknown to ants and bowerbirds becomes possible. So does intellectual and moral progress—the constant replacement of old norms with new, ever more complex and nuanced ones. Such progress does not require that people think of themselves as striving for Truth or for Goodness.

IV. PRICE: REPLY TO RORTY'S FURTHER REMARKS (FEBRUARY 2008)

Could We Be Mo'an?

Rorty begins his comments by pressing me on something that, as he notes, I flagged as a possible objection. As he says, I admitted "that one might doubt the possibility" of a Mo'an community—perhaps, as I put it, on the grounds that "a truth-like norm is essential to any practice which deserves to be called linguistic." But as Rorty also notes, I said that this possibility is "irrelevant to [my] thought experiment." "It seems relevant to me," Rorty replies, "I doubt that we can tell a plausible story about a Mo'an community. . . . Price's 'chatter of disengaged monologues' . . . is possible only as an enclave within a culture in which there is lots of engaged dialogue."

I'm happy to agree with Rorty about the underlying point here—the impossibility of a thoroughly Mo'an community. However, I think the objection backfires, from Rorty's point of view, because Rorty's alternative to my view of how language *actually* works is almost as implausible, on these same grounds, as the Mo'ans themselves. Rorty seems to think that my "third norm" (i.e., the norm I take to be made explicit in responses such as "That's true" and "That's false") is merely an *occasional* constraint on our assertoric practice, a local product of the need for cooperative action, on a case-by-case basis. Whereas I think that the kind of considerations that show the Mo'ans to be impossible show that nothing can count as an assertoric practice unless the third norm is at least "on" by default: a normative constraint that is always *presumed* to apply, unless canceled by agreement in particular cases.

More on this below, but first, to the issue of the relevance of the Mo'ans themselves, irenic but impossible creatures as Rorty and I agree in taking them to be. For my purposes, what mattered about the Mo'ans was that by seeing what their linguistic practice would lack, we see what truth adds to our own. What's missing for the Mo'ans—what the third norm provides for us—is (as I put it) "the automatic and quite unconscious sense of engagement in common purpose that distinguishes assertoric dialogue from a mere roll call of individual opinion." Let's agree, with Rorty, that when we consider the Mo'ans in the light of "arguments familiar from Wittgenstein, Davidson, and Brandom," we realize that there can be no such community. Removing that sense of engagement amounts to removing anything that might count as an assertion, or indeed as an expression of opinion, in the full-blown sense of the term. This is no reason to forget the lesson we learnt by trying to imagine the Mo'ans—on the contrary, as in many cases, the point of the thought experiment lies *precisely* in the fact that it leads us, in thought, to an impossible destination. (The lesson lies in the nature of the impossibility.)

The Mo'ans themselves aside, Rorty says that he doesn't see "why a radical interpreter would construe as *assertions* the noises made by organisms that *never* attempt to correct one another's behavioral dispositions—*never* try to get others to make the same noise they do." (The emphasis on "never" is mine.) As I've said, I don't see this, either. But nor do I see why "a radical interpreter would construe as *assertions* the noises made by organisms" who have a 'retail' (Rorty's term) or "opt-in" attitude to whether making different noises matter—who merely *sometimes* "try to get others to make the same noise they do." I claim that the "chatter of disengaged monologues is possible only as an enclave within a culture in which" the *default* is "engaged dialogue"—and that to recognize that is to see that there is a norm at work (a norm from whose constraints we need to opt out when we don't want disagreements to matter).

Before explaining this idea further, and contrasting it to what seems to be Rorty's view, I want to set aside some possible misinterpretations of my view, highlighted by Rorty's remarks.

Explicit and Implicit Norms

Rorty takes me to be committed to (what I agree to be) an implausible view of how this kind of norm could operate. However, we don't need to *state* the norm for it to be at work. I agree entirely with Rorty that "the use of 'true' in such contexts as 'What you have said is not true' would" suffice "to provoke conversational exchange." But that's *because* these expressions have a normative force—quite unlike, for example, my use of "Not for me!" when you have just asked the waiter for another beer, and I want to indicate that my preferences differ.

Quoting me again, Rorty notes that we "agree that 'truth is not a substantial property, about the nature of which there is an interesting philosophical issue.' . . . But since it is not," he says, "I cannot see what 'Seek the truth!' could add to something like 'Listen to other people's noises, figure out why they make different noises than you do, and try to find nonviolent ways of getting everybody to make roughly the same noises on the same occasions!'"

Well, I agree that "Seek the truth!" *by itself* doesn't do anything essential, any more than "Don't eat the ones that smell bad" adds anything essential, for creatures who already make the olfactory discrimination in question—or *could* add anything, for creatures who don't. In both cases, the explicit advice may enhance and sharpen the discriminative behavior of creatures who possess these abilities, but it doesn't produce such behavior where none existed before. In both cases, however, we can see the discrimination in question at work, in the behavior of creatures who don't have the vocabulary to make their own practices explicit, in the relevant respects. We see them discriminating by smell among otherwise similar pieces of food, eating some and rejecting others, and we see them discriminating among the utterances of their fellows, favoring some, disfavoring others. (It is true that in both cases we might be hard pressed to characterize the behavior if it wasn't something we do ourselves. But this is no objection to the claim that *they* make the discriminations without being able to say explicitly what it is that they are doing.)

My claim is that our practice of exposing our assertions to this kind of disfavor by other speakers is part and parcel of what *makes them* assertions—moves in a communal game, with a particular normative structure. The game doesn't require that its players can make its rules explicit, however. This is not to deny that the ability to state the rules explicitly (e.g., perhaps, by nominalizing "true") might enhance or refine the game, just as the availability of explicit evaluative language enables us to refine our judgments about many matters of taste and discrimination. But the *practice* of discrimination comes first—as it needs to, in the linguistic case, because the explicit step depends on the discriminatory practice in question.

Rorty's Alternative

In place of my "wholesale norm," as he calls it, Rorty wants to propose "lots of little retail norms," including " 'If it looks like a cloudless sky, call it blue!'; 'Don't all talk at once!'; and 'If you have asserted p and "if p, then q," do not deny q!' " I've already agreed on the last point. My wholesale norm doesn't need to be explicit. It is not clear whether the same is true of Rorty's "little retail norms," though apparently they could only be formulated in terms that presuppose assertion or related notions: "call," "describe," "deny," "asserted." If I'm right, asserting, calling, describing, and denying are all activities that already embody the third norm—not in the sense that they depend on an ability to make it explicit, to describe oneself as "seeking the Truth," but simply in the sense that to do any of these things is to engage in a social practice, one key feature of which turns on the default normative status of disagreements.

What alternative view can Rorty offer of these matters? He says: "The disagreement between Price and myself boils down to whether the practice of intersubjective justification is evidence of obedience to a big wholesale norm ('the norm of truth') or just to a recognition of the many concrete benefits resulting from social cooperation." Once more, I've already distanced myself from the view that we need *explicitly* to adopt a norm of truth. However—so long as "recognize" is *not* read as "can formulate explicitly" but merely as something like "feel the force of"—I do want to reiterate the claim that Rorty objects to here, that "unless individual speakers recognize such a norm, the idea that they might *improve* their views by consultation with the wider community is simply incoherent to them." It seems to me that the extra-subjective notion of improvement depends on the third norm—and, moreover, that this point is entirely in keeping with the insights of Wittgenstein, Davidson, and Brandom to which Rorty alludes.

In saying this, I'm not venturing any claim about how our ancestors came to play the game in the first place. Presumably the specifically linguistic norm arose on the back of other intersubjective norms and more basic inclinations to cooperative action. It has become something more than this, however—a norm with a life of its own, a norm that lives independently of particular, local projects of cooperation. Indeed, its effect is that *language itself* becomes a project of cooperation, a project that depends on the norm in question.

The difference between my position and Rorty's on this matter is illustrated by some comments from an exchange in 2003 (presented earlier in this chapter). Against Rorty, I think that his examples about chocolate and the Creed are exceptions. They are cases in which—not only "as genial, tolerant, easygoing types" but also, crucially, as types who know that food preferences and religion are special cases—we make a space for no-fault disagreement. The existence of such exceptions is no argument against my claim that the third norm applies by default (and that it is not, as Rorty here seems to suggest, a norm we "turn on" when resolving disagreement matters, by virtue of collaborative projects). There are countless ordinary cases that illustrate that this is so. For example, imagine that I hear a stranger advising a tourist to take the 378 bus to Bondi Beach. Believing (*correctly*, as it happens, though this doesn't matter to the example) that the 378 goes to Bronte Beach, not Bondi, I believe that the stranger is *mistaken* and has led the tourist into *error*. I have no common projects with the stranger, let alone with our foreign visitor, and I may be too lazy, too busy, too shy, or too constrained by a competing norm about talking to strangers or foreigners to step in and make their projects mine. Nevertheless, I take it that the stranger was *wrong*, and that the visitor is now *mistaken* about how to get to Bondi. I take examples such as bus routes to show Rorty's principle doesn't divide the cases anywhere near where we in fact draw the line.[1]

Am I in the grip of some philosopher's commitment to a Platonic ideal? Not at all: I'm just playing the common conversational game in the ordinary instinctive way, in which disagreements matter by default. As the case illustrates, they don't always matter *enough* to prompt the kind of behavior that might resolve them—perhaps they seldom do, in fact, in large communities—but by default, they always provide a normative pressure in that direction. Rorty's examples simply illustrate that the default can be cancelled fairly systematically for particular topics—but the fact that he had to choose religion and matters of taste is surely revealing. (Just try it with bus timetables or the location of ATMs.)

A rather different kind of response to Rorty's suggestion is provided by the case of philosophical disagreements themselves. Clearly, Rorty and I took the disagreement between us to "matter," in a sense in which our differing preferences concerning chocolate do not—even though, apparently, no cooperative action turned on the issue. We each took the other to be *mistaken* about something, in a sense not true of the chocolate case.

It seems to me that Rorty has two alternatives in response to this kind of example. He could say that our philosophical disagreements were a sign that even we pragmatists were still in the grip of the old Platonist mistake and that further therapy was necessary. Or he could say that we'd embarked on a common project, after all, albeit one with fewer immediate practical consequences than the number of rafters we put in the roof.

The first option wouldn't have called for an answer. It is an expression of a desire to leave this particular instantiation of the assertoric game, and the appropriate response is simply to wish the speaker well, in whatever alternative activities he chooses to pursue. As for the second option, I think it needs to be endorsed and returned with interest. I think that the case of philosophy merely illustrates (as an extreme and rarified case) the fact that language itself is a cooperative project—a project to which all normally functioning human beings are already signed up, long before they reach adulthood. Of course, many other factors determine to what extent, and in what ways, each of us chooses to participate in this common game, on particular occasions. But this variability is no challenge to the thesis that there is a fundamental game and a fundamental norm at the heart of the game. That was my claim, and I think that Rorty has not offered us any workable alternative.

After all, what's supposed to happen, in Rorty's view, when we simply *don't know* whether cooperative action turns on the issue in question? Or when we have different views about whether it does so? (Does the meta-issue get in or out, and by whose lights?) Moreover, if it was just *agreement* that mattered—where the need for cooperative action requires it—why not achieve it in some other way, such as deference to norms of social status?

In other words, I think that the only way that an assertoric, reason-eliciting practice can work is the way it actually does work: viz., for the application of the third norm to be positive presumptive. We take disagreements to matter by default and then cancel, hedge, or qualify that presupposition where necessary to avoid unproductive disputes about matters such as chocolate and religion and to bring some civility to the conflicts the norm induces.

"Realist" Truth Without Metaphysics

I want to close by emphasizing a more basic point from my paper, a point with which Rorty's comments did not engage. Even if it is true that conversation needs to work like this, there is nothing here to alarm a pragmatist. Truth is still explained in terms of its role in practice, not via metaphysics, even though the practice ends up looking more "realist" than pragmatists have usually recommended. As I say in the paper, I think that Rorty missed this option, of a pragmatist grounding for what has traditionally been thought of as a realist notion of truth. Here, as in other cases, the right course for a pragmatist is not to reject the practice but to reject the interpretation that the opposing, metaphysical tradition has placed on the practice. And the right way to do so is to show that we can account for the practice in homely, practical terms, without metaphysics. That's what I've tried to do by suggesting that we understand truth as a "convenient friction"—a norm that plays a particular vital and central role in our linguistic and cognitive lives.

NOTE

1. To the extent that there is a line, at any rate. I'm with Rorty in rejecting this "bifurcation thesis," as he calls it elsewhere, and I have myself argued that the possibility of no-fault disagreements is entirely a matter of degree. Different discourses admit them to different degrees and for different reasons, and no discourse is wholly free of them, for reasons related to the rule-following considerations. See my *Facts and the Function of Truth* (Oxford: Blackwell, 1988), chap. 8.

13

TWO DIRECTIONS FOR ANALYTIC KANTIANISM

NATURALISM AND IDEALISM

Paul Redding

Usually, analytic philosophy is thought of as standing firmly within the tradition of empiricism, but recently attention has been drawn to the strongly *Kantian* features that have characterized this philosophical movement throughout a considerable part of its history.[1] Those charting the history of early analytic philosophy sometimes point to a more Kantian stream of thought feeding it from both Frege and Wittgenstein, countering a quite different stream flowing from the early Russell and Moore.[2] In line with this general reassessment, Michael Friedman, for example, has pointed to the specifically Kantian features of the approach of Carnap and other members of the Vienna Circle.[3] For Friedman, the positivists should be seen as having emerged from the tradition of late-nineteenth-century neo-Kantianism. Although they had explicitly rejected Kant's analysis of geometric truth and his key concept of the "synthetic a priori" because of dramatic changes within science itself, this move should not be seen as any simple *abandonment* of Kantianism.[4] Rather, the positivists had *redefined* the nature of the Kantian a priori, by axiomatizing, relativizing, and historicizing it, so as to fit with the results of the contemporary sciences.

Whatever the exact history here, it would appear that such a broadly Kantian combination of a deflationary rejection of the type of substantive metaphysical knowledge from concepts alone—what Kant had alluded to as "dogmatic metaphysics"—together with a concern with those linguistic or cognitive frameworks that seem presupposed by any cognitively relevant human experience has become a common feature of much analytic philosophy. Moreover, what seemed promised here was a type of "naturalized" variant of Kantianism whose attraction is easy to feel. First, Kant's general claim concerning the "mediated" nature of human experience became something of a commonplace of much later-twentieth-century thought. *Some* version of the idea that our experiential contact with the world is mediated or conditioned by concepts—the concepts that form the nodes, as it were, of the "web" of those beliefs that we accumulate from experience and that in turn form the background to what can further be learned from experience—seems to be found in many parts of modern culture. Next, this "mediationalist" thesis that we lack direct or "unmediated" contact with reality seems to fit with a more general view of ourselves as limited, finite—indeed, *natural*—creatures, in contrast with which the very notion of some direct, unmediated contact with reality "as it is in itself" suggests a picture with distinctly *theological* overtones.

One move to a naturalized Kantianism has been described by Mark Sacks as involving separate steps with the resulting position being

> two steps removed from Kant, and one from the early Wittgenstein. It is not the *a priori* faculties of the mind, as in Kant, that are doing the work, but the structure of language; and, it is not the a priori structure of language, as in the early Wittgenstein, but the empirical linguistic frameworks in which in principle nothing need be held as immune to revision.[5]

Such a two-step move seems to capture the position of Sami Pihlström, for example, who conceives of a type of Kantianism that acknowledges the *circularity* of transcendental reasoning:

> Very simply, in order to examine the preconditions and limits of cognitive experience, the transcendental philosopher must already operate within the cognitive sphere s/he is examining. . . . [T]he transcendental is to be found—reflexively—within our natural practices themselves, not in any supposedly transcendent, metaphysically fixed point beyond our natural world.[6]

But such an approach must now face the question of the *type* of knowledge relevant to understanding these "natural practices" themselves. If they are "natural," will not the most appropriate form of investigation of them be just the type of investigation that has been so successful with respect to nature in general, that is, the modern natural sciences? And with this, naturalism threatens to simply consume the remnants of the desired "transcendental" approach in the way exemplified in Quine's classic attack on Carnap's transformation of Kant's synthetic a priori.[7] There, Carnap's conventionalized and relativized variant of Kant's transcendental structure ultimately became assimilated to a fully naturalized holistic "web of belief" constituting scientific knowledge. Historically, to many the appeal of Kant's transcendental approach had been its promised resistance to the perils of a "nihilistic" leveling of normativity, and it is just this threat that seems to return with Quine's naturalism.

In the following essay, after examining some general problems that accompany a naturalized Kantianism, I examine what I take to be an attractive recent version—Huw Price's "subject naturalism"—in terms of its capacity to respond to such threats. In the final section, I contrast it with another possible way beyond these problems, one based on the model of the approach of Kant's explicitly *idealist*, rather than *naturalist*, inheritors. This is the way of Robert Brandom and others who advocate the step from Kant to Hegel.

THE PROJECT OF NATURALIZING TRANSCENDENTAL PHILOSOPHY

In commenting on the analytic naturalization of the Kantian a priori, Mark Sacks has noted two general problems facing such an approach to philosophy. The first concerns the philosophical position's capacity to account for its own status:

> Although the given structures that cast the lines of individuation of our empirical reality are themselves not purported to be a priori or transcendental now, still the entire theoretical explanation itself clearly amounts to a synthetic *a priori* claim, and it is not clear how any such claim can be accommodated.[8]

While some might contest the claim that it is *clear* that this "entire theoretical explanation" has the status of a synthetic a priori claim, Sacks points to a definite problem here that I will call the problem of disciplinary identity. In short, if philosophy *does not* have synthetic a priori status, what then *is* the nature of the knowledge to which philosophers aspire? The common "naturalistic" answer that philosophy is "continuous" with the natural sciences just seems to be the admission that there *is* no basis for philosophy's distinct disciplinary status. "Continuous with" really just means "part of."

The second problem to which Sacks alludes is that

> once the Kantian transcendental psychology, or the Wittgensteinian *a priori* linguistic form, are each avoided, some of the safeguards of a transcendental idealist turn are undermined. . . . We are now faced with a relativism that goes beyond the mere instability of bare naturalism . . . different frameworks, or conceptual schemes, will result in different ontologies, different worlds, each of which can be the object of knowledge for those inhabiting the relevant framework, but none of them can lay claim to universality.[9]

These relativistic problems might in turn be seen as an expression of the process of denormativization that, as is commonly said, accompanies the naturalizing of an originally normative discipline. I will refer to this as the *nihilism* problem, a problem consequent on the collapse of some "foundational" discipline that is meant to assure the status of various normative claims. Historically, the model for this concerns the collapse of *religion*, Jacobi having coined the term when pointing to the consequences of a philosophy practiced free from the constraints of religion. In the early modern period, the type of theocentric *philosophy* that Kant came to criticize as "dogmatic metaphysics" had itself come to assume the foundational role otherwise played by *revelation*, and with Kant, it itself thereby became subject to the same critique. In turn, what we have referred to as the *naturalizing* of the Kantian a priori might itself be seen as representing a *further* stage in this process.

Jacobi's charge of nihilism had been made during the period of the early reception of Kant's philosophy, and it was a charge taken seriously by the early post-Kantian idealists like Friedrich von Schelling. In a youthful work, *Ideas for a Philosophy of Nature*, Schelling had speculated about the consequences of reflective thought, that process in which the knowing subject had, as he put it, "disentangled itself from the fetters of nature and her guardianship" such that that very nature came to be presented as an object knowable *for* that knower.[10]

Schelling held distinctly ambiguous, if not contradictory, attitudes toward this reflective disentanglement. On the one hand, reflection *is* the condition of freedom. It would be inconceivable how man had ever left the philosophical state of nature "if we did not know that his spirit, whose element is *freedom*, strives to make *itself* free, to disentangle itself from the fetters of Nature and her guardianship."[11] This was the attitude inherited from Kant and Fichte. But at the same time, reflection also seems to *undermine* the capacity for action. Reflection impedes action since the less [man] reflects upon himself, the more active he is. . . . Man is not born to waste his mental power in conflict against the fantasy of an imaginary world, but to exert all his powers upon a world which has influence upon him, lets him feel its forces, and upon which he can react . . . contact and reciprocal action must be possible between the two, for only so does man become man.[12]

This, with the theme of estrangement from some constituting community with nature, might be thought of as Schelling's *romantic* inheritance. From Schelling's point of view, "mere reflection" represents a kind of "spiritual sickness in mankind."[13] When humans break out of nature, as it were, they upset an original "absolute equilibrium of forces and consciousness," and although this is the precondition of their freedom and rationality, the point of such reflection must be to ultimately *reestablish* something like the original equilibrium "through freedom" as "only in equilibrium of forces is there health." Philosophy, therefore, must assign to reflection "only *negative* value."[14]

This problem of reflection, which, with the post-Kantian idealists, became one of the key problems to address, is still commonly commented on today. As Thomas Nagel has pointed out, while reflection purports or aspires to lead to a kind of aperspectival objectivity, a "view from nowhere" onto the world, it at the same time poses distinct problems for action *in* the world. In short, one can *act* in the world, only from *somewhere in particular*.[15] Among recent analytic philosophers, Bernard Williams has explored the dynamics of such a process in the context of ethics, a context in which, in his memorable phrase, "reflection can destroy knowledge."[16]

Like Nagel, Williams has extensively employed the Kantian metaphor of "perspective" to capture the mediated nature of our cognitive engagement with the world. In particular, he has stressed the necessarily perspectival nature of the *ethical* attitude. "I think about ethical and other goods," he claims, "*from* an ethical point of view that I have already acquired and that is part of what I am. In thinking about ethical and other goods, the agent thinks from a point of view that already places those goods, in general terms, in relation to one another and gives a special significance to ethical goods."[17] From such a "first-person" perspective of the agent, certain objects and states of affairs in the world will be invested with certain "action-guiding" qualities, the sorts of qualities that are expressed in normatively "thick concepts" such as "treachery," "brutality," or "courage."[18] However: "Looked at from the outside, this point of view belongs to someone in whom the ethical dispositions he has acquired lie deeper than other wants and preferences."[19] In the "reflective" move between these two points of view, the relationship of values to dispositions has changed. "It is not true from the point of view constituted by the ethical dispositions—the internal perspective—that the only things of value are people's dispositions; still less that only the agent's dispositions have value." Rather, from that point of view, the dispositions are dispositions to respond to the values *themselves*, which are conceived as "objective" and in the world, born by objects holding *normative* significance for the agent. But this realistic attitude does not withstand reflection, and from the point of view that "stands back" and reflects on this situation from "outside" this perspective, "there is a sense in which [the dispositions] are the ultimate supports of ethical value," not the objects themselves.

Williams formulates his critique in terms of the distinctness of "ethical" knowledge, but it is far from clear that theoretical knowledge *itself* escapes the corrosive effects of such reflection. Thus, for example, the idea that *epistemological reflection* can "destroy knowledge" in the *theoretical* context as well has been made by others such as David Lewis.[20] Here too, then, reflection, might destroy knowledge—such would seem to be the nihilistic situation of modern culture and its reflective propensities, predicted by Jacobi and popularized by Nietzsche. While few analytic philosophers have treated the topic as having such general significance, few would claim that the problems that can be grouped around the general notion of "nihilism" have been solved or have otherwise gone away.

HUW PRICE'S "SUBJECT NATURALISM"

In a series of papers Huw Price has suggested a novel way of conceiving the naturalization of philosophy that both problematizes the way of thinking of this process that has hitherto dominated analytic philosophy, and promises to avoid the problems of scientism and nihilism.[21] Moreover, Price's alternative, especially as recently expressed in the form of a "subject naturalism,"[22] does this in a way that manifests the distinctly *Kantian* features discussed earlier. Hitherto, naturalistic conceptions of philosophy, he claims, have regarded philosophy as taking as *its* object the world as science describes it. Such approaches then see their own task as one of finding a place for certain objects *in* this scientifically described world—objects that are not easily so located, such as those having to do with morality, meaning, or mathematics. Conceived in this way, philosophy typically addresses what he calls "placement problems." "A typical placement problem" he notes, "seeks to understand how some object, property, or fact can be a *natural* object, property, or fact."[23] One popular strategy is to *reduce* such objects, properties, or facts to the unproblematic ones of science. But there may be hidden assumptions implicit in this approach that are actually incompatible with a genuinely naturalistic view of *ourselves* as the subjects capable of such knowledge. Hence Price contrasts the "object naturalism" of such reductionist orientations with his own "subject naturalist" approach, which attempts to make explicit and hold on to a naturalistic approach to the knowing subject *prior* to the cutting in of the problematic hidden assumptions of object naturalism.

Price's deflationary critique of object naturalism starts with a distinction between two possible interpretations within which the placement problem might be posed. The "material conception" of the problem starts with a consideration of the problematic objects, properties, or facts themselves: "We are simply acquainted with X, and hence—in the light of a commitment to object naturalism—come to wonder how this thing-with-which-we-are-acquainted could be the kind of thing studied by science."[24] In contrast, the "linguistic conception" starts with our *talk* about these things, facts, and so on. Here, "we note that humans (ourselves or others) employ the term 'X' in language, or the concept *X*, in thought." But for the object naturalist, this linguistic framing of the placement problem quickly reduces to the material conception. "In the light of a commitment to object naturalism" he states, "we come to wonder how what these speakers are thereby talking or thinking *about* could be the kind of thing studied by science."[25] But this quick reversion to the material conception on the part of the object naturalist reveals a substantive assumption about the linguistic practices themselves— the assumption that they are to be understood *representationally*, that is, understood in terms of the things they are *about*. This representationalist assumption, Price says, "grounds our shift in focus from the *term* 'X' or concept *X* to its assumed *object*, *X*,"[26] and it is far from innocent. To bring out its problematic nature Price next invokes Quine's deflationary approach to semantic notions.

The representationalist move from linguistic practices to the objects they are about, he points out, should *not* be seen as a move down a Quinean deflationary semantic ladder, as deflationism is precisely meant to avoid the type of substantial metaphysical commitments implicit in the representationalist's approach. Thus, on Quine's own deflationary account, we start with talk about objects, and then any subsequent "ascent" to talk about the meanings of words or the truth of sentences is only an *apparent* one because, for Quine, talk of "the *truth* of the sentence 'X is F,' is just another way of talking about the *object*, X."[27] But conversely, if, reversing Quine's approach, we *actually* start with talk of our linguistic practices, neither should we then properly "descend" to talk of the objects these practices are purportedly about. The assumption that we *do*, the representationalist assumption, thus begs "substantial, non-deflationary semantic notions."[28] But these notions are typical of the problematic ones that reductionists hope to naturalize.

Kantian features indeed seem very close to the surface of Price's "subject naturalism."[29] One might say that Kant had adopted a stance toward traditional metaphysicians analogous to that of Price toward the object naturalist, in that it was central to Kant's philosophical project to bring into question untheorized assumptions about the conception of representation implicit in traditional metaphysics. The centrality of this semantic issue is clear in a letter by Kant during the period of the gestation of the *Critique of Pure Reason*. In 1772 Kant writes to Marcus Herz that up to that time his philosophy had "lacked something essential, something that in my long metaphysical studies I, as well as others, had failed to pay attention to and that, in fact, constitutes the key to the whole secret of hitherto still obscure metaphysics." This neglected topic concerned the nature of representation. "I asked myself," Kant goes on, "What is the ground of the relation of that in us which we call 'representation' to the object."[30]

More particularly, we might consider Kant's critical attitude to Locke in the first *Critique* as analogous to Price's deflationary, subject naturalist critique of the "object naturalist's" implicit semantic theory. In his "physiological" approach to the mind, Locke had given a causal explanation of the relation of mental representations to the world. However, it is clear from the combination of Kant's letter to Herz and his comments in the *Critique of Pure Reason* that Kant doesn't regard *this* as an answer to *his* question of the "ground" of the relation of representation to thing.[31] Rather, in giving a causal account of representation Locke had just begged the question of what it is that makes a representation of an object that caused it *a representation* of that object, and not just some *effect* of it.[32] It is crucial, of course, that for Locke such an effect *is* a representation, the possession of which will count as *knowledge*. Hence Locke's causal account seems to have begged what Price refers to as "substantial, non-deflationary semantic notions," and it is just *these* notions that Kant wanted to bring into question.

Much of the force and attraction of Kant's raising the question of the grounds of our capacity for representation comes from the mere fact of his doing so, rather than the particularities of his answer. Kant's point is that there must be *something* about we humans whereby our mental contents can purport to have representational status, and his approach of contrasting the human capacity here with some divine infinite capacity surely suggests a type of "subject naturalism." Once the question of the grounds of the representational relation is made an issue, it looks as if Locke has been by default committed to a mysterious, indeed magical, conception of his own mental powers. Lockean ideas just seem to be *intrinsically* representational.

In contrast, in his own complex account of our representational capacity understood in terms of the dual functions of intuitions and concepts, Kant contrasts our *finite* capacity for representation with God's purportedly *infinite* representational capacity. While God might have the capacity to directly intuit individual "things in themselves" we, restricted by our need to be sensuously affected by objects in the world, can only aspire to discursive representations of things that then appear to us relative to the specific conditions enabling our representational capacity. Indeed, the representational capacity that Locke assumes that humans have looks like the one Kant attributes to *God*. It is just this implicit, godly, or magical representational capacity, one that seems to presuppose the operation of what Hilary Putnam was later to capture with the phrase "metaphysical glue,"[33] that Kant is anxious to expose in the thought of his predecessors.[34]

CAN "SUBJECT NATURALISM" SAVE THE NATURALIZING PROJECT?

What might Price's subject naturalism have to say in face of the problems sketched by Sacks—the problems I have labeled as those of disciplinary identity and nihilism? Price is clearly aware of the latter *as* a problem: "The tide of naturalism has been rising since the seventeenth century," he notes, and "the regions under threat are some of the most central in human life."[35] The subject naturalist fights alongside the nonnaturalist to protect these regions by developing a critique of the scientistic excesses of earlier "object" naturalists, and, with this, Price finds himself on the same side of the barricades as advocates of a "broad" or "liberal" naturalism who want to rescue the idea of naturalism from its more "restrictive" and scientistic manifestations. But Price's response to the problem of disciplinary identity may create problems here. With regard to this problem, Price's strategy, I suggest, is to secure the disciplinary identity for *philosophy* by *assimilating* it to science. That is, it looks as if a "soft" or liberal naturalism is secured in the culture generally by the strategy of adopting a strict or scientistic naturalism *within* philosophy.

Subject naturalism aspires to achieve the metaphysical economy of naturalism without the expense of nihilism. While it is "naturalist in spirit," it "offers an olive branch to nonnaturalists"[36] because the nihilistic culprit was object naturalism:

> Object naturalism gives science not just center stage but the whole stage, taking scientific knowledge to be the only knowledge there is (at least in some sense). Subject naturalism suggests that science might properly take a more modest view of its own importance. It imagines a scientific discovery that science is not all there is—that science is just one thing among many that we do with 'representational' discourse.[37]

While this may seem good news for those hoping for a liberal philosophical naturalism, closer scrutiny reveals that subject naturalism may not offer support to liberal naturalists where they often want it—*within philosophy itself*. Like the object naturalist, the subject naturalist wants to assimilate *philosophy* to science—perhaps, as some of Price's formulations suggest, even to *natural* science. As he puts it in one place, naturalism per se is just "the view that the project of metaphysics can properly be conducted from the standpoint of natural science."[38] From this perspective, the problem with object naturalism is that it is *bad science*, and the strategy is to replace it by *better science*. "The story then has the following satisfying moral. If we do science better in philosophy, we'll be less inclined to think that science is all there is to do."[39] That is, we will be "less inclined to think that science is all there is to do" *outside* of philosophy.

The point at which Price's purportedly pluralistic naturalism threatens to slide into strict *philosophical* scientism is apparent when he shifts from the seemingly innocent idea that philosophy must start with the notion that "we humans are natural creatures" to the idea that it must start with "what *science* tells us about ourselves," because what science tells us is "that we humans are natural creatures."[40] But the critic may ask: Why *science* in particular—cannot nonscientific disciplines tell us important things about ourselves? And, in any case, if science, *which* science?

The idea that science is the place to start in the effort to learn about ourselves suggests that whatever the functions of nonscientific discourses, these cannot include that of telling us anything important about ourselves. But the idea that one can *only* learn about ourselves from science is, many will think, ludicrous. For starters, novels overwhelmingly portray human life in nonmagical, broadly naturalistic ways,[41] and the same could be said of many other literary and artistic genres. In fact, it could be argued (and has been) that different artistic forms, say painting and poetry, can "tell us something about ourselves" as natural material creatures—say, reveal to us otherwise unnoticed dimensions of the sensuous phenomenology of our lives, in ways that are very difficult if not impossible to convey in a maximally conceptualized medium like that to which the sciences aspire. Price wants to protect the nonscientific discourses from the rising tide of scientistic naturalism, but unless they have something important to tell us about ourselves, as humanists have traditionally thought, then it is unclear why this is worth the effort.

Viewed in terms of its positive aspect, Price's dictum seems to fare no better. For one, it is still not clear that the "science" that we are to do "better" in philosophy is a natural kind term. Rather than "science," a familiar objection goes, there are just many *sciences*, all of which tell us about ourselves as natural creatures in some particular way or another. Physics, for example, tells us about ourselves qua physical beings, biology about ourselves qua biological beings, sociology about ourselves qua social beings (social by our nature, as it were), and so on. Price is vague as to the identity of the science he recommends. The general idea that it studies the *functions* of different ways of talking suggests a science something like anthropology, say, or linguistics. Indeed, in one place, when singling out representatives of the type of appropriate form of approach within contemporary philosophy, he describes them as "addressing deep issues in socio-linguistic theory."[42] Naturalists are typically drawn to the "harder" sciences like physics and avoid "human sciences" like linguistics or anthropology or sociology, disciplines that are crossed by doctrinal disputes about the ultimate terms in which such human fields should be described. Moreover, naturalists in philosophy are likely to be worried about the presence of conflicting implicit *philosophical* theses underlying these disputes, justifying the "safer" model of the natural sciences.[43]

In any case, it is not clear, I suggest, that the nonnatural sciences are particularly helpful here. Consider the classic distinction between the natural sciences and the cultural or human sciences ("*Geisteswissenschaften*") such as history or ethnography. In very broad terms, conceived in the latter way, an *appropriately* "scientific" approach to humans is typically seen as interpretative rather than explanatory, or at least as having its explanations mediated by interpretations concerning the intentions attributed to agents, and so on. As such, the human sciences are seen as engaging not simply with "brute" facts that we take to be independent of any awareness of them but also with "institutional" facts that obtain only by virtue of the fact that they are recognized as so obtaining by the "bearers" of those institutions.[44] Linked to this, this approach tends to interpret regularities in human affairs in terms of patterns of rule following rather than the causal regularities of the contrasting natural sciences. Such a human-sciences approach can, of course, still treat humans as natural in some broad sense in which there is nothing godly or magical separating humans from the rest of nature, but once this is done, its advocates will typically stress the *discontinuities*.[45] But if the *Geisteswissenschaften* were allowed into the constitution of the philosophical standpoint, it is unclear as to how this would solve the problem of nihilism. In adopting the stance of an ethnographer to our own values and norms we will be supposed to assume a stance that is "value-free." This is just the reflective stance that, as Williams claims, "can destroy knowledge"—that is, destroy a *commitment* to those values.

Conceding the term "naturalism" to Price, we may ask whether his compelling points against "object naturalism" could be made within a *nonscientistic* and yet non-"supernaturalist" conception of *philosophy*? I think they can if the argument follows the route taken by one of Price's erstwhile allies in the struggle against the type of representationalist approach to language he opposes, Robert Brandom. This is the route of *idealism*.[46] Of course, many analytic philosophers may react to such a suggestion with horror or mirth. While the idealist tradition had been held in high regard in English-speaking philosophy in the last decades of the nineteenth century, it has rarely even been considered as a possible way of doing philosophy since. Idealism, "as everyone knows," was vanquished by Russell and Moore at the turn of the twentieth century.[47] This was, at least until recently, the generally unchallenged *official* story. Brandom suggests an alternative account of the recent trajectory of analytic philosophy that sees in it not the *naturalization* of Kantian transcendentalism but its *Hegelianization*.[48] After some brief but hopefully horror-allaying comments on the much misunderstood term "idealism," I will say in very general terms something of what I would see as an *idealistic* reinterpretation of Price's subject naturalism as involving. But rather than Brandom's self-consciously Hegelian orientation, I will appeal to the work of a philosopher who, on the face of it, shares Price's more Humean sympathies: Bernard Williams.

IDEALISM AND RECENT ANALYTIC NONNATURALISM

The term "idealism" is, in Anglophone circles, almost invariably associated with the figure of George Berkeley, but if Berkeley *can* intelligibly be referred to as an idealist, a characterization denied by a number of major interpreters,[49] then he is surely the *least* representative of the species. Berkeley himself didn't refer to his stance as idealism and used the much more descriptive "immaterialism," but perhaps a *better* name is the one used by his nineteenth-century editor, Alexander Campbell Frazer, who called it "spiritual realism."[50] It is the commonplace conflation of idealism with this immaterialist version of *realism* that has mired the interpretation of the movement of "German idealism" that immediately followed Kant and that is usually characterized by the three central figures, Fichte, Schelling, and Hegel.

From the time of the initial reception of the *Critique of Pure Reason*, Kant had objected to the claim of critics that his "transcendental" idealism was a species of Berkeley's doctrine, and much confusion here, I think, has been caused by the failure to attend to the *Aristotelian* dimension of Kant's own way of expressing his difference from the doctrine to which he referred as "material idealism." To be a "material idealist" is to be an idealist about "matter"—to reduce matter to the status of "ideas" in the mind of some subject, ultimately, for Berkeley, the mind of God. Thus "material idealism" or "immaterialism" was, as we have seen, just a consequence of Berkeley's spiritual *realism*. Kant was *not* an idealist about matter but rather an idealist about *form*,[51] and in this sense his philosophy was the exact *opposite* of Berkeley's.

The extent that Kantian *formal* "idealism" is alive in contemporary philosophical approaches to the *form* of experience and thought is consonant with the generally *non-supernaturalistic* character of Kant's idealism when contrasted with its Berkeleian opposite. Berkeley was, of course, a theist: the ultimate reality was the mind of God. Very crudely, it might be said that the idea of the "mind of God" was developed in the early Christian period by St. Augustine to designate the "place," as it were, where Platonic ideas were to be regarded as located. Here, the Pricean nature of Kant's *non-representationalist* strategy can be clearly seen in Kant's treatment of ideas like that of "God." Effectively, what Kant did was to point out that concepts like "God" do not *function* like ordinary referring empirical concepts such as "duck" or "electron." In his terminology, they are not "constitutive" concepts that are applied in judgments but "regulative" ones, regulating our theoretical and practical reasoning processes. We might say that for metaphysical notions like "God," Kant's slogan was something like the non-representationalist Wittgensteinian one that Price takes to heart: look to *the use*, not to the (supposed) referent.

But if Kant passes muster on this count, surely it will be objected, that *his* healthy and restraining sense of the finitude of humans was in turn overthrown by the later idealists who returned to just this type of "spiritualistic" thought? After all, didn't Hegel embrace the notion of "spirit" with a vengeance? Two things should be noted here. First, Hegel belonged to a tradition of post-Kantian thinkers who, coming on the scene after the "pantheism dispute" of the mid 1780s, tried in some way to reconcile the philosophies of Kant and Spinoza. While Hegel had a concept of "spirit," it was clearly not an *immaterialist one* like that of Berkeley.[52] Next, while it is still a relatively common view of Hegel that he reverted to the type of "dogmatic" metaphysical philosophy of which Kant was critical, according to many recent interpreters, this is radically mistaken: on their "post-Kantian" reading, rather than reverting to dogmatic metaphysics, Hegel had *extended* the type of Kantian critique, and retrospectively applied it to remnants of the dogmatic metaphysics that, he thought, afflicted Kant.[53] And just as Kant had shown that one does not need to be a philosophical theist to have a philosophical theology, Hegel showed that neither does one need to be a spiritual *realist* to have an account of *spirit*.

Here the notion of "objective spirit [*Geist*]" found in the human sciences (the *Geisteswissenschaften*) is relevant. Indeed, this very notion had been introduced into discussions of the human sciences by a philosopher, Wilhelm Dilthey, who had taken the idea from Hegel and tried to employ it for the project of a "human" alternative to natural science. The idea itself is not a particularly mysterious one, the realm of *objectiver Geist* as opposed to *Natur*, effectively appealing to the distinction between the realms of "institutional" and "brute" facts as described by analytic philosophers like Anscombe and Searle. Using a helpful locution of Bernard Williams, we might refer to the conception of "brute" facts as facts about "what is there anyway," effectively *nature* as it is revealed by science. Clearly, institutional facts are not easily thought of in this way, that is, as being there "anyway," independently of how they are conceived and talked about.[54] This is a commonplace of ethnographical methodology. If one were talking about a culture in which, say, there were *no* words that *in any way* functioned like our words, "marry," "spouse," "wedding," and so on, it would be very questionable to describe person X as *married* to person Y. In short, we might think of "institutional facts" as, along with those evaluatively "thick" facts of ethical knowledge, belonging to a realm accessible only to a person able to deploy all those relevant culturally transmitted perspectival concepts described by Williams, that is, by someone located within some collective "first-person" standpoint.

We might think of such "institutional facts" as accessed from a type of first-person-plural perspective as the sorts of facts of interest to those humanistic parts of our culture such as history or ethnography. This is the idea that holds out hope of a distinctly humanistic *science*. But as we have seen, for philosophy to appeal to this type of model may not help with the nihilism problem with respect to values. We might, with some ethnographic training, be able to describe the institutional facts constitutive of our own society, but this, as Williams's reflections on "reflection" point out, does not thereby *commit us* to those institutions. After all, an ethnographer might learn about what it is to be an Arunta tribesman without *becoming* one.[55] But while the idea of a "human *science*" may not be the appropriate model for philosophy, the related *looser* idea of a "humanistic discipline," as Williams presents it in the paper "Philosophy as a Humanistic Discipline,"[56] may be. Following Williams here, I suggest, gives us a sense of what an *idealistically* inflected version of subject naturalism might be like.

Williams responds to the charge of scientism made by Hilary Putnam. Williams is well known for having made a distinction akin to Price's between the functioning of concepts in the discourses of the natural sciences, on the one hand, and in evaluative or normative forms of talk, on the other. Moreover, his motivation here has features akin to Price's: while wanting to *deny* that, say, ethical judgments have the sort of objectivity characteristic of the natural sciences (they are not a part of the way the world is "anyway"), he wanted to avoid common naturalist strategies such as reductionism or eliminativism. Williams made this distinction by appealing to an "absolute conception" of the world on which the discourses of the natural sciences, but not those of the ethical realm, converge, and it was just this notion that was invoked by Putnam in his accusation of scientism.

But, Williams protests, he had never regarded *philosophy* as aiming at the absolute conception,[57] and he specifically rejects such an idea for the same reasons invoked by Putnam. Philosophy must account for relations such as semantic ones that are *normative*. While our account of semantics must, of course be *consistent* with physics, it is obvious, he thinks, that "any attempt to *reduce* semantic relations to concepts of physics is doomed."[58] Moreover, this is not a problem about physics per se, *no* science conceived in the natural explanatory mode, it would seem, could be adequate. While evolutionary explanations might, for example, seem better candidates in accounting for our linguistic practices, natural selection does not explain human cultural practices per se "but rather the universal human characteristic of having cultural practices, and humans' capacity to do so."[59] Presumably these same considerations would stretch to *any* empirical "sociolinguistics" of the type Price seems to suggest. Regardless of the identity of the science in question, if the normativity of semantic relations prevents them from being part of the "absolute conception of the world," this only brings out the fact that achieving such a view "would not be particularly serviceable to us for many of our purposes, such as making sense of our intellectual and other activities. For those purposes—in particular, in seeking to understand ourselves—we need concepts and explanations which are rooted in our more local practices, our culture, and our history."[60]

For Williams, then, philosophy should *not* aspire to an account of the way the world is "anyway" or, we might say, as it is "in itself." This we might describe as Williams's *Kantian* dimension. But we might regard his claim that our philosophical concepts and explanations are *necessarily* "rooted in our more local practices, our culture, and our history"—that is, rooted in our "objective spirit" whose norms must somehow *remain* normative for us—as manifesting his *Hegelian* dimension. Like Schelling, Hegel had criticized the idea that philosophy should be done entirely from the *reflective* position to which *natural* science aspires. What Schelling had treated as "disentanglement" Hegel treats in terms of the notion of "negation," and philosophy involves the *negation* of the initial "negation" of disentanglement: the philosopher needs somehow to return to that entanglement with the world that reflection denied.[61]

Like Hegel, Williams appeals to the necessarily *historical* dimension of philosophical reflection. Such a philosophical "memory," internal to the practice of philosophy, must address the relation between our present normative orientations and commitments (such as a commitment to science or certain liberal political values, for example) and the earlier configurations from which they arose and which they replaced. Here historical reflection cannot aspire to the type of "objectivity" that is the goal even of the empirical *Geisteswissenschaften*, as this is the type of historical narration that would lead to relativistic nihilism. Rather, such histories need to be recounted from the normative commitments *of the present point of view* from which our present commitments will be regarded as more rational than those they replaced: it will in some sense, therefore, be whiggish or, as Williams says, "vindicatory," lest "the history of our outlook . . . interfere with our commitment to it, and in particular with a philosophical attempt to work within it and develop its arguments."[62] Williams notes that the type of "wide-screen versions" of stories about the "unfolding of reason" or the "fuller realization of freedom and autonomy" as told by Hegel and Marx are not at all popular within contemporary philosophy. But "we *must* attend to it, if we are to know what reflective attitude to take to our own conceptions."[63]

Williams's gestures toward a type of Hegelianism here are, of course, not meant to suggest a commitment to the "wide-screen" version of the "unfolding of reason" that is associated with traditional pictures of Hegel, but *such a conception* of Hegel is not that found in the interpretations of the contemporary "post-Kantians" for whom Hegel has been traditionally *misconstrued* as a type of spiritual realist rather than an idealist.[64] In fact, many contemporary accounts of Hegel see in his theory of *Geist* an approach that has more in common with what, in the context of contemporary nonnaturalist accounts of practical reason, has been called, the "second-person" standpoint.[65] According to Stephen Darwall, this approach, which is seen in contemporary versions of contractualism such as that of T. M Scanlon,[66] finds an early exemplar in the post-Kantian idealist J. G. Fichte, who employed the idea of "reciprocal rec-ognition" in the context of his theory of "rights."[67] Indeed, many contemporary Hegelians see Hegel as having developed just such a "recognitive" approach to spirit on the basis of Fichte's starting point.[68]

According to Darwall, the second-person perspective is a version of the *first-person* perspective, the agentive perspective that grasps the world in terms of nonnatural normative concepts. In particular, it is the stance adopted in contexts of dialogical or reciprocal address, most obviously in the context of speech acts at the center of *practical reason* such as those of commanding or reproaching.[69] But Robert Brandom, following Wilfrid Sellars, also sees the paradigmatic act of *theoretical* reason—that of *asserting*—in these terms.[70] When I tell somebody something, he thinks, I undertake various types of commitment to which my interlocutor can hold me to account. For example, I implicitly undertake to give them *reasons* for accepting what I have told them, should they find reasons for doubting it, and I also commit myself to those assertions that logically follow from the original claim. In Brandom's "deontic scorekeeping" approach to reason, to assert a sentence is to place it in what Sellars called "the logical space of reasons,"[71] a justificatory "space" within which the "moves" of both theoretical and practical forms of reason-ing unfold.[72]

While there are differences among the various approaches in contempo-rary philosophy that appeal to something like the "second-person stand-point," there are a number of common features. First, approaching speech acts in this way tends to circumvent the tendency to think of their content "repre-sentationally" in relation to how the world is "in itself." This is most apparent in Brandom's "inferentialist" account of semantics that he draws from this approach, but it is also apparent, for example, in R. Jay Wallace's account of responsibility, in which there is a deflationary shift from more "metaphysical" questions such as "What is it to be a morally responsible agent?" to more prag-matic ones such as "What is it to treat someone as a morally responsible agent, or to hold a person morally responsible?"[73] Next, it becomes apparent that from this starting point even the *aspiration* to the aperspectival "absolute con-ception" is unlikely to come into view. Addressing another from the second-person standpoint presupposes that you are both committed to an at least overlapping set of norms that makes one's address intelligible. As Darwall makes clear, the picture that comes along with the second-person perspective is that of a *circle*: in particular a "circle of irreducibly second-personal con-cepts" into which "there is no way to break" from anywhere "outside."[74]

While the adoption of the second-person stance has become advocated as a way of thinking about the nature of moral philosophy that has a certain continuity with the Kantian approach, its extension to philosophy in general, as advocated by Brandom, also extends this in the direction of Hegel. We might now use this to answer William's metaphilosophical need when he insists that philosophy should not even aim at the aperspectival "absolute" point of view, as is implicit in contemporary naturalism. What can be seen emerging in these contemporary debates, I suggest, is a metaphilosophical analogue of what had been known in the nineteenth century as "idealism." To its advocates, such an "idealism," I suggest, with its insistence on the irreducibility of the normative, looks better equipped than contemporary philosophical naturalism to answer the two problems that have plagued much modern philosophy: its disciplinary identity, on the one hand, and its nihilistic tendencies, on the other.[75]

NOTES

1. See, for example, Mark Sacks, "Naturalism and the Transcendental Turn," *Ratio* 19 (2006): 92–106.
2. See, for example, Peter Hylton, *Russell, Idealism, and the Emergence of Analytic Philosophy* (New York: Oxford University Press, 1993), 223, and Michael Beaney, "Russell and Frege," in *The Cambridge Companion to Bertrand Russell*, ed. Nicholas Griffin (Cambridge: Cambridge University Press, 2003), section 6.
3. Michael Friedman, *Reconsidering Logical Positivism* (Cambridge: Cambridge University Press, 1999). Similarly, Sacks notes: "For both Carnap and Quine questions of existence, and the structures we can have knowledge of, are settled by the linguistic apparatus with which we approach them," and this linguistic or conceptual framework "plays the role of transcendental psychology in Kant: it determines or individuates the world which we approach by means of it" ("Naturalism and the Transcendental Turn," 97–98).
4. Friedman, *Reconsidering Logical Positivism*, 7.
5. Sacks, "Naturalism and the Transcendental Turn," 98.
6. Sami Pihlström, "Recent Reinterpretations of the Transcendental," *Inquiry* 47 (2004): 294.
7. W. V. Quine, "Two Dogmas of Empiricism," in *From a Logical Point of View* (Cambridge, Mass.: Harvard University Press, 1953).
8. Sacks, "Naturalism and the Transcendental Turn," 99.
9. Ibid.
10. F. W. J. Schelling, *Ideas for a Philosophy of Nature*, trans. E. E. Harris and Peter Heath (New York: Cambridge University Press, 1988), 10. Schelling adds: "As soon as man sets himself in opposition to the external world . . . the first step of philosophy has been taken. With that separation, reflection first begins, he separates from now on what Nature had always united, separates the object from the intuition, the

concept from the image, finally (in that he becomes his own *object*) himself from himself" (10).

11. Ibid.
12. Ibid., 10–11.
13. Ibid., 11.
14. Ibid.
15. Thomas Nagel, *The View from Nowhere* (Oxford: Oxford University Press, 1984).
16. Bernard Williams, *Ethics and the Limits of Philosophy* (Cambridge, Mass.: Harvard University Press, 1985), 148.
17. Ibid., 51.
18. Ibid., 129. Thick concepts combine descriptive "world-guided" and evaluative "action-guiding" aspects.
19. Ibid., 51.
20. Lewis wonders whether "when we look hard at our knowledge, it goes away." David Lewis, "Elusive Knowledge," in *Papers in Metaphysics and Epistemology* (Cambridge: Cambridge University Press, 1999), especially 434.
21. See, for example, Huw Price, "Metaphysical Pluralism," *The Journal of Philosophy* 89, (1992): 387–409; "Naturalism and the Fate of the M-Worlds," *Proceedings of the Aristotelian Society*, supp. vol. 7 (1997): 247–67; and "Naturalism Without Representationalism," in *Naturalism in Question*, ed. Mario De Caro and David Macarthur (Cambridge, Mass.: Harvard University Press, 2004), 71–88.
22. Price, "Naturalism without Representationalism."
23. Ibid., 75.
24. Ibid.
25. Ibid., 76.
26. Ibid., 77.
27. Ibid.
28. In short, an object naturalism that starts from the linguistic realm and descends to the level of objects and facts "rests on substantial theoretical assumptions about what we humans do with language—roughly, the assumption that substantial 'word-world' semantic relations are part of the best scientific account of our use of the relevant terms" (ibid., 78).
29. These Kantian features are further apparent in Price's writings on time in that Price rejects a realistic interpretation of the asymmetry of past and future and the notion of cause that accompanies it. See, for example, *Time's Arrow and Archimedes' Point: New Directions for the Physics of Time* (New York: Oxford University Press, 1996). Such asymmetry is, in his account, a feature that is projected onto time from certain features of our own practices.
30. Kant to Marcus Herz, 21 February 1772, in Immanuel Kant, *Correspondence*, ed. Arnulf Zweig (Cambridge: Cambridge University Press, 1999), 132–33.
31. Kant frames his comments on Locke in terms of the impossibility of an empirical deduction of the pure a priori concepts, that is, the concepts to which an empirical concept must conform in order to count as having a representational function.

Immanuel Kant, *Critique of Pure Reason*, ed. and trans. Paul Guyer and A. W. Wood (Cambridge: Cambridge University Press, 1998), A86–7/B120.

32. Using a legal analogy, Kant accuses Locke of confusing a question of *fact* with a question of *right* or *law*. Locke's causal explanation cannot give an account of the normative status that adheres to the ideal of knowledge or representation. The "right" of some mental state, as it were, to be called a "representation." "Since this attempted physiological derivation concerns a *quaestio facti*, it cannot strictly be called deduction" (Kant, *Critique of Pure Reason*, A87/B120). A few pages earlier Kant had introduced the distinction between "questions of right (*quid juris*)" and "questions of fact (*quid facti*)" in relation to the transcendental "deduction" he uses to establish the normative status of the a priori concepts that are the necessary conditions of the *representational function* of any empirical concept (ibid., A84–5/B116–7).

33. Hilary Putnam, *Realism and Reason: Philosophical Papers*, vol. 3 (Cambridge: Cambridge University Press, 1983), 18.

34. Putnam links his own critique of "magical" theories of reference to Kant's critical philosophy in *Reason, Truth, and History* (Cambridge: Cambridge University Press, 1981), 60–64.

35. Price, "Naturalism and the Fate of the M-Worlds," 247.

36. Ibid., 248.

37. Ibid., 88.

38. Ibid., 247n. 1.

39. Ibid., 88.

40. Price, "Naturalism Without Representationalism," 73, emphasis added.

41. That novels are fictional in no way impugns this if we take the Aristotelian point that such representations novels are about *possible* human lives rather than *actual* ones.

42. John O'Leary-Hawthorne and Huw Price, "How To Stand Up for Non-Cognitivists," *The Australasian Journal of Philosophy* 74 (1996): 290n. 21.

43. I am grateful to James Chase for having pointed this out.

44. That Melinda is married to Jack is an institutional fact. That Melinda is in bed with Ralph is a brute (and perhaps, for Jack, brutal) fact.

45. Here, the culture-nature distinction is made internal to some more inclusive sense of nature.

46. See my *Analytic Philosophy and the Return of Hegelian Thought* (Cambridge: Cambridge University Press, 2007).

47. For a telling reassessment of the usual story here, see Stewart Candlish, *The Russell/Bradley Dispute and Its Significance for Twentieth-Century Philosophy* (Houndsmills: Palgrave Macmillan, 2007).

48. Robert B. Brandom, *Making It Explicit* (Cambridge, Mass.: Harvard University Press, 1994), and *Tales of the Mighty Dead: Historical Essays in the Metaphysics of Intentionality* (Cambridge, Mass.: Harvard University Press, 2002). Richard Rorty

describes Brandom himself as attempting to "usher analytic philosophy from its Kantian to its Hegelian stage." Richard Rorty, introduction to Wilfrid Sellars, *Empiricism and the Philosophy of Mind* (Cambridge, Mass.: Harvard University Press, 1997), 8–9. Significantly, Brandom treats Quine's critique of Carnap as a *Hegelian* critique of Carnap's Kant. See Brandom, *Tales of the Mighty Dead*, chap. 7.

49. A. A. Luce, for example, denies that Berkeley was "an idealist" if by that term is meant "the type of philosophical doctrine found in the recognized idealists—Kant, Hegel, and Bradley. If those three are idealists, then Berkeley is not" (*Berkeley's Immaterialism: A Commentary on his "A Treatise Concerning the Principles of Human Knowledge"* [London: T. Nelson , 1945], 25). See also J. R. Roberts, *A Metaphysics for the Mob—the Philosophy of George Berkeley* (New York: Oxford University Press, 2007).

50. A. C. Frazer, *Berkeley and Spiritual Realism* (London: Constable, 1908).

51. In a letter to J. S. Beck, from 4 December 1792, we find Kant helpfully clarifying the relation of his idealism to Berkeley's immaterialism. Countering the claim of those who had identified his "Critical Idealism" with the philosophy of Berkeley, Kant explains: "For I speak of ideality in reference to the form of representation while they construe it as ideality with respect to the matter, i.e., ideality of the object and its existence itself" (Kant, *Correspondence*, 445).

52. See, for example, the history of this movement as given by Frederick Beiser, *German Idealism: The Struggle Against Subjectivism, 1781–1801* (Cambridge, Mass.: Harvard University Press, 2002). I give a different account but one that equally contests the construal of idealism as a "subjectivist" doctrine in *Continental Idealism: Leibniz to Nietzsche* (London: Routledge, 2009).

53. Two well-known advocates of this view are Robert Pippin, *Hegel's Idealism: The Satisfactions of Self-Consciousness* (Cambridge: Cambridge University Press, 1989), and Terry Pinkard, *Hegel's Phenomenology: The Sociality of Reason* (Cambridge: Cambridge University Press, 1994).

54. In Hegel's terms, institutional facts would be thought of independently of creatures with *subjective* spirit—what we might call "intentionality."

55. Of course, I do not mean to deny the stance of the "participant-observer" in ethnography. But this idea can be misleadingly overliteralized. Ethnographers cannot study societies and *cease to be* ethnographers in the process, which would typically be the case if they literally *became* proper members of the studied society. Most societies just don't have a role for "ethnography."

56. Bernard Williams, "Philosophy as a Humanistic Discipline," *Philosophy* 75 (2000): 477–96. I mean by "humanistic discipline" here the humanities broadly conceived, and by the "human sciences," the more "social science" orientation that strives for something like the "objectivity" and explanatory stance of the natural sciences while acknowledging the nonreducibility of its objects to "nature." Many actual "human sciences" are, I think, pulled in these different directions, my point simply being that there *are* different directions in which to be pulled. Were Price's "natu-

ralism" to be construed broadly and "science" extended to the range of the human *sciences*, it would, I believe, given the spirit of his "naturalism," still stop short of the humanities as conceived here.

57. Williams asks, "Should the idea that science and only science describes the world as it is in itself, independent of perspective, mean that there is no independent philosophical enterprise? That would follow only on the assumption that if there is an independent philosophical enterprise, its aim is to describe the world as it is in itself, independent of perspective. And why should we accept that?" (Williams, "Philosophy as a Humanistic Discipline," 481). Williams first introduced the idea of an "absolute conception of the world" in *Descartes: The Project of Pure Enquiry* (Harmondsworth: Penguin, 1978), 64. Hilary Putnam attacked this is *Renewing Philosophy* (Cambridge, Mass.: Harvard University Press, 1992), chap. 5.

58. Williams, "Philosophy as a Humanistic Discipline," 484. This loss of normativity is just another example of what I have referred to as the nihilism problem facing philosophical naturalism.

59. Ibid., 485.

60. Ibid., 484.

61. As with the idealists, for Williams, it could be said that the appropriate way of conceiving of philosophy is not on the "reflective" model of "consciousness"—with its abstract opposition between the knower and the known—but on that of a collective "*self*-consciousness" in which the idea of such an ultimate separation does not make sense (ibid., 488).

62. Ibid., 489.

63. Ibid., 488. On the extent of Williams's Hegelian tendencies, especially in relation to his critique of Kantian moral philosophy, see Mark P. Jenkins, *Bernard Williams* (Chesham: Acumen, 2006), 84–86.

64. In earlier work, Williams himself had drawn attention to the incoherence of those usual conceptions of any type of socialized Cartesian mind—what he calls "aggregated solipsism"—that might be invoked in relation to idealism. See Bernard Williams, "Wittgenstein and Idealism," in *Moral Luck: Philosophical Papers, 1973–1980* (Cambridge: Cambridge University Press, 1981).

65. Stephen Darwall, *The Second-Person Standpoint: Morality, Respect, and Accountability* (Cambridge, Mass.: Harvard University Press, 1996).

66. T. M. Scanlon, *What We Owe Each Other* (Cambridge, Mass.: Harvard University Press, 1998).

67. J. G. Fichte, *Foundations of Natural Right: According to the Principles of the Wissenschaftslehre*, ed. Frederick Neuhouser, trans. Michael Baur (Cambridge: Cambridge University Press, 2000); Darwall, *The Second-Person Standpoint*, 20–22. Scanlon characterizes the social relation underlying moral obligation in terms of the notion of "reciprocal recognition" in *What We Owe Each Other*, 162.

68. See, for example, Robert Williams, *Recognition: Fichte and Hegel on the Other* (Albany: State University of New York Press, 1992); and my *Hegel's Hermeneutics* (Ithaca, N.Y.: Cornell University Press, 1996).

69. These are acts that, as Darwall points out, are "grounded in (*de jure*) authority relations that an addresser takes to hold between him and his addressee" (Darwall, *The Second-Person Standpoint*, 2).

70. See, especially, Robert B. Brandom, *Making It Explicit* (Cambridge, Mass.: Harvard University Press, 1994), and *Articulating Reasons* (Cambridge, Mass.: Harvard University Press, 2000). In contrast, Darwall does not see theoretical reason as ultimately resting on the same dialogical grounds (*The Second-Person Standpoint*, 55–60. This seems a key distinction in separating the more Hegelian approach of Brandom from the more Kantian one of Darwall.

71. Sellars, *Empiricism and the Philosophy of Mind*, 76.

72. See, in particular, Wilfrid Sellars, "Language Rules and Behavior," in *Pure Pragmatics and Possible Worlds: The Early Essays of Wilfrid Sellars*, ed. J. F. Sicha (Atascadero, Calif.: Ridgeview, 1980).

73. R. J. Wallace, *Responsibility and the Moral Sentiments* (Cambridge, Mass.: Harvard University Press, 1994), 1. Wallace claims that "metaphysical interpretations" that "postulate facts about responsibility that are completely prior to and independent of our practice of holding people responsible" as ones that "are to be avoided, if possible" (84–85). Analogously, Robert Brandom advocates a "pragmatic phenomenalist" approach to knowledge. Rather than pursue the question what it is *to be* a knower, the pragmatic phenomenalist approach will pursue the different question of what is entailed by *taking* another to be a knower (*Making It Explicit*, 297).

74. Darwall, *The Second-Person Standpoint*, 11–12. "What the second-person stance excludes is the third-person perspective, that is, regarding, for practical purposes, others (and oneself), not in relation to oneself, but as they are (or one is) 'objectively' or 'agent-neutrally'" (9).

75. I am most grateful to David Macarthur and Huw Price for helpful criticisms of earlier versions of this chapter.

PART VI

NATURALISM AND HUMAN NATURE

14

HOW TO BE NATURALISTIC WITHOUT BEING SIMPLISTIC IN THE STUDY OF HUMAN NATURE

John Dupré

A good start for understanding naturalism is the idea that in describing the world we should appeal to Nature rather than, for instance, Pure Reason or the interpretation of Holy Writ. Of course, appealing to Nature isn't such an easy matter. We can't just ask her, because she doesn't speak. Perhaps Francis Bacon was nearer to the mark when he suggested that we needed to torture her. But even apart from the unpleasantness of the metaphor, and the genuine possibility that it may legitimate unpleasant actions, the messages elicited from Baconian acts of torture require a good deal of interpretation. At any rate, one point I share with all contemporary philosophers who describe themselves as naturalists is that I assume that the best ways of interrogating nature can be found by looking at the empirical sciences. Though due caveats should be entered about the importance of a critical attitude to the interpretation of the deliverances of science, these will not be my main concern here.

The preceding might be thought of as the most minimal epistemological account of naturalism. But there is also an ontological aspect. Though I have been critical of many interpretations of this doctrine, I do subscribe to a very modest kind of physicalism. To begin with, this is the view that everything that exists is made of the sort of stuff described by physics and chemistry. Of course, this idea does not apply to things that, if they exist in some sense, are surely not made of anything. So, for example, I do not want to enter the sometimes arcane debates about whether mathematical objects exist. My own version of the ontological aspect of naturalism might better be described as "anti-supernaturalism." That is to say, we should try to understand the world, insofar as we can at all, without appealing to anything that is entirely beyond the reach of empirical knowledge. Excluded by anti-supernaturalism are such things as immaterial minds or souls, vital forces, and divine beings. Should such things prove to be amenable to standard methods of empirical investigation they would, I take it, to that extent cease to be properly super-natural. An obvious topical example of an excluded explanation would be a special creationist account of the universe or the origin of humans.

Though this sort of naturalism is far from trivial, it has less specific content than many ideas that are currently associated with the thesis by contemporary philosophers. For example, it is common for philosophers to state that as naturalists they are committed to explaining everything solely in terms of physics and chemistry. Such philosophers thereby embrace some kind of physicalist reductionism. For me, physics and chemistry are naturalistic accounts of the stuff of which the world is made, but it is no part of my naturalism to assume that everything in the world can be explained in terms of theories of material stuff. And in fact I find this thesis highly implausible.

I cannot explain my reservations about reductionism here in any detail, though in the past I have done so at length.[1] Here I can only summarize my position. First, it can be said with some confidence that empirical science has not substantiated any robust version of reductionism by its success in implementing actual reductions. The first stage in such an implementation would be the reduction of chemistry to physics—the explanation of the empirical facts of chemistry in terms of appropriate elements of the theory of elementary particles. In fact, it appears that there is some debate as to whether the hydrogen atom, the simplest possible chemical object, might be understood fully in terms of physical law, and few suppose that even a large atom, let alone a complex macromolecule, is likely to be susceptible to such complete reduction.[2] Traditional reductionism then requires a sequence of further reductions up the hierarchy of structural complexity. Other traditional reductionist projects along this route, such as the reduction of Mendelian transmission genetics to molecular genetics, or society to the behavior of constituent individuals, are generally agreed to have been misconceived. One might well argue that reductionism in this sense, since quite unsupported by empirical science, is strongly contrary to the spirit of naturalism.

In the face of this difficulty various responses can be distinguished. Some philosophers have urged an eliminativism that denies the real existence of the objects of higher-level sciences to the extent that these cannot be subsumed under the laws of physics. Others have argued that despite the minimal successes of reductionism in the strong sense, a commitment to such a view is so deeply implicated in the ambitions and methods of the sciences that the naturalist is committed to some kind of reductionism in principle though one unattainable in practice, as part of a commitment to the epistemic authority of the sciences. In what follows I hope to show why such a conclusion would be quite unwarranted. But first I shall consider a response that is common among devotees of a variety of responses to reductionism, the idea of supervenience. According to adherents of various versions of this doctrine, although it may be impossible to reduce a realm of phenomena to its subordinate structural level, nonetheless once the lower level is fixed, so is the upper level. The lower level does, then, in a clear sense determine the upper level. This doctrine will provide a convenient way to approach what I take to be the most fundamental errors in the association of naturalism and reductionism.

Supervenientists propose that although we cannot in practice explain the behavior of complex objects by appeal to physical laws operating on their constituent parts, this is for contingent reasons; such behavior is nonetheless ultimately determined by these constituent parts. One relevant practical obstacle might simply be the complexity of the objects we study, and the consequent computational intractability of the applications to them of physical laws. This is commonly taken to be the problem with the reduction of chemistry. Another obstacle, widely cited in explanation of the irreducibility of objects in biology or the social sciences, is the problem of multiple realizability. The idea here is that the kinds of things these higher-level sciences address are highly diverse from the perspective of physics. Hence physical laws could not, in principle, be applied to them in a general way. For any particular realization of the higher-level property or thing, however, supervenience guarantees that that physical realization will be sufficient to determine matters at the higher level.

As I have mentioned, supervenience is generally a response to the admitted failure of actual reduction. Again, therefore, it seems in obvious respects to contravene the spirit of naturalism. I want to suggest, more strongly, that reflection on empirical knowledge indicates why it is untrue. Consider, for example, the question whether the properties of biological macromolecules could, in principle, be fully understood in terms of a sufficient knowledge of their constituents (atoms, let us say) and sufficient computational ability.

Consider protein molecules. The important property of proteins in biological cells is their ability to interact with other molecules. An important subclass, enzymes, catalyze chemical changes to such other molecules. But, first, being a particular kind of enzyme, a protease or a DNA polymerase, say, is a relational property among the enzyme, the substrate to which it binds, and the transformation that it catalyses. As it happens, a particular protein may have several kinds of enzymatic activity, and what it actually does depends on where it is in the cell and what else is in its vicinity. Proteins are known with significant numbers of quite different functions, and there is surely no reason to suppose that there is any particular limit to how many such functions they might have in different imaginable contexts.[3] It is interesting that this phenomenon has been widely referred to as protein "moonlighting," presumably reflecting the assumption that any protein has a day job—perhaps even the job to which it is most naturally suited. But of course such an assumption of a natural or primary task is entirely gratuitous, reflecting only the function that the molecule was first observed to serve.

Of course, I do not mean to deny that the ability of a particular molecule to function as a particular kinase or protease, say, is explicable in terms of simpler chemical knowledge. As a matter of fact, such explanations can often be given in considerable detail, in terms, for example, of the electrochemical topography of the site to which the substrate binds, changing the shape of that substrate in ways that lower energy barriers to chemical transformations. But the fact that the molecule acts as one kind of enzyme rather than as one of many other kinds—one might say, indeed, the fact that it is, in this context a kinase rather than a protease—depends on the cellular context in which it currently resides. Thus, finally, the property of *being a kinase* does not supervene on the chemical features of the molecule but only, at most, on the chemical features of the molecule and a much larger part of the cellular context in which it is found. This example, finally, exemplifies a central general claim I want to make against reductionist positions: reductionist methods explain how it is possible for an entity to have a particular capacity, but to understand what capacities it exercises, and even, I want to say, what capacities it actually rather than merely possibly has, requires seeing the entity in a larger context. A different way of reaching a similar conclusion is to note that in Nature there are no closed systems. Any bounded part of the world will, for this reason, only be partially describable without appeal to interactions that occur beyond the boundary. Almost reductionistically intelligible behavior can be generated by making efforts to approximate closed systems, as is standard procedure when setting up laboratory experiments or building machines. The reason it is difficult to make reliable machines or successful laboratory experiments, one might say, is precisely because nature is not composed of machines, though biological organisms have features that in certain respects resemble machines.

One final point should be stressed on the topic of supervenience. A natural response to the problems I have raised about the nonclosure of real systems is to suggest that the system on which I have proposed to ground the higher level property is too limited. If we want to describe the full supervenient base of a particular protease, for example, we need to include not only the relevant protein model, but some part of the surrounding contents of the cell. But of course the problem may recur. The further features of the cell, or entities within it, which we need to add to the supervenient base may also obtain their properties or nature in part from their surroundings. Though it would take more detailed argument to establish that this problem really does occur generally and frequently, I am confident that it does, and, of course, the general failure of supervenience on any predeterminable base requires only that it occurs sometimes. So we shall eventually be driven to some such thesis as that everything supervenes on the microphysical state of the entire universe. This is a thesis about which I am happy to remain agnostic. I note only that whatever its grounding, it is hardly an empirical discovery. So, finally, it has no place in the kind of naturalist philosophy I espouse.

The contrast between my own position and the reductionist position just indicated can be illuminated by considering so-called downward causation: the possibility that the behavior of the parts of a system is to be causally explained by the capacities and consequent behavior of the whole. For reductionists, this idea is anathema. We should explain what complex things do solely by looking at the capacities and behavior of their physical parts. In my view this is an epistemological prejudice. When I decide to move my arm, say, to reach for a cup of coffee, it is my coffee-seeking behavior that explains the movement of the physical parts—bones, muscles, or the atoms and smaller things of which these are made. My ability to move my arm should certainly be explained in terms of the properties or articulations of these smaller things, but their actual movement at this moment is to be explained in terms of the goals or intentions of the larger system they compose, myself.[4]

This brings us, then, to the problem of understanding human capacities and behaviour. The introduction of downward causation points to one way in which what humans can do depends on much more than merely aspects of their structure, even the structure of their brains. The clearest illustration of this point is in the context of language. As Wittgenstein most famously argued, language is not something that could be the property of an isolated individual organism. Language requires rules for its correct and incorrect uses, and rules can only exist as a feature of the practice of a collective. Since most of the distinctively human things we do depend on language—language, for example, makes possible the intricate division of labor that drives complex human societies—distinctively human capacities emerge only when humans are embedded in complex societies. My ability to write (or better, perhaps, to present publicly) this paper depends on my ability to carry out countless routine social transactions, none of which would be possible without my membership in a language-using society. Chimpanzees and dolphins seldom write or deliver philosophy papers.

Banal though the preceding point may be, it seems to me that generally reductionist approaches to human behavior, of which evolutionary psychology is a notable example on which I have focused in the past, fail to grasp it.[5] Evolutionary psychology suffers from reductionist prejudices at three main points: first, in its exaggeration of the role of genes in the development of brains; second, in its account of the relation between brains and behavior; and third, and partly consequent on a simplistic view of what genes do, in assuming a simplistic and unsustainable view of evolution.

How do genes build brains? Of course, they do nothing of the kind. Brains develop through a long sequence of cell divisions and differentiations. Just as in the case of proteins particular genes have particular capacities as a result of the enormously complex molecular and structural environment in which they reside. This point was less obvious when it was imagined that genes were discrete elements of the nuclear DNA from which a particular protein product could be read off with a knowledge of the genetic code. But the understanding that some 23,000 "genes" are involved in the production of perhaps a million final protein products is sufficient to show that this assumption is untenable. In fact, contemporary knowledge of alternative reading frames, alternative splicing, post-transcriptional editing, post-translational processing of various kinds, and so on has given us considerable insight into how this proliferation of proteins is achieved. The genome is, of course, an essential resource for the functioning of cells, but what it does at any moment is massively underdetermined by its intrinsic properties. The sequence structure explains its capacity to direct the production of countless possible RNA sequences. Which sequences are actually generated, and in what quantities, is determined by the cellular, or often wider, context; and similarly this wider context determines what proteins are finally produced from the great quantity of RNA sequence in the cell. The growth in our understanding of this matter, incidentally, explains the seriousness of the debate as to whether it is really useful to maintain the concept of the gene as a discrete element of the genome. One reason in favor of eliminating it is that it remains intimately connected with the idea that there is some developmental trajectory, or even developmental outcome, inherent in the nature of the sequence. And the all too frequent talk of genes *for* everything from brown eyes and double-jointed thumbs to aggression, homosexuality, or language learning is surely very widely understood in just this way.

Abandonment of the idea of a telos somehow inherent in a sequence of nucleotides makes room for a much richer picture of the process of ontogeny, of which the development of the brain is just one part. First, the development even of the basic structure of the organism, including its brain, is not something implicit from the start in the genome sequence but a process that is reliably reproduced through the repeated assembly of a wide range of developmental resources. The chemistry of the cell, but also a variety of external inputs into the system from physical deformations consequent on cell division to the provision by parent birds of nests and model songs, have effects on the process, often including the ways in which genomic resources are exploited.[6] In the human case, it is quite obvious that the capacities for behavior that depend on the development of the human brain are generated by a process to which environmental influences provide a substantial part of the input. Here, the speaking of a particular language (or any language) is only one of a multitude of possible examples. Of course, evolutionary psychologists do not quite deny this and are indeed liable to considerable annoyance when it is suggested that they do. The fact, however, is that they see environmental influences as of quite secondary importance, as mere triggers for the exercise of capacities or for the particular mode of exercise (as in speaking one language rather than another). The basic capacities thus triggered are "wired in" to the brain as they are earlier "encoded" in the genome.

The lack of appreciation of the fundamentally interactive nature of development becomes clearest in evolutionary psychologists' essential appeal to a thoroughly simplistic account of evolution. Appreciation of the diversity of developmental resources allows us to see that evolutionary change in the characteristic features of a population of organisms may be explained in many ways besides the traditional one of looking for changes in the nucleotide sequence of the genome. So, with respect to the case in point, provisions of different cultural resources—schools, scientific knowledge, reality television, or whatever—may certainly produce characteristically different brains. But very importantly, it is increasingly clear that change in nucleotide sequence, traditionally conceived as producing the template for novel proteins, is not the only possible way in which genomes can evolve, and probably far from the most important. As just noted, the same genomic resources can be used to produce a vast range of possible products. There are also important possibilities in duplication of parts of genomes, for example, polyploidy in many plants, transpositions, inversions, deletions, and so on.[7]

That evolutionary psychology is committed to denying, or at least ignoring, this point is evident in its foundational argument that modern human brains are essentially the same as Stone Age human brains. Why should not human brains have changed in fundamental ways in, say, the last 1,000 years? The answer offered by evolutionary psychologists is that evolution, by which they mean the accumulation of genetic changes in a population, is much too slow. As a matter of fact, I strongly suspect that if the simple population genetics models for the spread of point mutations generally implicitly or explicitly invoked here were the only available resources for evolution, even the million or two years allowed for human evolution would hardly be long enough for the development of brand new mental organs of the kind evolutionary psychologists commonly postulate. But this is not the place to go into the possible refinement of evolutionary theory generally.[8] All I want to say here is that it is an entirely open question, a priori, how much change in the capacities of brains can be generated by changes in the external inputs to them. The empirical evidence is that there can be quite a lot. It seems possible, for example, that the recently acquired capacity to use mobile phones, notoriously far more richly developed among those growing up with this technological addition to their culture, will bring about deep changes in human functioning. The relation between social contact and spatial proximity, arguably, is being fundamentally altered, and this could have major effects on the nature of human life. I don't want to assert that this is the case but only to suggest that it is entirely possible that the presence of mobile phones is becoming a developmental resource that will generate significantly different kinds of people.

I earlier separated the question of the relation of genes to brains from the question of the relation of brains to behavior. But it should by now be clear that the distinction is not clear cut. Humans begin to behave at birth or earlier, and their behavior elicits responses from the environment that, in turn, affect the development of the brain. Humans develop capacities in interaction with their environments, and what those capacities are depends both etiologically on the kinds of environment with which they have interacted and occurrently on the environment in which they find themselves. What needs to be added here are some critical comments on the very idea of the brain causing behavior. Just as the genome is part of a system that interacts with the surrounding world, so is the brain part of a much larger system. People behave; many things cause their behavior—the behavior of others, opportunities or dangers perceived in the environment, long-term goals, and so on. In this proper sense in which we look for the causes of behavior, brains are unlikely causes unless I am a neuroscientist, a brain surgeon, or perhaps a chef unimpressed by the dangers of prions. Brains, it is much better to say, embody some of the central capacities that make behavior possible. Peoples' brains vary in the capacities they embody, and this partly explains the differences in behavioral capacities. But distinguishing the brain from the general nexus of environment, social context, and action, as in general the thing that causes behavior, is no more than a curious metaphysical prejudice. Ironically, in view of the materialistic and naturalistic avowals of those most liable to espouse this kind of talk, it is best seen as a hangover from Cartesianism—a dualism of body and brain replacing the dualism of body and mind.

This leads me to repeat an earlier point: those who think of themselves as hard-nosed materialist naturalists often end up ironically committed to philosophical positions that have little grounding in empirical knowledge of the world. Our empirical knowledge of humans is drawn from a wide variety of sources, and it is philosophical argument not experience that leads so many philosophers to suppose that only a privileged few should be accorded any credence. This suggests some more general thoughts on what I take a properly naturalistic philosophy to be. One central aspect of a naturalistic philosophy is simply the attempt to make broader use of what seem the best grounded aspects of our empirical knowledge of the world including, centrally but not exclusively, science.

As I have argued for many years, there have been many successful scientific projects (and many unsuccessful ones, too, of course) and the attempt to weave the former into one overarching narrative has been unsuccessful. Part of the task of naturalistic philosophy, then, is to make sense of these facts. Of course this could be done in many ways. As indicated in my preliminary discussion of reductionism, one possibility is to see the lack of a unified account of the world as merely a reflection of the inadequate state of our knowledge. Of course I accept that the state of our knowledge is in many ways woefully inadequate, but I prefer a different view of the diversity of knowledge.

The concepts in terms of which we develop scientific knowledge are, of course, abstractions. We explore reality by developing particular representations, and sometimes these representations are powerful tools. But such representations mirror only a few carefully selected aspects of the thing in view. Consider, as an illustrative example, our representation of DNA molecules in terms of the familiar four letters A, C, G, T. This has been an enormously useful mode of representation both in facilitating deeper insights into the molecules in question and also in further applications. For example, it is superbly adapted to the comparison of genomes which is central to the investigation of both genomic function and phylogeny. But it isn't always fully appreciated what a minimal representation of the chemical itself it is. One should note in passing, that this is at most a representation of the DNA molecule rather than the whole genome construed as the set of chromosomes replete with structural relations to a range of protein molecules and so on. But even as a representation of the DNA molecule it is highly simplified. Some twenty nucleotides can be found in DNA molecules, but we choose to mention only four. The best known illustration of this point is that many of the "cytosine" molecules in human and other genomes at any one time have been converted into 5-methyl-cytosine, a crucial epigenetic change known as methylation that inhibits gene expression. It would be perfectly possible to use a five-letter code including a letter representing 5-methyl cytosine. Instead we choose— for both historical and pragmatic reasons—to see this merely as a modification of the cytosine molecule.[9]

It is hardly news that scientific concepts embody abstractions. But it seems less often noticed that this makes it entirely intelligible, even to be expected, that we should have a pluralistic science. The more complex the objects we are trying to understand, the less likely it becomes that a particular abstract representation will be sufficient for understanding every aspect of its behavior. Note that this remark is entirely independent of questions about reference. Whether one believes that a sequence of Cs, Gs, Ts, and As refers rigidly to a DNA molecule or, say, refers to whatever best fits the description embodied in this representation is quite irrelevant to how adequate this representation is for providing a comprehensive account of DNA molecules. Note also that there is nothing antirealist implied by anything I have said. I take it that the scientific representations that I have been considering do, indeed, refer to objects in the world. It is merely that they do not fully describe those objects. And for that reason, the classes of objects picked out by these representations will not always perfectly coincide.

At any rate, it is widely agreed that humans are among the most complex objects—perhaps the most complex objects—we attempt to understand, so it should be no surprise that doing so will require a range of quite distinct types of representation. This was the central argument of my book *Human Nature and the Limits of Science*, in which I criticized the imperialistic claims of evolutionary psychologists and rational choice theorists to have preeminent conceptual resources for addressing human nature.[10] As a matter of fact, I also claimed that these perspectives were in some respects deeply flawed, though certainly evolutionary and economic approaches to human nature, developed with appropriate modesty in a pluralistic context, have important potential insights to offer.

Let me conclude with a quite different philosophical point, but one that emphasizes the inevitability of the kind of pluralism I advocate. The reductionist approach to science, at least insofar as any attempt is made to implement it in practice, is based on a hierarchy of entities in which each level comprises entities composed of the entities described in the level below. I have already agreed that there is a central place in science for explanatory ventures based on this model. The capacities of entities are, in part, explained by mechanical models describing the properties and relations of their parts. However, I want to suggest that the arrangement of entities into such a hierarchy is not something unambiguously determined by Nature. In fact, this is itself an abstraction, and different abstractions of entities as constituting the structural hierarchy may serve different investigative purposes. Let me illustrate with two examples.

Although many central concepts in biology have come under sustained philosophical criticism, notably the species and the gene, it is widely assumed that the concept of an organism is unproblematic. If an elephant wanders into the room we're pretty confident that exactly one organism has come onto the scene. But this assumption is much more problematic than it seems.

Restricting our attention to cellular organisms (excluding, that is, viruses and the like) it is standard to divide organisms into two main categories, unicellular and multicellular. Multicellular organisms comprise animals, plants, and fungi with fruiting bodies; unicellular organisms include everything else—bacteria, archaea, and a diverse set of eukaryotic organisms (protists, yeasts, and so on). Note first that almost everything in biology belongs in the second category. For 80 percent of the history of life that is all there was; even today the overwhelming majority of organisms are single-celled in this categorization—indeed, even half of current biomass. For this reason it is strange that we have a term, "microbe," that covers the latter category but no such term for the former. I have advocated elsewhere that we call them "macrobes."[11]

At any rate, this distinction is highly problematic. Cells form a great diversity of cooperative multicelled enterprises, and the forms of organization we think of as plants and animals are only two examples of such enterprises. Most microbes, especially bacteria, live in biofilms, complex communities usually including many different kinds of microbes, which adhere to almost any moist surface. The biofilms on our teeth that we know as dental plaque, for example, include around 500 varieties of cell. These cells exhibit complex structural and functional differentiation, and most microbes appear only to thrive in such communities. This is a central part of the reason why only about 1 percent of microbes can be cultured in isolated clones in the laboratory. There are very good reasons for treating such communities as a kind of organism. From an ecological perspective, at any rate, these are the entities that interact with the surrounding environment, and their constituent parts exhibit a cooperative division of labor much like the cell types in a more familiar multicellular organism. From a phylogenetic perspective, on the other hand, we will need to distinguish the various kinds of cells that constitute these cooperative wholes. Similar questions, though without the further problem of the genomic diversity of cell types, have long been raised about the status of social insects; many experts have proposed that the colony should be treated as a superorganism.

Having presented the suggestion that biofilms are a kind of multicelled organism then suggests a new perspective on macrobes. It is standard to think of a human, say, as a highly differentiated set of cells with more or less identical genomes, all originating from a single zygote. While there is nothing wrong with this perspective, and there are questions that it is well suited to address, there is another quite different way of looking at the biological entity. Ninety percent of the cells that make up a functioning human are in fact not "human" cells at all but vast assemblies of microbes, most especially in the digestive tract but also covering the skin and colonizing all the bodily orifices. These are not merely passengers but are increasingly recognized as essential constituents of a symbiotic whole. Organisms with their symbiotic microbes removed function very poorly, if at all, and develop abnormally. The expression of genes in "animal" cells is often modulated by the interacting microbes. In fact, 99 percent of the genes to be found within the human body reside in these microbes rather than in the "human" cells, and some scientists are working toward an expansion of the human genome project into the human biome project, a full catalogue of the genes that make up a functioning human.

Similar things can be said about plants. The roots of a plant, apart from the essential function of keeping the plant in place and, in many cases, upright, serve as the home to exceedingly complex consortia of fungi and bacteria that serve to process nutrients for the use of the system. Fungi live within the roots and send hyphae into the surrounding soil where they extract nutrients. The fungi, in turn, derive nutrients from the photosynthetic activities of the plant. There is even growing evidence that the mycelial networks that connect with plant roots may serve as a conduit for transferring nutrients between plants, even plants of different species, suggesting that there may be grounds for seeing entire plant communities as constituting superorganisms.[12]

The last point raises the second case that I want to mention, though only in the briefest way. Given, as I have argued, that a vast range of human capacities are contingent on the relation of humans to larger social groups of which they are part, it seems plausible that these social groups themselves should be accorded an ontological standing comparable with that of the isolated individual. My suggestion is just that the recognition of the general indeterminacy within biology of the boundaries of a system should make such a move much more acceptable. If this is right, it provides a very strong rebuttal of Margaret Thatcher's famous assertion that "There is no such thing as society."

The examples I have just sketched point to a way in which I propose to expand my earlier thesis of promiscuous realism.[13] It is not just, as that thesis asserted, that there is no unique way of dividing things into kinds but also that there is no unique way of dividing stuff into things. The latter thesis might be dubbed "promiscuous individualism." As with my understanding of promiscuous realism about kinds, I don't at all want to deny that the various individuals that might be carved, for instance, out of the total terrestrial assembly of cells, are real. It is just that they are not uniquely real. I take it that if this thesis were accepted it would provide another significant nail in the coffin of traditional materialist reductionism, or anyhow the version that assumes that there is an objective and unique hierarchy of objects, the behavior of each of which should be understood by reference to properties of its parts.

Let me sum up. What I am advocating is a more committed naturalism. Traditional naturalisms have tended to be contaminated by metaphysical or epistemological commitments that were no part of our empirical knowledge of the world. The family of varyingly powerful reductionist theses are the outstanding examples of such theses. My argument here is that a truly committed naturalism not only should eschew such a priori commitments but actually provides excellent reasons for thinking they are false. The reality that actually confronts empirical enquiry takes many forms from many different perspectives, and its investigation can only be hindered by the commitment to a single authoritative, hierarchical perspective.

I have also argued that this philosophical position has profound implications for our approach to the study of the human. We should not expect there to be a unique and foundational approach to human nature. Human nature is a philosophical abstraction that provides an umbrella for many possible empirical abstractions. Many systems of such abstractions can be found across the range of the humanities and the human sciences and many other possible approaches are very likely possible. Of course, I do not want to say that all have equal value. But each should be valued individually according to the sorts of insights into the human it can provide. Knowledge of human nature tout court, I believe, is only likely to be achieved by someone who is able to assimilate insights from a variety of perspectives. This, I suppose, is a philosophical achievement rather than a scientific one, but then, as I have said, Human Nature is not a first-level scientific abstraction but a philosophical abstraction grounded on as wide as possible a range of scientific as well as more traditionally humanistic perspectives. Naturalism, or anti-supernaturalism, provides weak but fundamentally important constraints on which such perspectives may prove the most illuminating.

NOTES

1. John Dupré, *The Disorder of Things: Metaphysical Foundations of the Disunity of Science* (Cambridge, Mass.: Harvard University Press, 1993).

2. It must be said that the extent to which chemistry is reducible to physics remains somewhat controversial. For various discussions, see Davis Baird, Eric Scerri, and Lee McIntyre, eds., *Philosophy of Chemistry: Synthesis of a New Discipline*, Boston Studies in the Philosophy of Science 242 (Dordrecht: Springer, 2005). The failure of higher level reductions, on the other hand, is not controversial.

3. C.J. Jeffery, "Moonlighting Proteins," *Trends in Biochemical Sciences* 24 (1999): 8–11.

4. In an interesting recent paper ("Top-down Causation Without Top-down Causes," *Biology and Philosophy* 22 [2007]: 547–63), C. F. Craver and William Bechtel, though they object to the expression "downward causation," preferring to conceptualize it in terms of mechanistically mediated effects, offer strong support for a similar general idea.

5. See J. L. Barkow, Leda Cosmides, and John Tooby Barkow, eds., *The Adapted Mind* (New York: Oxford University Press, 1992); and John Dupré, *Human Nature and the Limits of Science* (Oxford: Oxford University Press, 2001).

6. Here I summarize the perspective known as Developmental Systems Theory. See Susan Oyama, *The Ontogeny of Information* (Durham, N.C.: Duke University Press, 2000); or, for an extremely helpful overview, P. E. Griffiths and R. D. Gray, "Developmental Systems and Evolutionary Explanation," *Journal of Philosophy* 91 (1994): 277–304.

7. For extensive details, see T. R. Gregory, ed., *The Evolution of the Genome* (San Diego: Elsevier, 2004).

8. Outstanding sources for more up-to-date discussions of evolution than the antiquated views generally assumed by evolutionary psychologists are Eva Jablonka and M. J. Lamb, *Evolution in Four Dimensions* (Cambridge, Mass.: MIT Press, 2005); and M. J. West Eberhard, *Developmental Plasticity and Evolution* (New York: Oxford University Press, 2003).

9. For much more detailed discussions on the points surveyed in this paragraph, see Barry Barnes and John Dupré, *Genomes and What to Make of Them* (Chicago: University of Chicago Press, 2008).

10. Dupré, *Human Nature and the Limits of Science*.

11. More generally on the neglect of microbes in the philosophy of biology, see John Dupré and Maureen O'Malley, "Size Doesn't Matter: Towards a More Inclusive Philosophy of Biology," *Biology and Philosophy* 22 (2007): 155–91. For more on the ensuing difficulties in defining an organism, see John Dupré and Maureen O'Malley, "Metagenomics and Biological Ontology," *Studies in the History and Philosophy of the Biological and Biomedical Sciences* (2007): 834–46.

12. John Whitfield, "Underground Networking," *Nature* 449 (2007): 136–38.

13. Dupré, *The Disorder of Things*.

15

DEWEY, CONTINUITY, AND MCDOWELL

Peter Godfrey-Smith

John Dewey's voluminous output includes substantial work in meta-physics, epistemology, philosophy of mind, and philosophy of science. Within analytic philosophy, the implicit conclusion drawn for some decades now has been that Dewey is not especially relevant or useful in these areas. He is seen, I suspect, as well-meaning but lacking rigor and as not grappling effectively with the issues that are most alive to us now. I argue here for the contemporary relevance, as well as the intrinsic merit, of some of Dewey's ideas on these topics. The area I will look at is general discussion of the relation between "mind and world." My main focus will be on questions about belief and knowledge, as opposed to the overtly metaphysical side of the mind-body problem.

Dewey sought to assert "continuity" between mind and world. This was his response to a number of famous "dualisms" in philosophy. As the term suggests, Dewey's is a "no gulf" view, a denial of problematic separation. But this does not mean that he refused to see a problem and baldly asserted that the mind is in straightforward contact with the world at large. There is more to his treatment than that. In this case and others, Dewey's analysis took the following form. Where we see a traditional philosophical "gulf" (mind and matter, thought and world, fact and value), there tends to be a real distinction present that has a kind of functional significance. But events in the history of philosophy, in interaction with the history of science and politics, will have led to this distinction acquiring a distorted philosophical role, one in which it poses a problem in the form of a gulf or dualism. Attention to the functional significance of the distinction, and its history, enable the problem be partly solved and partly dissolved.

Some parts of recent philosophy have also sought to develop views that break with the usual forms taken by philosophical theory in this area. One example is John McDowell's *Mind and World*.[1] McDowell makes philosophical claims about perception, content, reasons, and norms. But he also engages in criticism of the way these problems are usually set up and addressed. This includes criticism of familiar forms of philosophical naturalism. McDowell's positive view is naturalistic in a sense—it involves a reconceptualization of what is natural to humans. But much of the mindset associated with the naturalistic tradition in recent philosophy is rejected.

Dewey is normally associated now with "pragmatism." But from about 1925 onward, his preferred term for his outlook was "naturalism." Dewey's naturalism is not mainstream contemporary naturalism, the naturalism of people like Philip Kitcher, Alvin Goldman, David Papineau, and Kim Sterelny.[2] It is a distinctive view. Though I think that mainstream contemporary naturalism is philosophically viable, one criticism of this approach that should be taken seriously is the claim that it is too comfortable with the terms of discussion and problem framings that originate in more traditional philosophy. In that area, some of the anxieties seen in contemporary antinaturalism may well be onto something. But rejecting naturalism itself is not the right response to the problem, and some of Dewey's ideas provide useful resources in developing a better approach. So the aim of this chapter is to present parts of Dewey's view as a *useable response* to present-day concerns, including some of those that motivate McDowell.

McDowell will be my sole representative of dissatisfaction with mainstream naturalism. Aside from the significance and influence of his work, McDowell is a good choice because the points of contact between Dewey and McDowell are rich and informative. (For this point, and for the stimulus to write this chapter, I am indebted to discussion with Bharath Vallabha.) Even the relation between the titles of the two key works is provocative: *Experience and Nature* on Dewey's side and *Mind and World* on McDowell's. A case could be made that the two books would be more accurately named with the titles switched, or at least that both books would best be called *Experience and Nature*.[3] We see in both works the attempt to dispose of "dualistic" conceptions of the relation between thought and world. Both, however, regard the familiar philosophical format for systematic treatment of the problem as misguided. Both trace various philosophical dualisms to a source or root, an ur-Dualism that generates others. In each case we see a careful treatment of experience but also a reconceptualization of nature, without which the traditional oppositions cannot be overcome. We also see a careful treatment of relation between human and nonhuman cognition and the role of culture and language in marking the difference.[4]

The next section will outline some of Dewey's central ideas, in highly compressed form. (The "Dewey" of this paper is Dewey of the 1920s and 1930s, his later but not final period.) The section after that will mount a criticism of some of McDowell's views as presented in *Mind and World*. The final section will look at a Deweyan response to some of the specific issues that motivate McDowell.

DEWEY ON MIND AND NATURE

The place to start, with Dewey, is with the idea of a problem-solving account of cognition. But we must then extend "outward" our usual picture of what that involves. We don't see problem solving as something that begins with a thinker receiving some input, some "recalcitrant experience," and terminates in cognitive adjustments that resolve the internal puzzle, or even as something that terminates in action. Instead, our treatment starts with the environmental conditions that characteristically *prompt* thought, the conditions that generate what agents experience as problematic situations. For Dewey, these are typically conditions of variability, uncertainty, and change. At the "output" end, for Dewey, we typically have action, but also the effects of action on the agent's environment. This involves the transformation of the conditions that prompted thought, the conditions that posed the problem. In tracking the state of the world and arriving at effective forms of action, agents use stable features of the world as *resources*; these make it possible to get purchase on the unstable features.

The simplest version of this picture is one in which generalizations can be made about both the "before" and "after" conditions. Problems are posed by change and instability, and intelligent action characteristically restores stability in organism-environment relations. That highly structured view, which is more marked in Dewey's *Logic* than *Experience and Nature*, seems suspiciously simple.[5] In *Experience and Nature*, we see definite generalizations on the input side, but not so much on the output side. Perhaps that is more reasonable. What is more important here is the overall "shape" of the account, where it begins and ends, and especially the idea that action is directed at transforming the prior state of external conditions and that we might give a philosophical account of the characteristic form of those changes.

Dewey gave his problem-solving account of intelligence a naturalistic basis in a combination of biology and psychology, especially nascent social psychology. He also drew in detailed ways on anthropology. From biology, Dewey took the idea of continuity between human cognition and a larger class of organic responses in animals. This larger class of adaptive responses displays, as a shallower precursor, some features of the general pattern seen in problem solving. The link to biology is also used to support Dewey's insistence that our epistemic commerce with the world develops out of various kinds of nonepistemic commerce with it and remains embedded within this larger context of interaction.

From social psychology, Dewey took the idea of a theory of communicative and cooperative action and argued for the primacy of this viewpoint in developing a theory of meaning. If we want to know what meanings are—what sort of involvement representations have with their objects, how finely meanings should be divided, whether meaning is a unified phenomenon—we should look to what sorts of properties figure in a theory of communicative action. If we marry this perspective with the idea that meaning and representation are essential to genuine thought, the combination yields a social theory of mind.

So if we imagine looking down on a world, and isolating the mental parts of it, for Dewey these are the parts of the world in which living agents engage in problem solving with the aid of a socially sustained linguistic medium of representation. Such activity has, for Dewey, a distinctive causal profile that features, as a significant variable, a sort of "extendedness" or "reach." The precursors of thought in nonhuman animals involve a low-level and comparatively "shallow" pattern of adjustment and transformation of environmental conditions. With the advent of language, the causal reach of these capacities is greatly extended and focused. What results, human intelligence, is not just something we can see as useful for us. Viewed as part of nature at large, it is a causal factor of an unprecedented kind.

That concludes my outline of Dewey's naturalistic view of mind-world relations. Dewey also gives a theory of how traditional philosophy has responded to these facts. This is a charting of a long historical process of, roughly speaking, understandable distortion of those facts, via their interaction with the development of social and economic conditions in the West from ancient times to the present. This process yields the familiar problems of modern philosophy.

A large proportion of Dewey's work is concerned with charting this history, over and over, from many angles. His method involves a particular kind of historical narrative, one with little explicit role for individuals as opposed to the general social and economic factors. Much of the narrative itself is inessential here. But the contribution that a story of this general kind makes to the overall package of ideas is important. So I will give a highly compressed sketch of the narrative.

The story starts in classical Greek thought and its relation to social conditions. Knowledge is seen as the domain of a leisured class. Accordingly, knowledge is seen as a matter of contemplation; there is a devaluation of practical experimentation and craft. For Dewey, however, it is in some ways an *inescapable* fact that knowledge, as discussed above, involves use of the stable to deal with the unstable. This fact makes its way into Greek thought in the form of a distinction in kinds of being. The unchanging is regarded as superior and more real.

As I understand the argument, Dewey describes two factors at work here. One is the treatment of knowledge in isolation from its context within our larger nonepistemic traffic with the world. That makes it possible to treat the epistemic roles of various things as reflecting their degree and kind of reality. The second is a forcing of the different epistemic roles of the changing and the stable into a framework that treats knowledge as matter of contemplation. Knowledge becomes contemplation of the unchanging, and the unchanging acquires a superior kind of reality.

We then jump forward to a profound shift in social conditions, associated with the rise of capitalism and early modern science. This produces a transformation in actual epistemic practice—in knowledge and methods. In particular, we see a new emphasis on individuality, experiment, and methods continuous with craft. We also see the discovery of the enormous power of a focus on the mechanical structure of nature—for Dewey, on mechanical *aspects* of natural affairs. But there is a failure to make corresponding shifts on the philosophical side, which would properly accommodate the new practices of knowledge gathering. The result is a retention of the idea that the objects of knowledge are the genuinely real, in combination with a new conception of what is known and knowable. What has turned out to be knowable is the mechanical order, so the things apparent in ordinary experience (colors, everyday objects, values) become philosophically problematic. The problem of mind-world relations is also exacerbated by the subjectivist side of modern philosophy, which is an exuberant distortion of the genuinely new and important role for individuality and innovation in the modern period.

So for Dewey, the standard problems of modern philosophy are largely products of a philosophical inheritance from the Greeks, combined with a new context of actual epistemic practice. In particular, the Greeks were responsible for

> the introduction into nature of a split in Being itself, its division into some things which are inherently defective, changing, relational, and other things which inherently perfect, permanent, self-possessed. Other dualisms such as that between sensuous appetite and rational thought, between the particular and universal, between the mechanical and the telic, between experience and science, and between matter and mind, are but the reflections of this primary metaphysical dualism.[6]

The argument is not that all these *distinctions* stem from the primary dualism. They are all decent, free-standing distinctions, properly understood. They have their place in real aspects of experience and our ongoing attempts to deal with it. The argument is that the appearance of a philosophical *problem* around them stems in each case from the "primary dualism."

Dewey's narrative and the associated arguments are often interesting and plausible. But to accept the diagnosis as a whole is to deny that at least many of those philosophical problems have a "life of their own," originating in genuinely puzzling aspects of experience and how the world works. It is also important that even if one has a deflationary attitude to these problems, this attitude does not *require* a "primary dualism" at the root of them all. So although some parts of it have particular force and relevance, I won't try to make Dewey's millennium-spanning diagnosis do much philosophical work in this paper. The point that Dewey thinks we have *reached* via this road is important, however. In the case of the relation between the knowing mind and the external world, we are left with a picture in which the aspiring knower is supposed to hope for the existence of a strange sui generis relation of copying, linking a private inner state to an external condition. The mental state at one end seems to be not of the natural world in both its composition and activities; both the constitution and the *point* of the prized copying relation are obscure.

What does restoring the right perspective involve? For Dewey, it does *not* involve fashioning a direct answer to the questions bequeathed to us by the philosophical tradition. That would be doomed. Further, we cannot hope that the deflationary historical account will clear away all the inherited confusions by itself, leaving us able to assert the relevant pieces of science in raw form, expecting them to give us a philosophical understanding of the situation. For Dewey—especially the Dewey of *Experience and Nature*—the way out of the problem also involves a *casting* of the naturalistic account in a particular way, a way that involves moves within metaphysics and epistemology.

My emphasis here is quite different from that of Dewey's main champion in recent decades, Richard Rorty.[7] For Rorty, Dewey's diagnostic narrative enables a simple dissolving of a great mass of unwanted philosophic matter. Basic naturalistic facts about people also contribute something of a successor picture, but little of this is actually needed. And what is not just unnecessary but downright unhelpful is Dewey's foray into systematic metaphysics and epistemology. This, for Rorty, is a lapse into a kind of theorizing that Dewey has completely undermined.

Rorty and I agree on three components that can be isolated in Dewey: the diagnostic narrative, the introduction of information from biology and psychology, and the "M&E" moves that cast the naturalistic material in a particular form. Rorty regards the third as regrettable. Sometimes he writes as if it was simply not there. But in at least one essay, "Dewey's Metaphysics," Rorty departs from what I see as a misleading interpretive idealization and instead singles out and rejects the substantive metaphysical side of Dewey's later thought.[8] My approach, in contrast, uses all three parts, and trusts in the sheer dissolving power of Dewey's historical narrative less than Rorty does.

As with the other two components, I will briefly sketch the metaphysical and epistemological framework that Dewey offers in *Experience and Nature*. This material will seem especially archaic, and it is in various ways overstructured. But I think it includes some ideas of value and contemporary relevance.

Dewey recognizes both "intrinsic qualities" and relations, and each is assigned distinct roles in his account. In ordinary experience, intrinsic qualities of things can be encountered in a non-epistemic fashion. They are "had" without being known. The business of knowledge is the tracking of relations—sequences, coexistences, patterns.[9] This because of the overall function of intelligence, which is tracking and utilizing stability to deal with uncertainty. Dewey rejects attempts to deflate or reduce one or the other metaphysical category; he rejects both the Aristotelian downgrading of relations and modern structuralist suspicions of intrinsic qualities.

The role of relations in Dewey's account is significant in several ways here. First, Dewey uses them to resolve some standard questions about where the objects of everyday experience can be located in the physical world. For Dewey, the description of the world given to us by modern physical science is a description of relational features of nature, in very pure form.[10] That has the consequence that the physical description is *designed to omit* much of what is real.[11] Science's focus on one set of natural features does not entail rejection of the other. Indeed, Dewey accepts the simple metaphysical argument that relations need positive relata.[12] So such things as aesthetic properties, colors, and so on receive a comparatively easy ride in this kind of naturalism— probably too easy a ride. Dewey's view of physical description also enables him to adopt (what I interpret as) a form of neutral monism about the mind-body problem in its narrowly metaphysical form.[13]

The idea that knowledge has a special concern with relations also has consequences regarding what sort of connection with the external world is *sought* by a knower. As a first pass, we might say that a knower wants to represent facts about relations. But this brings the knower himself into a new set of relations with external things. So the knower joins and transforms the relational structure, and this is so even before the new knowledge has issued in actions that result in changes to external affairs by a commonsense causal standard. This move is not supposed to imply a collapse into an undifferentiated holistic metaphysics. Most of the possible ways in which a knower might come into a new set of relations with external conditions are empirically idle, insignificant. The ones that are not idle are those that involve genuine new potentialities for action and causal impact on the world.

This completes my sketch of Dewey's account of mind and knowledge, how traditional philosophical problems have arisen, and the proper philosophical framing of a naturalistic view. I do not endorse anything like the whole of this package. I would reject the simple way in which the relational and intrinsic are assigned distinct roles in epistemology, where the intrinsic qualities are "had" and the relations are known. Second, the idea that physics' focus on relations makes available a neutral monist view of the mind is intriguing but seems somehow too easy. Third, Dewey struggles with the need to assert but not overinflate the idea that changes to the content of knowledge imply changes to the objects known. But Dewey's ideas, even where outdated and imperfect, offer valuable vantage points from which to view philosophical problems, and in many cases offer resources with considerable power. Dewey did succeed in thinking his way outside of frameworks, pictures, and habits whose influence is strong and continuing.

For example, Dewey's naturalistic treatment of the impact of mind and intelligence on the world provides him with the basis for a striking diagnosis and critique of "idealist" philosophies, which I would also apply to some contemporary "social constructivist" views. For Dewey, these are philosophies that recognize, in distorted form, the active function of intelligence but try to make sense of this role within a view that neglects the causal role that the mind has in the control of action and consequent transformation of environmental conditions.[14] For Dewey, questions of whether reality is "found" or "made" can be answered by saying that we find the world in one state and, via action, we make it enter another.

Another part of his picture is especially relevant here. For Dewey, ordinary distinctions become philosophical problems not solely through mishandling. Both ordinary distinctions and their bloated philosophical analogues tend to have a kind of functional significance. One example is the gulf between mind and world. Early in *Experience and Nature* Dewey notes that everyday experience is *of* objects in the world. It takes a definite mental operation to sever experience from nature and treat the mind as a self-contained domain. This is the conception of experience that troubles philosophy, of course, but Dewey does not merely see it as a misstep. The drawing of a line between mind and nature in the modern period reflects a new pattern in actual epistemic affairs. Thinking of the mind and the physical world as self-sufficient domains turned out to have enormous practical power. "To distinguish in reflection the physical and to hold it in temporary detachment is to be set upon the road that conducts to tools and technologies, to construction of mechanisms, to the arts that ensue in the wake of the sciences."[15] The problem in philosophy has arisen not by according the distinction between mind and world a special status but by allowing the distinction to "harden" in a particular way.

CRITICISM OF *MIND AND WORLD*

In this section I turn from Dewey to McDowell. I focus primarily on some crucial moves made toward the end of *Mind and World*. But I sketch some background first.

In his first chapters McDowell argues that there are aspects of human reason and belief that cannot apparently be captured by a description of humans in ordinary naturalistic terms. It is assumed that the modes of description and understanding available to a mainstream naturalist involve locating things in a "realm of law," where this is distinct from the idea of a realm of cause-effect relations. Via Kant, McDowell sees thought as a combination of receptivity and spontaneity. The understanding is a faculty of spontaneity— an active power and one linked to freedom. This is because belief involves taking responsibility, taking a stand on how things are. So the overall problem can be summarized by saying: "we cannot capture what it is to possess and employ the understanding, a faculty of spontaneity, in terms of concepts that place things in the realm of law" (*MW*, 87).

McDowell's solution to the problem has two elements. One is a detailed treatment of experience as a passive capacity but one linked to understanding (and hence to spontaneity) in a crucial way. McDowell argues that experiences have conceptual content, a kind of content that gives them genuine justificatory capacities. Much of the subsequent debate has concerned the viability of this idea, but it will not figure much in my discussion.

The second aspect of the solution is a set of metatheoretic claims. I will break these also into two. First, we have McDowell's isolation of the true source of the problem. "I have now introduced my candidate for that role: the naturalism that leaves nature disenchanted" (*MW*, 85). A genuine resolution of the problem includes a reconceptualization of what counts as natural, for philosophical purposes. This involves the recognition of what McDowell calls "second nature," which is a certain kind of human involvement in practices of judgment, interpretation, and reason giving, acquired by enculturation and immersion in a tradition. Enculturation or *Bildung* brings us into a condition where we can recognize and work within the "space of reasons."

Last, we have the claim that our philosophical treatment of this situation should not take the form of "normal philosophy," featuring a familiar kind of bridging operation between two sets of facts that seem problematically distinct. Second nature is properly understood from a standpoint *within* the set of practices it involves, as opposed to a "sideways-on" standpoint. If a challenge is raised that alleges a "spookiness" or "rampant Platonism" in this picture, the right response is not to try to rebuild the space of reasons with the minimal tools that mainstream naturalism allows. That is the kind of operation that is no longer needed. "We need not connect this natural history to nature as the realm of law any more tightly than by simply affirming our right to the notion of second nature" (*MW*, 95).

My focus will be on these abstract and partially metatheoretic aspects of McDowell's view in this section. In the next section I will grapple with some of the particular issues that motivate McDowell's claims about meaning and justification.

There is a point in *Mind and World* of usefully close contact with Dewey. McDowell notes that modern philosophy is characterized by problems that take the form of dualisms. One view of Wittgenstein sees him as uncovering a fundamental dualism, a "dualism of norm and nature," that is the "source of the familiar dualisms of modern philosophy" (*MW*, 93–94). McDowell says this fits the picture he has been developing, but he then urges that we reject the bridging operations that are the usual response to problems of this kind. These are operations that start out with the materials on one side of the divide, and erect something that more or less passes for the facts or features associated with the other side. McDowell thinks he has given us warrant for treating any such operation as unnecessary.

Here is a key passage:

> The naturalism of second nature that I have been describing is precisely a shape for our thinking that would leave even the last dualism not seeming to call for constructive philosophy. The bare idea of *Bildung* ensures that the autonomy of meaning is not inhuman, and that should eliminate the tendency to be spooked by the very idea of norms or demands of reason. This leaves no genuine questions about norms, apart from those we address in reflective thinking about specific norms, an activity that is not particularly philosophical. There is no need for constructive philosophy, directed at the very idea of norms of reason, or the structure within which meaning comes into view, from the standpoint of the naturalism that threatens to disenchant nature. . . . We need not connect this natural history to nature as the realm of law any more tightly than by simply affirming our right to the notion of second nature.
>
> (MW, 94–95)

To assess these claims, let us first ask what McDowell's theory of second nature is intended to be a theory *of*. It is a description of *our* second nature, a description of a set of skills and habits that are natural to properly enculturated humans. As I read McDowell, these might be described as habits of interpretation, judgment, reflective epistemic responsibility, and sensitivity to relations of entailment and support. In a low-key way, I will refer to this as a "framework" that we use in judgment, reflection, and interpretation. If "framework" seems problematic then "habits of thought and action" can be substituted in its place (*MW*, 84). I will argue for the viability and importance of a form of investigation of these features that McDowell seeks to resist.

For McDowell, the way in which second nature counts as natural to us makes it unnecessary to engage in a philosophical project of locating these features in the world as conceived by science. "The bare idea of *Bildung* ensures that the autonomy of meaning is not inhuman." But what does the bare idea of *Bildung* really do? For McDowell, "*Bildung*" is the best available term for a kind of enculturation process in which a person is made sensitive to reasons, and to features of the world that depend on the notion of reason, such as meaning and inquiry. Surely all the *bare* fact of Bildung establishes is that normal processes of enculturation lead to us *finding it natural* to apply the framework, to our exhibiting certain habits. So one observation that can be made immediately is that the bare fact of *Bildung* does not preclude the framework we acquire by normal enculturation from embodying factual commitments that are false. A process of enculturation *could* surely yield such a thing.

That possibility may not arise in this case; it may be an error to treat the framework that McDowell has in mind as one that embodies factual commitments. The set of habits and capacities that constitutes second nature might be quite different in character. But this difference does not affect the fundamental point because an analogous question can then be asked about second nature: What, as a matter of fact, does the framework do for us, what sort of coordination does it give us with one another and with the world at large. This coordination may not go via factual commitments, true or false, but it has some character or other. What difference does it make to us as organisms that we have this feature, and does how it affect our modes of interaction with the rest of the natural world?

To ask this is to take a "sideways-on" perspective toward our second nature, something that McDowell wants to resist. But what grounds has he to resist it? First, we should deflect the possible claim that taking such an approach lands us back in the familiar gulf-bridging projects of modern philosophy. The investigation may be carried out in this way, but it need not. It could be combined with a deep critique of the standard ways in which these problems have usually been generated and addressed in the philosophical tradition. That is, it could be undertaken within an unorthodox naturalism of the kind we see in Dewey. McDowell sometimes associates the idea of a sideways-on perspective with the idea of an "outer boundary" separating thought from reality at large. But the idea of such a boundary can receive a critical treatment within a perspective that is nonetheless comfortable with a sideways-on point of view. Here Dewey's relevance is vivid. Sideways-on need not mean boilerplate-dualistic.

For McDowell, however, there is enough on the table to head off any attempt to undertake a further philosophical investigation of second nature. Drawing still from the passage above:

> This leaves no genuine questions about norms, apart from those we address in reflective thinking about specific norms, an activity that is not particularly philosophical. There is no need for constructive philosophy, directed at the very idea of norms of reason, or the structure within which meaning comes into view, from the standpoint of the naturalism that threatens to disenchant nature.

One way to read this passage is to see it as directed specifically against a form of naturalism that is organized around the notion of natural law. If this was so, it would be easy to reply that many naturalists—not just unorthodox ones like Dewey—would deny that laws are central to the scientific outlook, either in general or in the context of understanding human capacities. However, I think it is clear that McDowell aims these points at a bigger target than the specific combination of naturalism plus a law-based philosophy of science. Indeed, later in the book it is not only a philosophical treatment of human enculturation that is resisted by McDowell. He seems averse to *any* kind of systematic theoretical investigation of the matter, of the sort that could threaten to generate philosophical consequences: "We can regard the culture a human being is initiated into as a going concern; there is no particular reason why we should need to uncover or speculate about its history, let alone the origins of culture as such" (*MW*, 123). A scientific treatment of the evolution of culture is possible, but for McDowell it is not very "pressing" (*MW*, 123). And it must not become the basis for any sort of systematic third-person account "of what responsiveness to reason is" (*MW*, 124)

This aversion to theoretical investigation, spreading here from systematic philosophy even to neighboring disciplines, is surely a weak spot in McDowell account. It aims to deter, for example, the following kind of inquiry: Suppose we have a philosopher who has taken full heed of the missteps in the tradition, especially the erecting of boundaries between thought and nature, also someone who does not think of science as obsessed with locking events into laws. What the philosopher wants to do is ask general questions about how the "habits of thought and action" involved in our use of normative concepts relate to other facts about us and how these habits function as human cognitive tools. When this philosopher says that such an investigation should mesh with what we learn from science, do not think "physics" when he says "science." Instead, think social psychology, a field that overflows with the most startling results almost untapped by philosophy. Think comparative psychology, which has recently become intensely concerned with the ways in which various nonhuman animals have *partial* analogues of the key human characteristic of cultural learning and is intensely concerned with how and why the human lineage took an extra step.[16]

McDowell, or others, may think that even within an obsessively nondualistic framework there is something impossible or incoherent about a sideways-on investigation of our basic normative and semantic concepts and habits. It is true that some of the features of second nature that McDowell wants to place off-limits from systematic investigation might be said to be truly ground-level, in a way that makes it hard to imagine asking what human life would be like without them. But whether this is granted or not, other elements of what McDowell regards as second nature are far more specific and cannot receive this protection.

In the key passage quoted earlier, for example, McDowell says that the bare idea of *Bildung* "should eliminate the tendency to be spooked by the very idea of norms or demands of reason." But at this point we should insist on a distinction between the "very idea of norms," of any kind at all, and the particular normative framework that includes "demands of reason." Even if it is somehow misconceived and fruitless to try to ask sideways-on questions about what difference is made to our lives by the existence of any normative structure whatsoever, it is a much more specific question to ask what difference is made to our lives by an acceptance of the idea of a *demand* of reason. Here and elsewhere, the structure that McDowell sees us merely asserting our "right" to is richer and more specific than a minimal skeleton of normative orientation. It includes a particular *way* of configuring our norms of judgment, reflection, and interpretation. Indeed, it is clear that enculturation does in fact give people a framework with this kind of richness.

Another illustration of the substantive character of the inheritance that McDowell would have us affirm our "right" to is the framework of psychological description and interpretation that McDowell recognizes. This is especially vivid in McDowell's handling of the cognitive attributes of nonhuman animals.[17] As McDowell denies that animals have understanding, he denies that they have experience of the world or beliefs about it. But he is also quick to deny that this forces him to treat animals as automata. He makes what I read as two distinct moves here. First, he says that animals have a kind of "proto-subjectivity." I assume that this entitles them to weakened analogues of propositional attitudes. A different concession is made, however, with respect to states like pain and fear. Animals can be attributed the ordinary forms of these states, not weakened or scare-quoted analogues. Animals can feel pain and fear because "nothing in the concepts of pain and fear implies that they can get a grip only where there is understanding, and thus full-fledged subjectivity" (*MW*, 120). This two-pronged treatment of animal cognition contrasts with a one-pronged treatment, according to which *all* ordinary concepts of psychological states, including pain and fear, are similarly enmeshed in the rich normative structure in which the concepts of belief and reason live.

Suppose for a moment that the one-pronged option is true. That is, suppose the one-pronged view is a true expression of some of the constraints that govern the habits of thought and action that constitute second nature. (Indeed, it has been suggested to me that this is a better expression of McDowell's overall picture, as reflected especially in his other works.) Whatever it is that bars the attribution of belief to animals also bars the attribution of fear and pain. On this second view, it is clear that our second nature is shaping the contours of coherent psychological description in quite definite ways, and in ways that have downstream consequences. The same is true, though not quite as vividly, of the two-pronged view as well. My point is that while it may be true that one or other of these frameworks "feels natural" to us, given our enculturation, as a way of navigating the normative and semantic domain, that surely should not be the end of the story. We can and should ask about how the contours generated by this framework of psychological description relate to the contours revealed by empirical psychology, and we can also ask whether our usual framework of psychological description is the best one for us to use, given our goals.

So McDowell's second nature includes a mix of more skeletal and more fleshed-out elements. And once we focus on the substantive elements, it is evident that asserting our "right" to them cannot be the end of the matter. McDowell might not be interested in sideways-on questions about how our usual normative framework operates in social life, and where various elements of the framework come from, but he cannot claim that any attempt to ask these questions is doomed to devolve into a pointless dualistic oscillation between unacceptable options.

ABOUTNESS AND IDEALS

I have sketched Dewey's view of the relation between mind and world, and I have criticized some parts of McDowell's. But of course, Dewey and McDowell are focusing their attention on different parts of this huge topic. It is not the case that they are trying to answer exactly the same questions and giving different answers. McDowell's project is guided by subtle features of what he takes to be our core semantic and epistemic concepts, especially normative features of those concepts. This sort of material plays little role in Dewey's treatment. In part this is because Dewey is more "zoomed out" than McDowell is at this part of his story; in part it is because Dewey is not thinking about meaning with anything like the sophisticated armamentarium of late-twentieth-century philosophy. In any case, it can sometimes seem that Dewey is just steamrolling over the subtle features of the landscape that are guiding McDowell's project. So in this section I will say more about what an unorthodox naturalism of Dewey's kind would make of some phenomena central to the argument of *Mind and World*.

For Dewey, the way to approach the phenomenon of meaning is through a theory of communicative behavior. Meaning enters the world when first behaviors and then persisting artifacts acquire a role in guiding and stabilizing joint action, especially action directed at other things. The promissory notes Dewey leaves here are huge, but Brian Skyrms's recent work includes a more rigorous treatment of this basic idea.[18] In particular, Skyrms's work includes a careful treatment of the relation between cooperative and competitive aspects of the process.

As noted earlier, Dewey committed himself also to the view that communicative interaction of this kind is essential to the existence of genuine thought. This is another point of interesting contact with McDowell, but I will not address it here. My focus instead will be on how a Deweyan approach would treat the distinction between "genuine" meaning and representation, as opposed to more rudimentary kinds of involvement between inner states (or other quasi-representational structures) and the world.

A distinction of this kind is a central feature of McDowell's treatment. McDowell insists on the importance of a particular "demanding" sense of belief, which he thinks is tied to the idea of genuine "aboutness." As McDowell emphasizes in his "Afterword," he does not claim that no simpler forms of directedness or involvement between inner states and conditions in the world could exist and figure in theories of various kinds. But he insists that the richer senses of belief and aboutness are real and philosophically important. In particular, these semantic concepts are bound up with epistemic concepts of responsibility—with the idea of "taking a stand" on how the world is—and hence with the idea of freedom.

I have already criticized parts of the view that McDowell reaches by following this road. But how do these distinctions and connections, that seem so vivid and important to McDowell, appear from a Deweyan point of view? Are they to be swept aside, treated as unwanted creatures of philosophical fantasy?

For Dewey, the richest and most fine-grained semantic phenomena are seen as arising out of simpler kinds of behavioral coordination and sign use. Verbal behaviors and representational artifacts go from having one-time or haphazard roles in particular episodes of behavioral coordination to having stable and persisting roles in a community of agents. This is a process in which objects acquire, via their embedding in a behavioral context, a novel kind of causal role: functioning as a representation. For Dewey this is an enriching of the causal powers of ordinary objects; acquiring meaning is acquiring the capacity to affect the course of events in a particular kind of way. This change in causal power is a matter of degree. The new causal roles of objects used as representations begin as diffuse and unreliable ones, but once stabilized they can have both longer "reach" and greater focus.

While it is hard to make these ideas precise, the core point is that, for Dewey, there can be more and less "demanding" senses of semantic concepts like meaning and aboutness, but these should correspond to varieties of semantic phenomena that are empirically different. They should be linked to more and less fine-grained ways in which signs are used in social life. Further, this need not involve merely a continual "ramping up" of the same *kinds* of causal powers. Innovations and qualitative changes can be recognized as real in such a framework.

Here are two examples of such qualitative changes. First, in humans we find not just first-order representation use, but thought and talk *about* representations. The empirical phenomena of language and thought include the existence of a framework that we use to describe, predict, influence, and manage the representation use of ourselves and others. Second, we find the partial entanglement of semantic concepts with epistemic concepts and perhaps also with those of responsibility and freedom. Given the causal importance of stabilized representations, it is natural for us to hold people accountable for what they say, for what they overtly "take a stand" on. With this may also come an encouragement of internal attitudes of epistemic responsibility. Thus it may happen that our concepts of belief and aboutness become bound up with broader ideals of responsibility and with scripts of justification and normative assessment.

I suggest that McDowell overstates the extent of these entanglements, and they are probably much more open and disorderly than McDowell makes them appear. McDowell binds up our network of semantic and epistemic concepts as tightly as he possibly can, snipping off an array of mixed and intermediate uses, in the hope of isolating an inescapable structure. But from a Deweyan point of view, there is no need to deny that there is *something*, a real phenomenon, here. If real, this phenomenon will have its own empirical features, discernible from sideways-on.

So much of the philosophical heavy lifting that McDowell wants to engage in with the aid of this material is rejected. But the idea that there is a link between semantic concepts and ideals of justification and responsibility need not be simply steamrolled or ignored by a naturalism of this kind. These phenomena acquire, when real, a different role in the overall story.

NOTES

This paper was given as the keynote address at the 2006 Harvard/MIT Graduate Philosophy Conference. I am grateful to those present for valuable criticisms. I am also grateful to Liz Camp, Richard Rorty, Jane Sheldon, Dmitri Tymoczko, Bharath Vallabha, and Kritika Yegnashankaran for comments on an earlier draft.

1. John McDowell, *Mind and World* (hereafter *MW*) (Cambridge, Mass.: Harvard University Press, 1994).

2. Philip Kitcher "The Naturalist's Return," *Philosophical Review* 101 (1992): 53–114; Alvin Goldman, *Epistemology and Cognition* (Cambridge, Mass.: Harvard University Press, 1986); David Papineau, *Philosophical Naturalism* (Oxford: Blackwell, 1993); Kim Sterelny, *Thought in a Hostile World* (Oxford: Blackwell 2003).

3. Dewey eventually said that it might have been better to call his book something like "Nature and Culture," which also would have been as good or better as a title for McDowell's book.

4. Dewey's name does not appear at all in McDowell's book, although in his "Afterword" McDowell is willing to be called one kind of "pragmatist" (*MW*, 155). One facet of Dewey's thought appears indirectly in McDowell's discussion, filtered through Rorty. The differences between Rorty's treatment of Dewey and mine are discussed below.

5. John Dewey, *Logic: The Theory of Inquiry* (New York: Henry Holt, 1938); reprinted in Jo-Ann Boydston, ed., *John Dewey: The Later Works, 1925–1953*, vol. 12 (Carbondale: Southern Illinois University Press, 1986); John Dewey, *Experience and Nature*, rev. ed. (New York: Henry Holt, 1929).

6. Dewey, *Experience and Nature*, 123–24.

7. Richard Rorty, *Consequences of Pragmatism* (Minneapolis: University of Minnesota Press, 1982).

8. Richard Rorty, "Dewey's Metaphysics," in *New Studies in the Philosophy of John Dewey*, ed. S.M. Cahn (Hanover, N.H.: University Press of New England, 1977), 45–74; reprinted in *Consequences of Pragmatism* (Minneapolis: University of Minnesota Press, 1982), 72–89.

9. "Things in their immediacy are unknown and unknowable, not because they are remote or behind some impenetrable veil of sensation of ideas, but because knowledge has no concern with them. For knowledge is a memorandum of conditions of their appearance, concerned, that is, with sequences, coexistence, relations" (Dewey, *Experience and Nature*, 86).

10. See also Peter Godfrey-Smith, "Dewey on Naturalism, Realism, and Science," *Philosophy of Science* 69 (2002): S1–S11.

11. "Physical science does not set up another and rival realm of antithetical existence" (Dewey, *Experience and Nature*, 136). "It is only with respect to the function of instituting connection that the objects of physics can be said to be more 'real'" (139).

12. Ibid., 87.

13. Dewey in *Experience and Nature* accepts the term "emergentist."

14. "Hence an office of transformation was converted into an act of original and final creation" (Dewey, *Experience and Nature*, 158).

15. Ibid, 10.

16. Michael Tomasello, *Cultural Origins of Human Cognition* (Cambridge, Mass.: Harvard University Press, 1999).

17. See, especially, *MW*, 114–21.

18. Brian Skyrms, *Evolution of the Social Contract* (Cambridge: Cambridge University Press, 1996).

16

WITTGENSTEIN AND NATURALISM

Marie McGinn

The concepts of human nature, natural human reactions and interests, and human natural history are pervasive in Wittgenstein's later philosophy. The idea that our language-games rest on "very general facts of nature,"[1] or that if certain things were different from what they are, "this would make our normal language-games lose their point" (*PI*, 142), is an important theme of his later work. The notions of *form of life* and *language-game* and the emphasis on our everyday practice, our education and training, and the application of linguistic techniques all serve to draw the reader's attention to "the spatial and temporal phenomenon of language" (*PI*, 108), to our life with language, to language as it is woven in with a multitude of human activities. The emphases on description, attention to particular cases, and coming to command a clear view of our use of expressions, and the suspicion of abstraction and idealization, are fundamental to the later Wittgenstein's conception of philosophical method. In all these ways, and for all these reasons, Wittgenstein's later philosophy gives a strong impression of embodying some form of philosophical naturalism: it is concerned not with questions of justification or with foundations but with a realistic understanding of what we do. However, the questions of exactly what form Wittgenstein's naturalism takes and of exactly what role it plays in his philosophy are not easy to answer. Yet the issue is central for the interpretation of Wittgenstein's work. It is, for example, key to the interpretation of his remarks on rule following and his treatment of psychological concepts, the topics I will focus on in this paper.

◎ ◎ ◎

Exploiting the naturalistic strand in the later philosophy has been central to attempts to reply to Kripke's claim that Wittgenstein's work contains the ingredients for the construction of a skeptical argument whose conclusion is that there are no facts about meaning, that there is nothing that makes the attribution of meaning, either to oneself or to another, true.[2] However, it is possible to discern two quite different conceptions of the form of naturalism that Wittgenstein is held to believe provides the means to escape from the skeptical paradox that he articulates: "No course of action could be determined by a rule, because every course of action can be made out to accord with the rule" (PI, 201). In both conceptions, Wittgenstein is seen as intending to provide genuine relief from the paradox. He is held to show that there is a viable conception of a rule and of acting in accordance with a rule that survives his undermining of both the idea that there is such a thing as a rule that "forces an application on us" (PI, 140) and the idea that understanding is a state "which is the *source* of correct use" (PI, 146). The two conceptions disagree, however, about both the conceptual resources that are employed in Wittgenstein's form of naturalism and the philosophical objectives that his appeal to some form of naturalism is intended to accomplish.

On the one hand, there is the view, put forward, for example, by Crispin Wright,[3] that the central notion in Wittgenstein's naturalism is that of *primitive dispositions* and that this notion is employed in a constructive account of what constitutes meaning or of what rule following consists in, which is, in some respects at least, at odds with our prephilosophical understanding of what meaning and going by a rule amount to. On the other hand, there is the view, put forward, for example, by John McDowell,[4] that Wittgenstein's naturalism does not retreat to a set of nonnormative concepts and proceed to construct an account of what meaning or going by a rule consists in. His naturalism, rather, is merely intended to show that our ordinary concepts of meaning and understanding, or intentionality generally, depend for their sense upon a background of normative practices, which are part of the natural history of human beings and which we are brought by our training to participate in. The difference between these two conceptions of Wittgenstein's naturalism can, in part at least, be traced to the different concepts of nature that inform the interpreters' understanding of what Wittgenstein means when he appeals to our natural responses or suggests that "there is a way of grasping a rule which is *not* an *interpretation*" (PI, 201), that "when I obey a rule, I do not choose. I obey a rule blindly" (PI, 219).

Wright is implicitly operating with what might be called the "standard" concept of nature, on which to describe a response as "natural" is essentially to conceive of it nonnormatively, as the expression of a "primitive disposition." In this understanding, the aim of Wittgenstein's naturalism is philosophically substantial: to escape the apparently catastrophic consequences of the realization that there is no such thing as grasping a rule that cannot be interpreted by providing an account of what meaning consists in that starts from a purely naturalistic base, with "naturalistic" construed in a way that conforms to this standard, or restrictive, conception of nature. The aim is to show how normative talk of meaning and intentional states can be grounded in the mere contingent fact of agreement in primitive dispositions to respond. McDowell, by contrast, is operating with a concept of nature that is broad enough both to accommodate those activities of human beings that are essentially described in normative terms and to recognize that a human being initiated into these practices comes to inhabit a normative space, so that a subject's natural responses, as such, can be understood as an expression of his recognition of what a rule requires of him. In this understanding, the aim of Wittgenstein's naturalism is essentially therapeutic: to show how human performances that are in themselves "nothing but 'blind' reaction[s] to a situation" can also, in themselves, be "case[s] of going by a rule" (MVR, 242). The aim is to show that we can acknowledge normative relations between an individual's intentional states and what he goes on to say and do without thereby committing ourselves to the Platonist mythology of rules as rails. The question is which concept of nature, and thus which form of naturalism, is operative in Wittgenstein's thought. Before we answer it, however, we need a firmer grip on the two ways of understanding Wittgensteinian naturalism.

◉ ◉ ◉

Wright believes that Kripke is mistaken in thinking that Wittgenstein's aim is—problematically—to call into question the reality of one's meanings. Wittgenstein's response to the paradox that he articulates (PI, 201) is, Wright points out, not to offer an accommodation with it but to reject it on the grounds that it depends upon a faulty premise: "the idea that determinacy of meaning depends upon *interpretation*" (RI, 85). The fact that we follow rules—or mean addition by "plus"—is not in question; what is in question is what following a rule—or meaning addition by "plus"—actually consists in. However, Wright also believes that Kripke is correct in thinking that Wittgenstein takes his reflections on rule following to uncover genuine, philosophical difficulties. According to Wright, these difficulties concern our prephilosophical understanding of the nature of intentional states, and he believes that Wittgenstein's response to them leads him to ideas about the nature of such states that are, to some extent at least, revisionary of ordinary understanding of them. The difficulties that Wittgenstein uncovers arise in connection with the need to reconcile the fact that we have noninferential, authoritative knowledge of our own intentions—and are thus in a position to reply authoritatively to Kripke's skeptic—with the fact that the identity of our intentional states "is constitutively answerable to . . . (subsequent) capabilities and behaviour in a fashion which is broadly analogous to that of dispositional states" (*RI*, 86).

The difficult question that Wittgenstein is alleged to pose is how there can be quasi-dispositional states, whose identity depends upon what a subject subsequently says or does and that are noninferentially accessible to the subject: "How could there be . . . a state, available immediately to the subject, apt for authoritative avowal and non-inferential recall, yet possessing determinate, potentially infinite content?" (*RI*, 125).

In this interpretation, Wittgenstein identifies a real philosophical problem—the problem of understanding how there can be states with the properties of intentional states—to which his naturalistic account of the nature of intentional states is intended to provide a satisfactory solution. With the naturalistic solution to the problem in place, we are in a position to give a straightforward reply to the skeptic: "Since I can know my present intentions non-inferentially, it is not question-begging to respond to the sceptic's challenge to my knowledge of my past intentions to reply that I may simply remember them" (*RI*, 126). For Wright, the real problem is "seeing how and why the correct answer [—that we noninferentially know of our present meanings and intentions, and may later noninferentially recall them—] can *be* correct" (*RI*, 177).

On Wright's view, the central concern of the *Philosophical Investigations* is with the nature of intentional states, and its principal target is a picture of the mind that Wittgenstein believes comes very naturally to us and that is deeply embedded in our ordinary thinking. Wright characterizes this picture as "roughly Cartesian." According to the picture, the mind constitutes a realm of objects—sensations, moods, emotions, and intentional states—which can be brought under concepts and made the objects of introspective awareness.

We are concerned here specifically with intentional states. Wright believes that the intuitive appeal of the Cartesian picture arises in part from its apparent ability to explain one of the distinctive marks of concepts of such states: that their fist-person use is constituted by avowals, which are standardly accepted as groundlessly authoritative. Our prephilosophical idea is that our noninferential knowledge of our own intentional states is based on privileged introspective access to states that already anticipate what will, in the future, constitute acting in accord with them.[5]

Our prephilosophical idea is thus held to presuppose what Wright calls a "Platonized" conception of intentions. According to Wright, Wittgenstein undermines this Platonistic, prephilosophical idea by observing, first, that there is, in general, no distinctive phenomenology associated with being in an intentional state with a specific content and, more important, that whatever phenomenological features such states do have they cannot "sustain the kinds of internal connection with aspects of a subject's (subsequent) doings and reactions which mental states of this kind essentially sustain" (RI, 296). Nothing that could plausibly count as an event in consciousness is either necessary or sufficient for a subject to count as suddenly coming to understand a word, "plus," as meaning addition or as intending his interlocutor to continue "1,002" after "1,000" in response to the order "add 2." That is to say, nothing that is before a subject's mind guarantees that he will go on in any particular way or that the concepts he applies to himself actually apply to him. The question is: "How, if nothing happening in the subject's consciousness is uniquely distinctive of the concept that comes to apply to him, can he be in *position* to apply it, usually with complete confidence?" (RI, 133). Wittgenstein's naturalism, it is held, provides the answer to this question.

According to Wright's understanding, Wittgenstein's naturalistic response to the question has two parts. First of all, there is the move away from thinking of understanding as an occurrent mental state and toward thinking of it in more "functional" terms. We need to think of coming to understand a word as "the onset of some kind of capacity or complex of capacities, or of a range of dispositions" (RI, 135). Wittgenstein's idea "that there is a way of grasping a rule which is not an interpretation" is to be understood as an appeal to "certain subrational propensities towards conformity of response, towards 'going on in the same way,'" "a whole plethora of naturalistic classificatory dispositions," (RI, 124) where "naturalistic" denotes that these classificatory dispositions are not to be understood in terms of mastery of a rule of classification, but as brute dispositions to respond in a certain way—the way that "seems right"—in new cases. The idea of normative constraints on what a subject says and does are absent from this naturalistic base. In order for an individual's classificatory responses to come under a normative constraint, and thereby qualify for assessment as correct or incorrect, some account of what constitutes correctness or incorrectness, which is independent of his natural responses, needs to be given. However, the immediate problem that concerns Wittgenstein, Wright believes, is that his naturalistic conception of understanding, or of intentional states, as grounded in primitive (nonnormative) dispositions to respond in certain ways in the future, appears to make "a mystery out of the phenomenon of first-person *avowal*" (*RI*, 136). In what, according to Wittgenstein's naturalistic conception of intentional states as brute dispositions to respond to future situations in certain ways, does a subject's special authority concerning his own intentional states reside? How can a subject be in a position authoritatively to report what the content of his current intentional state is, given that there is nothing that puts him in a privileged position to know or predict how he will respond in the future? This takes us to what Wright sees as the second part of Wittgenstein's response.

Wright argues that Wittgenstein believes it is hopeless to try to explain the first-person authority of avowals of intentions in terms of a subject's cognitive access to a determinate state of affairs. He believes that Wittgenstein's fundamental insight is to abandon what he calls a "detectivist" conception of avowals of intentional states and to recognize that the authoritative status of first-person avowals is not a reflection of any special authority that a speaker has concerning the content of his own intentional states. The authority of avowals is, rather, a reflection of the fact that in our language-game, other things being equal, avowals are treated as constitutive of the subject's being in the intentional state that he thereby ascribes to himself. A subject's cognition of a state of affairs, which in itself constitutes something with which subsequent behavior may either accord or fail to accord, no longer comes into the picture. A subject is simply "moved" to make an avowal of an intention, and it is part of the language-game that we play with concepts of intentional states that first-person authority is "unofficially granted to anyone whom [we] take seriously as a rational subject" (*RI*, 138).

Thus, Wright's Wittgenstein holds that "it is part of regarding human beings as persons, rational reflective agents, that we are prepared to ascribe intentional states to them, to try to explain and anticipate their behaviour in terms of the concepts of desire, belief, decision and intention," (*RI*, 138) and that it is "a fundamental anthropological fact" that, as a result of our initiation into language, we are moved to make avowals in such a way that the most satisfying framework for interpretation and prediction of behavior is one that accords authority to first-person ascriptions. For Wright's Wittgenstein, "the roots of first-personal authority for the self-ascription of [intentional] states reside not in cognitive achievement, based on cognitive privilege, but in the success of the practices informed by [our] cooperative interpretational scheme" (*RI*, 138). A corollary of this non-detectivist view of avowals is that the pattern of performance imposed on a subject by an intention he has declared himself to have "is not settled independently of his judgement of the matter" (*RI*, 142). The subject's judgment as to whether a subsequent performance accords with a former intention "serve[s] to determine, rather than objectively accord with or violate the content of his anterior intention" (*RI*, 142). The unexplained fact that the judgments that a subject is inclined to make concerning what fulfils his former intentions tend to accord with the appraisals that others, similarly trained, are inclined to make is what permits mutual interpretation.

In Wright's understanding, it is this demand for mutual interpretability that provides the normative constraints that are missing from the naturalistic base and that are needed to make an individual's performances assessable for correctness or incorrectness. The constraint of having one's sincere avowals cohere with subsequent outward performances in a way that allows one's overall behavior to be interpreted as the behavior of a rational agent is what supplies the standard of correctness, against which the question of accord between one's avowals and what one goes on to do can be judged. There is a standard of correctness against which, given, say, my avowal that I mean addition by "plus," both my avowal and my subsequent performances will be judged for correctness. However, in Wright's account, it comes, not from any normative constraint imposed by my mental state of intending to use the plus sign in accord with the rule for addition but from the demand that I go on to give answers to addition problems that others call "giving the sum," if they are to accept both my avowal as the avowal of an intention to mean addition by "+" and my subsequent behavior as acting in accord with it. This is, he concedes, "a step in the direction of a broadly Communitarian response" (*RI*, 87) to the problem of how the normative requirements of a rule are constituted. For, on his view, Wittgenstein is committed to the view that "the validity of [my] self-impressions is . . . constitutively constrained by their contribution to my ability to make sense of myself to others in my speech-community" (*RI*, 88).[6]

Wright concedes that one consequence of the form of naturalism that he finds in Wittgenstein is that "the notion we tend to have of the objectivity of meaning is untenable" (*RI*, 142). The idea that meaning can be constituted once and for all, either for an individual speaker or for a community, has been shown to be empty. We have to come to see "ourselves as the perennial creators of our concepts, not in the style of conscious architects but just by doing what comes naturally" (*RI*, 78). Wright recognizes the threat that this presents for anything worth regarding as genuine normativity. It may seem that the account reduces the requirements of a particular rule, in any particular case, to whatever the majority of speakers deem it to be. This would clearly be to abandon altogether the notion that a rule imposes requirements on our subsequent performances. The position Wright attributes to Wittgenstein would, in that case, be indistinguishable from Kripke's skeptical solution; talk of what is "correct" or "incorrect" would be a mere *façon de parler* to which nothing in reality corresponds. It seems essential, therefore, that some way be found for Wittgenstein to deny that the requirements of a rule in any particular case are actually constituted by a majority of speakers' agreeing that that is how the rule is to be applied.

However, while Wright believes that Wittgenstein does indeed deny that majority agreement determines whether a judgment is correct, he believes that there is nothing in his text that provides, from within the sort of naturalistic account he is seen as offering, anything like a full account of how the notion of the requirements imposed by a rule are constituted independently of community agreement in particular cases. We do, on occasion, judge that a consensus judgment was mistaken, and so the distinction between what is right and majority judgment is one that appears to have content within our ordinary practice. Nothing more than that, Wright suggests, can be said. What is crucial for Wittgenstein, he argues, is that we have seen through both the mythological conception of rules as rails and the Cartesian conception of intentional states as determinate states of affairs to which subjects have privileged access and which in themselves determine a standard of correctness for subsequent behavior. Having rid ourselves of these philosophical misunderstandings and grounded our talk of meaning, understanding, and going by a rule in the natural phenomenon of actual, widespread human agreement in judgment, we must reject any further demands for explanation and simply assert, with Wittgenstein, that "this language-game is played" (*PI*, 654). At this point, Wright suggests, we have to learn "one of the hardest lessons which the pursuit of Wittgenstein's later thought may have to teach us": "to know when philosophy can tell us nothing further" (*RI*, 318).

◎ ◎ ◎

John McDowell has been highly critical of Wright's interpretation of Wittgenstein. He argues that Wright's picture of how meaning and understanding are constituted "is not recognizable as a picture of meaning and understanding at all" (*MVR*, 223). Insofar as Wright's picture abandons the idea of ratification-independent patterns of application, McDowell believes that it puts our intuitive notion of objectivity under threat. We are no longer entitled to "the idea of things being thus and so anyway, whether or not we choose to investigate the matter in question, and whatever the outcome of any such investigation" (*MVR*, 222). It is not merely that the revisionary implications of Wright's picture are at odds with Wittgenstein's insistence that philosophy "leaves everything as it is" (*PI*, 124), but that the picture Wright attributes to Wittgenstein is indistinguishable from "one according to which the possibility of going out of step with our fellows gives the *illusion* of being subject to norms, and consequently the *illusion* of entertaining and expressing meanings" (*MVR*, 235). As McDowell sees it, in holding that all there is at "the basic level" is brute dispositions to respond, Wright's picture is no more successful in accommodating a recognizable conception of meaning than Kripke's skeptical solution.[7]

In the same way, he believes that Wright's claim that "there is nothing for an intention, conceived as determining subsequent conformity and non-conformity to it autonomously and independently of its author's judgements on the matter, to be"[8] is destructive of our intuitive notion of intentions, according to which "they are fully identifiable by their subject in advance of being acted on" (*MVR*, 317). Wright's fundamental error, McDowell believes, is to embrace "the picture of a basic level at which there are no norms" (*MVR*, 242). It is, in other words, Wright's conception of the naturalism that Wittgenstein intends to provide relief from the paradox that is at fault (*PI*, 201). Wright is mistaken in taking Wittgenstein to be operating with a conception of nature that is devoid of norms or with a notion of "natural response" that is entirely nonnormative, and thus mistaken in supposing that Wittgenstein's aim is to try to reconstruct normative notions from this nonnormative base. In opposing Wright's interpretation, McDowell offers a completely different understanding of the form of naturalism that is to be found in Wittgenstein's later philosophy.

The idea that McDowell believes goes missing in Wright is that an individual subject can be in a state of mind that "can fully encompass . . . within itself a determination of what counts as conformity with it" (*MVR*, 320). As Wright sees it, the only possible way of making sense of that idea is by embracing the form of Platonism that Wittgenstein has undermined.[9] For Wright, rejecting Platonism leaves us, as far as the individual subject is concerned, with a picture in which a subject's future responses constitutively determine what counts as conformity with a former intention and in which the notion of normative constraint only enters in with the idea of a context of mutual interpretation. McDowell believes that this is a complete misunderstanding of Wittgenstein. As McDowell reads him, the naturalism that provides relief from the paradox (*PI*, 201) does not involve abandoning the idea that meanings can come to mind or the idea that intentions, in themselves, impose normative constraints on future action; it merely rejects a tempting misconception of what these things amount to.

Thus, we do indeed avoid the disastrous idea that understanding is always interpretation "by stressing that, say, calling something 'green' can be like crying 'Help!' when one is drowning—simply how one has learned to react in this situation" (*MVR*, 242), but the point is to see that this natural response can, just as such, be a case of going by a rule. The fact that a speaker responds immediately, without guidance, when asked to describe the color of a given object does not mean that the description he gives is nothing more than a "subrational propensity" to react. What makes his immediate response a case of going by a rule is that the speaker has been initiated into certain normatively structured practices of employing color words, that he has mastered the relevant techniques, and that he now applies these techniques, without guidance, in accordance with the rules that he has mastered. Thus, McDowell finds in Wittgenstein what he calls "a naturalism of second nature," or a form of "naturalized Platonism,"[10] in which normatively structured activities and the capacity to be in mental states with normative implications for future behavior are seen as part of the natural history of human beings.[11]

According to McDowell's understanding, the aim of Wittgenstein's naturalism is precisely to show how to preserve the autonomy of meaning—the idea that grasp of the meaning of a word imposes normative constraints on future responses—once it is recognized that the idea of rules as rails is a myth. The autonomy of meaning never depended on our being able to ground the notions of correctness and incorrectness outside human practices or on the idea of the "super-rigid," ethereal, "logical machine": "it is wrong to suppose platonism is implicit in the very idea that meaning and intention contain within themselves a determination of what counts as accord with them" (*MVR*, 320). Realizing that the idea of rules as rails is empty is a threat neither to the idea of the autonomy of meaning nor to the idea of objectivity that depends upon it. By coming to see that "obeying a rule is a practice"—that the notion of a rule is internally linked with the idea of a customary practice of employing expressions—we are free to recognize that what makes my application of the word "green," on a particular occasion a case of going by a rule is not that my use of the word is guided by a rule (or an interpretation of it) but the fact that my response is the result of my initiation into a customary way of employing color terms. If my response is not seen in the context of my participation in a customary practice of employing expressions, then there is nothing we can point to that distinguishes it from a brute response, a piece of verbal behavior. It is only when we see it in its context, as an action undertaken by someone who has been initiated into an existing, normatively structured, linguistic practice, that we can see it as a case of applying a rule.

This account of Wittgenstein's response looks either question begging or in danger of collapsing into the kind of communitarian reading that Wright offers *only*, McDowell argues, if we presuppose a restrictive conception of nature that is unable to embrace the idea that our initiation into the normative practices that are part of human natural history results in an individual whose life has acquired a normatively structured shape. Once we recognize that by "our natural history" Wittgenstein means "the natural history of creatures whose nature is largely second nature," then we can allow him to embrace the idea that "human life, our natural way of being, is already shaped by meaning."[12] Wittgenstein's aim is to get us to see that, from within the perspective of a life that has been so shaped, through the initiation into communal linguistic practices, there are perceived normative constraints on what can be said or done, to which we respond immediately, without the intermediary of an interpretation. At the bottom of our practice lies not brute dispositions to verbal behavior but the ability of human beings to respond, in the way the practice of employing expressions requires, without guidance from a rule. Thus, when Wittgenstein says that justifications for my following a rule in the way I do run out and "then I am inclined to say: 'This is simply what I do'" (*PI*, 217), McDowell doesn't hear this as a retreat to a level where no norms operate. The point is, rather, that we apply rules without guidance, immediately, in the way we have been trained. According to this reading, Wittgenstein is never in the business of formulating a genuine philosophical problem whose solution depends upon a constructive account of what meaning and understanding consist in. Wittgenstein's naturalism is merely a series of reminders of the background against which our ordinary concepts of meaning, understanding, and going by a rule actually function; the point of the reminders is that they are "an attempt to recall our thinking from running in grooves that make it look as if we needed constructive philosophy."[13]

How does all this help us to make sense of the idea of subject's intentional states as states with which his subsequent performances may or may not accord. On McDowell's interpretation, Wittgenstein's remarks on intentional states—and on understanding, in particular—are not to be understood as putting the idea of a normative connection between a subject's intentional states and his subsequent performances in doubt, but as rejecting various candidate conceptions of what such states amount to. Wittgenstein's concern is with the danger of our falling into a problematic form of Platonism when we try to make sense of such states. This, in McDowell's view, is exactly what Wright does when he takes it for granted that there is nothing but Platonistic mythology in the idea that an intention determines independently of its author's subsequent judgments what counts as conformity with it. Wright is therefore led to suppose that Wittgenstein, in rejecting Platonism, denies that there is any normative connection between a subject's mental state and his subsequent behavior.

As McDowell sees it, Wittgenstein's aim is merely to reject a conception of intentional states as "queer" processes, processes that "in a queer way" anticipate the future. This leaves room for Wittgenstein to reclaim the notion a meaning's coming to mind, of our coming to understand a word in a flash, while getting rid of the idea that these states are uncomprehended processes that lie behind the "more or less characteristic *accompaniments* or manifestations of understanding" (*PI*, 152). The idea is never under threat that if I form the intention to multiply 25 by 25, then, given the content that my intention determinately has, only my coming up with the answer 625 will count as executing that intention. Wittgenstein's aim is merely to get us to recognize that I am capable of forming that intention only because I am party to the practices of calculation that are constitutive of the concept of multiplying. Given that I have been initiated into these practices and that I am participating in them, nothing more than the formation of the intention itself is needed to determine what counts as conformity; it does not, in particular, await upon my own later judgments about what constitutes conformity with that intention to determine its content.

According to McDowell's interpretation of Wittgenstein's naturalism, there is no problem with the idea that intentions are fully identifiable by their subject in advance of being acted on. McDowell concedes, however, that there might still seem to be a question how identification in advance is possible, given that the concepts that specify the intention's content internally connect it with the future performance that counts as executing it. One might, in other words, still be puzzled by the question: How can the subject know that *multiplying* the numbers is what he intends to do? But for McDowell, this puzzlement is no longer an expression of an urgent philosophical need to explain how there can *be* states with the properties of intentions. It is, rather, a call to bring an aspect of our language-game into clearer focus, and he thinks that here an appeal to the default status of avowals does have a role to play in Wittgenstein's thought. Moreover, he acknowledges that Wittgenstein recognizes that the default status that first-person avowals have in our language-game depends upon certain deep contingencies. However, his understanding of the nature of these contingencies and the role they play in Wittgenstein's thought is completely different from that of Wright.

For Wright, as we saw, "the contingency [is] that taking the self-conceptions of others seriously . . . almost always tends to result in an overall picture of their psychology which is more illuminating . . . than anything which might be gleaned by respecting all the data except the subject's testimony."[14] For McDowell, on the other hand, there is simply no question of a creature's vocalizations so much as counting as expressions of self-conceptions if taking them at face value does not serve the attempt to understand them: "The words (if that is what they are) cannot convey self-conceptions at all unless they can be understood in such a way as to help make sense of the subject, by being taken to express something that is available to the subject 'just like that'" (*MVR*, 318). The contingency comes in at quite another place. The contingency lies in human beings being such that they are capable of acquiring mastery of concepts—for instance the concept of multiplication—that figure in the expression of their intentions.

Insofar as the default status of avowals is partly constitutive of our language-game, mastery of the relevant concepts includes an ability to apply these concepts in the expression of intentions "in a way that meshes with, for instance, one's subsequent performances in the manner required for the self-conception in question to help to make sense of one, but without one's needing to wait and make sure of the mesh before one can know that the concept applies" (*MVR*, 318–19). It is not, of course, by observing anything that one knows one's own intentions; it would, as McDowell points out, "be more nearly right to say that one knows them by forming them" (*MVR*, 319). The deep contingency that, in McDowell's understanding of Wittgenstein's naturalism, underlies this possibility for knowledge of one's intentions in advance of acting on them "is the contingency that human beings can be initiated into the capacity to place themselves within a 'normatively' structured space of possibilities" (*MVR*, 319). The contingency is that we can master the concept—that is, the practice—of multiplication; having mastered the concept we can use it to express our intentions and go on to apply it, without guidance, in the way we have been trained to, in the way that accords (or fails to accord) with the intention we have expressed.

◎ ◎ ◎

It is clear that these two interpretations raise a question not simply about how the remarks on rule following are to be understood but also about the form of naturalism that is to be found in Wittgenstein's philosophy as a whole. The question which of the two interpretations of his remarks on rule following is correct is closely intertwined with the question what concept of nature Wittgenstein should be understood to be operating with when he remarks, for example, that "what we are supplying are really remarks on the natural history of human beings" (*PI*, 415). Is he, as Wright suggests, operating with a restrictive concept of nature that means he is describing mere non-normative contingencies out of which a revisionary, philosophical account of what meaning and rule following consist in must be constructed? Or is he, as McDowell suggests, operating with a concept of nature that embraces normative human practices and the capacities of individuals to master and employ rules, and intent merely on clarification undertaken in response to philosophical misconceptions of what meaning, intending, inferring, and so on amount to? The aim of this section is to look at the nature of Wittgenstein's naturalism and see if a case can be made for one of the above conceptions over the other. The question is, clearly, a large one that can't possibly be treated fully in the space of a short chapter. However, I hope that enough of the background can be sketched to make an informed judgment possible.

McDowell and Wright are agreed that Wittgenstein's naturalism is, on the one hand, intended to provide relief from the paradox given in *Philosophical Investigations* (201), and on the other hand, developed in opposition to a certain form of Platonism. In part 1 of *Remarks on the Foundations of Mathematics*, Wittgenstein raises the question of "what inferring really consists in."[15] He is concerned with our tendency to be "misled by the special use of the verb 'infer'" into imaging "that inferring is a peculiar activity, a process in the medium of the understanding" (*RFM*, 1:6). His aim is to get us to "look at what happens" when we infer and to recognize that "there is nothing occult about the process; it is a derivation of one sentence from another according to a rule; a comparison of both with some paradigm or other, which represents the schema of transition; or something of the kind." This is, of course, not something that goes on "in the medium of the understanding, but "may go on on paper, orally or 'in the head.'" Nor is it essential that we consult a paradigm. It may be that all that occurs is "our saying 'Therefore' or 'It follows from this' or something of the kind." He goes on: "We call it a 'conclusion' when the inferred proposition *can* in fact be derived from the premise." And he asks: "Now what does it mean to say that one proposition *can* be derived from another by means of a rule?" (1:7). Does it mean "going by the rules of inference"? Or does it mean "going by such rules of inference as somehow agree with some (sort of) reality" (1:8)? The latter idea expresses the Platonism he rejects; the former expresses a view that is opposed to this

Platonism and reflects his own naturalistic understanding of what the process of inferring consists in. The question is: How are we to understand this naturalized, non-Platonist conception of "the rules of inference"?

In the remarks that follow, Wittgenstein makes a connection between his naturalized conception of a rule and the existence of a convention. He says that "what we call 'logical inference' is a transformation of our expression" (*RFM*, 1:9): we pass, for example, from one proposition ($(x)Fx$) to another (Fa) in accord with a rule. He gives as an example of such a transformation of one expression into another: "the translation of one measure into another," for example, the translation of inches into centimeters. In this case, too, we pass from one expression to another according to a rule. But what makes a particular rule of translation "right"? Wittgenstein writes: "And of course there is such a thing as right and wrong in passing from one measure to the other; but what is the reality that 'right' accords with here? Presumably a *convention*, or a *use*, and perhaps our practical requirements" (1:9).

In this case, it is clear that "the reality that 'right' accords with" is not something abstract or ethereal but the actual use of these methods of measurement within our practical lives, the practice of using rulers marked in inches and centimeters to obtain and state results of measurement in the course of carrying out a host of different activities: building, dressmaking, carpentry, and so on. "The reality that 'right' accords with," when we transform expressions from one system of measurement into expressions of another, is our practice of employing the relevant techniques or methods of measurement in various activities within our everyday lives. It is insofar as this practice exists and is woven in with our activities that "there is such a thing as right and wrong in passing from one measure to another" (*RFM*, 1:9).

Wittgenstein's use of this example makes clear one of the principal themes of his naturalism: it is not the case that everything that can be expressed in the form of a rule for transformation of expressions has an application, and it is the application—its use within our human activities—that transforms its status from something merely empty to a rule by which we proceed.[16] What makes the translation rule "1 inch = 2.54 centimetres" a rule of inference, that is, a rule by which we *can* infer from the statement that a table is 36 inches long the statement that it is 91.44 centimeters long, is not merely that 36 inches and 91.44 centimeters coincide on a ruler; taken by itself, a ruler with marks in inches on one side and centimeters on the other means nothing. It is the fact that rulers marked in inches and centimeters have a certain employment within our practical activities and that the rule of transformation is one that we can employ within these practices that makes it a rule of inference, a rule that we can use to draw conclusions on which we can act.[17] These practices themselves are part of human natural history, they reflect not only human interests but the world that we inhabit. If we imagine these differently, then our practice of measurement loses its point. Without the background of these practices, he suggests, the rule "1 inch = 2.54 centimeters" means nothing; there is nothing corresponding to it.

The example makes clear that, in the attempt to clarify what the process of correct inference amounts to, Wittgenstein's naturalizing aim is to redirect our thoughts away from the mythology of "the logical machine" and toward the human practices in which inferring or calculating are actually employed. We need to look at "what kind of *procedure* in the language-game inferring is" (*RFM*, 1:17). It is in this sense that he tries to get us to see inferring and calculating as part of our natural history: they are procedures that human beings carry out in the course of the countless practical activities in which they engage; they are "technique[s] that [are] employed daily in the most various operations of our lives" (1:4).[18] It is against the background of these practices—our actual, concrete engagement in these procedures in the course of our practical lives—that it makes sense to speak of the existence of a rule, which in turn gives sense to talk of what *can* be inferred or of carrying out a calculation *correctly*.

There is clearly no sense in any of this that Wittgenstein's naturalism takes the form of a constructive account of what going by a rule consists in. So far, at least, the naturalizing move appears to be, as McDowell argues, aimed at overcoming philosophical misconceptions—the idea of inferring as an "occult process," the idea that rules of inference must correspond with some abstract, "very general, very rigid," reality—and getting us to see "that nothing out of the ordinary is involved" (*PI*, 94). Insofar as it makes sense to speak of the reality that corresponds to a rule, that reality is the fact that the rule is employed, or can be employed, within the activities of our everyday lives. However, this is clearly not the end of the story. For if inferring is a matter of making such transitions between propositions that accord with the rules of inferring that are employed in the various activities of our lives, how do we learn these rules? How do we know what we are to do in order to follow them? How is it determined what counts, in new cases, as being in accord with them? These questions, which Wittgenstein's naturalizing move clearly invites, once again open up the prospect of the sort of constructive account of what rule following consists in that Wright finds in his remarks.

It is central both to Wittgenstein's opposition to Platonism and to the naturalism that characterizes his whole approach to the topic of rule following that he recognizes that a "pupil's capacity to learn may come to an end" (*PI*,

143). There is nothing that *guarantees* that a pupil will acquire the capacity to go on independently in the way we all do. This raises two questions. First of all, what would happen if it became generally the case that people no longer agreed in the results of measurement or in the result that is obtained when one number is multiplied by another? The implication of Wittgenstein's naturalism is that the background that is needed to make sense of talk of correct inference, or correct calculation, would be lacking in these circumstances. That this is indeed his response to the question is clear: "our normal language-games would lose their point" (*PI*, 142); "calculating would lose its point" (*RFM*, 3:75). This may seem to bring his naturalism much closer to ideas that figure in Wright's interpretation: Our agreement in the results that we get when we measure an object, or multiply one number by another, is a happy, inexplicable contingency without which there is no criterion that determines what counts as correct measurement or being in accord with the rule of multiplication. No judgment as to the result of a given measurement or calculation would be correct if in general human beings did not agree in the answers they are inclined to give: correctness is (possibly with some qualification) constituted by the result that most people would give, and without this agreement no one's judgment can be considered correct.

It is, however, precisely this view—that human agreement plays some form of constitutive role in determining which judgments count as correct— that Wittgenstein appears to distance himself from in the following remark: "The proposition that this room is 16 foot long would not become *false*, if rulers and measuring fell into confusion. Its sense, not its truth, is founded on the regular working of measurements" (*RFM*, 3:75). That is to say, that we are expressing judgments as to the length of an object or the result of a calculation *at all* depends upon the existence of a certain consensus in the results of measuring and calculating. The agreement that forms the essential background to our talk of the length of a given object, of what follows from a given proposition, or of carrying out a calculation does not determine the *truth* of the judgments we express but their *sense*. The possibility of our expressing senseful judgments, and thus of our judgements being assessed for correctness or incorrectness, truth or falsity, presupposes a stable background of consensus in the results of measuring, inferring, and calculating. However, correctness or truth of our judgments is not itself, Wittgenstein seems anxious to stress, to be defined in terms of the notion of consensus.

Thus, the form of consensus that Wittgenstein appears to be concerned with here is not, as Wright's account implies, one that is needed in order to determine what constitutes the correct application of a rule but rather one that is essential so that our words—for example, "This room is 16 foot long"—have a sense that can be assessed against reality for truth or falsity. It is the existence of senseful judgment *as such* that depends upon our agreeing in a practice of measuring, inferring, and calculating, that is, on the fact that we all get the same results when we apply our methods of measurement, the rules of inference, or the rules of arithmetic; there is nothing to assess for correctness or incorrectness without this background of agreement.[19] This, as Wittgenstein observes, "is not agreement in opinions but in form of life" (*PI*, 241), that is to say, it is prior to questions of truth or falsity. That we can predict, on the basis of our own calculations or knowledge of mathematics, the results that someone else will get belongs to the essence of calculating: "we should not call something 'calculating' if we could not make such a prophecy with certainty" (*RFM*, 3:66). In the same way, "what we call 'measuring' is partly determined by a certain constancy in the results of measurement" (*PI*, 242). This is what it means to have a practice of using a technique: it "pertains to the essence of a technique" (*RFM*, 3:66), and thus to the phenomenon of rule following as such: "Consensus belongs to the essence of *calculation*, so much is certain. I.e.: this consensus is part of the phenomenon of our calculating" (*RFM*, 3:67). It is against the background of this natural consensus in how the methods of measurement, the rules of inference, or the rules of arithmetic are applied that it makes sense to speak of our measuring, inferring, or calculating, or of our expressing senseful judgments that are, in virtue of their sense and how the world is, true or false.

However, all this now leads to a second question. If the possibility of our utterances expressing judgments of measurement or mathematical judgments presupposes a certain consistency in results, then are the rules that define these techniques mere empirical statements about what human beings generally do? The question seems once again to open the way for something like Wright's thought that Wittgenstein provides an account of what constitutes going by a given rule in terms of how most people respond once they have received a suitable training with the rule. Does Wittgenstein's naturalizing move mean that he equates the proposition $25 \times 25 = 625$ with the proposition, Most people, when asked to multiply 25 by itself, give the answer 625? If so, then isn't there, once again, the suggestion that it is human agreement that decides what is true and what is false. The idea the $25 \times 25 = 625$ is the *right* result, the one that the rule of multiplication *requires*, or which *must* be given, seems to evaporate, and we seem, once again, to be left defining what is right in terms of human consensus. Wittgenstein seems to come very close to saying something like this in the following remark from *Lectures on the*

Foundations of Mathematics. He imagines a case in which we ask someone to multiply two very large numbers and, having arrived at a result, says "This is what I got." He goes on:

> This is not [a] mathematical proposition. How do we pass from this to the mathematical proposition: "So-and-so times so-and-so *is* so-and-so"?
>
>   It has been said: "It's a question of general consensus." There is something true in this.

However, he then goes on to ask: "Only—what is it we agree to? Do we agree to the mathematical proposition, or do we agree in *getting* this result?" And he insists that "these are entirely different" (*LFM*, 106–7). The question is what does he understand the difference to be, and what significance does he think it has.

What we must agree in, Wittgenstein goes on to claim, is "in *getting* this," "in what [we] do." He goes on: "Mathematical truth isn't established by all [of us] agreeing that it's true—as if [we] were witnesses of it. *Because* [we] all agree in what [we] do, we lay it down as a rule, and put it in the archives. Not until we do that have we got to mathematics" (*LFM*, 107). The idea seems to be that the rule of multiplication is defined by its results, so that every new calculation—every proof—is an extension of the definition, and hence of mathematics. Wittgenstein's rejection of Platonism means that there is no internal relation between 25 × 25 and 625 "unless you have both terms already" (*LFM*, 108), that is to say, unless the connection has been made, the calculation carried out. Moreover, our adopting the result of a new calculation as a rule is, he suggests, essentially grounded in the fact that this is "the natural way to do it, the natural way to go—for all these people" (*LFM*, 107). This certainly sounds very close to Wright's thought that Wittgenstein sees us as "the perennial creators of our concepts, not in the style of conscious architects but just by doing what comes naturally" (*RI*, 78). However, this would be to neglect the significance of Wittgenstein's observation that we must agree not on the result of the calculation but on getting this result and that the result that, after suitable training in the technique of multiplication, we agree on in getting, is then laid down as a rule or put in the archives. This suggests that Wittgenstein does not, as Wright claims, reject the idea that our practice is one in which we establish paradigms or rules with which future actions—for example, my deriving from the proposition that there are 25 groups of 25 children the proposition that there are 625 children altogether—may accord or fail to accord. It isn't a later majority judgment that this is what counts as accord with the established rule that determines the matter, but the rule itself, as it is laid down in the system of mathematics: "4 + 1 = 5 is now itself a rule, by which we judge proceedings. This rule is the result of a proceeding that we assume as *decisive* for the judgement of other proceedings. The rule-grounding proceeding is the proof of the rule" (*RFM*, 6:16).

Our practice of establishing and employing rules of this kind is grounded in our capacity to be brought by our training to a point where we all do the same when asked to multiply two numbers, but it is the rules themselves that we appeal to in justification of a judgment or an inference in any particular case. And it is, of course, crucial to Wittgenstein's naturalism that although these rules are, in some sense, arbitrary, they are employed in a thousand different human activities, which gives them a significance and reality that they would otherwise lack. Not everything that could be put forward as a rule has a use in our lives: "Thinking and inferring (like counting) is of course bounded for us, not by an arbitrary definition, but by the natural limits corresponding to the body of what can be called the role of thinking and inferring in our life" (*RFM*, 1:116).

The theme of our capacity to use particular samples or examples as rules or paradigms, which we go on to apply in countless different kinds of circumstance, is an important one in Wittgenstein's later thought. We could, he suggests, learn to do mathematics entirely by counting apples and proving statements of the form "20 apples plus 30 apples is 50 apples." The point is whether we have only proved that 20 apples plus 30 apples is 50 apples, or even only that *this* group of 20 apples plus *this* group of 30 apples is 50 apples, or whether we "have thereby proved also that 20 chairs + 30 chairs = 50 chairs." He asks: "What is the difference between proving it for apples alone and proving it for chairs, tables, etc? Does it lie in what I write down?" That is to say, does it matter that I write down "20 apples plus 30 apples is 50 apples," rather than "20 + 30 = 50"? He goes on: "Obviously not—nor in what I think as I write it. But in the use I make of it" (*LFM*, 113).

It is a fact of our natural history that we do use examples or samples in a way that, in a certain sense, goes beyond the sample itself. It is insofar as this is the case—insofar as we can see particular, concrete demonstrations and calculations under the aspect of generality—that human natural history includes normative practices in which the idea of proof plays a role and in which the concepts of calculation, inference, measurement, and so on have meaning. Although the proof of "20 apples plus 30 apples is 50 apples" may be based on a procedure carried out on particular, concrete objects, or with particular marks or scratches, we regard it "as demonstrating an *internal property* (a property of the *essence*) of the structures" (*RFM*, 1:99). And this means that we treat the proposition that we've established "nontemporally," as something that isn't subject to further, empirical test but that can be given a general application, or used in inference, and against which it can be judged, in particular circumstances, whether I have counted a set of objects correctly, calculated the area of a floor correctly, drawn a correct inference from a given empirical proposition, and so on. It is our naturally agreeing in how we respond to and employ examples, samples, pictures, formulae, rulers, and so on, how we are taught to use them in a host of different human activities, that is the essential background against which it makes sense to speak of norms against which particular, concrete performances can be judged for correctness or incorrectness, or to speak of what *can* be inferred or to describe a rule for translating between units of measurement as right or wrong. Someone counts as participating in our practice only if he employs these norms or paradigms or samples in the way that we do, that is, if his life with them agrees with ours. Thus:

> The laws of inference can be said to compel us; in the same sense, that is to say, as other laws of human society. The clerk who infers [as follows: a regulation says "All who are taller than five foot six are to join the . . . section." A clerk reads out the men's names and heights. Another allots them to such-and-such sections.—"N.N. five foot nine." "So N.N. to the . . . section"] *must* do it like that; he would be punished if he inferred differently. If you draw different conclusions you do indeed get into conflict, e.g. with society; and also with other practical consequences.
>
> (RFM, I:17 AND 116).

All this suggests that Wittgenstein's naturalism never leads him to abandon the commonsense idea that there are rules that determine whether we have calculated or drawn an inference correctly. A recurrent theme of his remarks is that the naturalizing move, which he offers the reader as relief from the paradoxes that seem to threaten when we realize the emptiness of the idea of logical or mathematical reality, does not come down to defining truth, or what following a rule correctly consists in, in terms of brute consensus. The aim, as McDowell argues, is to "leave everything as it is," while showing that "nothing out of the ordinary is involved." What we are dealing with are the distinctive capacities of human beings to govern their activities, in ways that agree, by reference to norms that they mutually recognize, and with the distinctive capacity of individuals to respond to a certain characteristic training with examples in a way that constitutes mastery of the way a rule or norm is employed, of how it is applied: "The rule-governed nature of our language permeates our life";[20] for us "the order 'add 3' completely determines every step from one number to the next" (*PI*, 189). This mastery, which nothing guarantees, is shown in the individual's going on to apply the rule in different circumstances in the way that we all do. It is in this sense that what the pupil grasps goes beyond the particular examples he has been given: he *sees* the use that he can make of them. There is nothing occult here, yet it is a remarkable fact about human beings that they can be trained with rules in this way, that they do see, that is, treat, particular concrete examples or demonstrations as establishing paradigms that can be employed in a host of different circumstances. However, this may give rise to one final question: How does someone know how he is to apply the rule in a new case? How does he know what counts as acting in accord, in a new case, with a norm or rule that he's employing?

This is a question that Wittgenstein returns to repeatedly throughout the remarks on rule following. It arises first in the opening paragraph of the *Investigations*, when the interlocutor asks: "But how does [the shopkeeper] know where and how he is to look up the word 'red' and what he is to do with the word 'five'?" And Wittgenstein responds here as he will respond throughout these remarks: "Well, I assume that he *acts* as I have described. Explanations come to an end somewhere." The idea that understanding consists in an *unguided* capacity to *do* something is recurrent:

Don't you understand the call "Slab!" if you act on it in such-and-such a way?

(*PI*, 6)

The term "language-*game*" is meant to bring into prominence the fact that the *speaking* of language is part of an activity, or of a form of life.

(*PI*, 23)

We may say: only someone who already knows how to do something with it can significantly ask a name.

(*PI*, 31)

One gives examples (of a game) and intends them to be taken in a particular way.—I do not, however, mean by this that he is supposed to see in those examples that common thing which I—for some reason—was unable to express; but that he is now to *employ* those examples in a particular way. . . . The point is that *this* is how we play the game. (I mean the language-game with the word "game.")

(*PI*, 71)

To understand a sentence means to understand a language. To understand a language means to be master of a technique.

(*PI*, 199)

There is a way of grasping a rule which is *not* an *interpretation*, but which is exhibited in what we call "obeying the rule" and "going against it" in actual cases.

How can he *know* how he is to continue a pattern by himself—whatever instruction you give him?—Well, how do I know?—If that means "Have I reasons?" the answer is: my reasons will soon give out. And then I shall act, without reasons.

(*PI*, 211)

When someone whom I'm afraid of orders me to continue the series, I act quickly, with perfect certainty, and lack of reasons does not trouble me.

(*PI*, 212)

"How am I able to obey a rule?"—if this is not a question about causes, then it is about the justification for my following the rule in the way I do.

If I have exhausted the justification I have reached bedrock, and my spade is turned. Then I am inclined to say: "This is simply what I do."

(*PI*, 217)

When I obey a rule, I do not choose. I obey the rule *blindly*.

(*PI*, 219)

One does not feel that one has always got to wait upon the nod (the whisper) of the rule.

(*PI*, 223)

We look to the rule for instruction and *do something*, without appealing to anything else for guidance.

(PI, 228)

Don't always think that you read off what you say from the facts; that you portray these in words according to rules. For even so you would have to apply the rule in the particular case without guidance.

(PI, 292)

There is no sense in these remarks that the lack of guidance renders the notion of going by a rule inapplicable to an individual. The point is, rather, that even in those cases in which a rule or a norm is explicitly consulted (as it is in *PI*, 1) there is, in the end, the unguided action of someone who has mastered a technique. And if we ask where the connection between what someone does now and a particular technique is made, then Wittgenstein's answer is clear: "a person goes by a sign-post (follows a particular rule, applies a given technique) only insofar as there exists a regular use of sign-posts (the rule, the technique), a custom" (*PI*, 198). It is the background circumstances, not something accompanying the action, that constitutes it as a case of going by a rule. However, given that these background circumstances exist, it is the unguided action itself that is constituted as a case of obeying an order, going by a signpost, continuing a pattern, calling something a game, and so on; it doesn't wait upon the interpretation by others to transform what is in itself a brute response, a "subrational" propensity to vocalize, into a significant act.[21] In the same way, "insofar as I do[, for example,] intend the construction of a sentence in advance, that is made possible by the fact that I can speak the language in question" (*PI*, 337). There is nothing—no experience, no feeling—by which I know what my intention is: "I don't read [my intention] off from some other process. . . . Nor am I *interpreting* [the] situation and its antecendents" (*PI*, 637). As in the case of my applying a rule with whose use I am familiar, I can say, without guidance from anything, "just like that," what I intend to do.

◎ ◎ ◎

In the dispute between Wright and McDowell concerning the nature of Wittgenstein's naturalism, the evidence seems clearly on McDowell's side. On the evidence of the remarks we've considered, the characterization of Wittgenstein's view as a form of "naturalized platonism" appears well justified.

However, it is also the case that there are important aspects of Wittgenstein's naturalism that are, at the very least, not central to McDowell's account of his remarks on rule following. In particular, the emphasis on aspects of what McDowell calls "first nature"—our prerational, natural propensities—is fundamental, for example, in Wittgenstein's attempt to clarify what is involved in our capacity to respond to proofs and demonstrations *as* proofs and demonstrations, or to treat a particular, concrete object as a sample or paradigm that we employ in different circumstances. There is, in general, a great emphasis in Wittgenstein's remarks on the role of preconceptual, untrained responses and reactions as the root of our ordinary language-games.[22]

Lars Hertzberg has drawn attention to the importance of this theme in Wittgenstein's remarks on both our concepts of cause and effect and our psychological concepts.[23] In these cases, too, Wittgenstein sees our early training in our language-games as exploiting, and in a certain sense rooted in, our natural—that is, uneducated—reactions to events (e.g., our intuitive looking for a cause), our natural expression of sensations and emotions, and our natural responses to the actions and emotions of others. It is not merely that our words are woven into our lives, and are given their life thereby, but that we can see our being drawn into our life with language as itself something that has grown out of, or is a development of, a life that we have, as it were, primitively, simply by virtue of being the kind of creature we are. Wittgenstein expresses the point, in "Cause and Effect: Intuitive Awareness," as follows:

> The origin and the primitive form of the language game is a reaction; only from this can more complicated forms develop.
>
> Language—I want to say—is a refinement. "In the beginning was the deed."
>
> I want to say: it is characteristic of our language that the foundation on which it grows consists in steady ways of living, regular ways of acting.
>
> Its function is determined *above all* by action, which it accompanies.
>
> We have an idea of which ways of living are primitive, and which could only have developed out of these.[24]

Wittgenstein is careful to make clear that remarks of this kind are not intended to serve as explanatory hypotheses concerning the actual origins of human language.[25] They are, rather, intended to draw our attention to a "formal connection" or a "formal ordering," one which, by placing a particular phenomenon in a certain context, makes it less puzzling than it appears at first sight. He does this, for example, in response to our puzzlement about how the connection between a name and a sensation is set up (*PI*, 244). He writes: "Here is one possibility: words are connected with the primitive, the natural, expressions of sensations and used in their place. A child has hurt himself and he cries; and then the adults talk to him and teach him exclamations and, later, sentences. They teach the child new pain-behaviour."

Wittgenstein's naturalism is as central to his rejection of the myth of the inner as it is to his rejection of the myth of logical or mathematical reality. And in this case, too, his concern is to show that nothing of significance has been lost in abandoning the myth. In the first instance, his ordering suggests, everything takes place in the public realm of our primitive life with others.[26] It is through a gradual refinement of this life with others, a refinement in which the acquisition of language plays a crucial role, that possibilities for both refined feelings, complex intentions, and so on, and also for deception and hiddenness develop: lying, dissembling, mocking, mimicking, acting, and so on, are also part of the natural history of human beings. This emphasis on primitive reactions is no doubt misrepresented in Wright's interpretation, but it is clearly central to Wittgenstein's reflections.

All this suggests that Wittgenstein's naturalism is a much more complex and significant element in his thought than McDowell's characterization of him as a naturalized Platonist brings out. His naturalism represents not merely an escape from the paradoxes of rule following but also a fundamental and all-pervasive approach to philosophical perplexity. One way to sum this up might be to suggest that what is absent from McDowell's account is the way in which naturalism is fundamental to Wittgenstein's philosophical method. Wittgenstein's highly distinctive form of naturalism is clearly influenced by Goethe. There is only space left to list some of its principal, naturalistic features: the importance of seeing things in context, of looking at particular cases, of seeing connections, of looking at how something develops or unfolds in time and of recognizing patterns; the rejection of explanation in favor of description; the use of analogies and comparisons; the suspicion of abstractions, hypostatizations, and idealizations; the avoidance of dogma; the appeal to the reader's full sensuous awareness of phenomena and the attempt to make phenomena present to the imagination; and finally, the consistent emphasis on doing over knowing, on the application or employment of linguistic techniques in everyday, human activities and on the roots of our language-games in primitive responses and reactions. It is by means of remarks that exploit these methodological principles that Wittgenstein tries to overcome the intellectual temptation, which he regards as the root cause of philosophical perplexity, to idealize, create abstractions, and hypostatize objects, and to get us to see that "nothing out of the ordinary is involved."

NOTES

1. Ludwig Wittgenstein, *Philosophical Investigations*, 2nd ed. (hereafter *PI*) (Oxford: Blackwell, 1998), 230.

2. See Saul Kripke, *Wittgenstein on Rules and Private Language* (Oxford: Blackwell, 1982).

3. The papers by Crispin Wright in which he develops these ideas and which are referred to in this paper are all collected in *Rails to Infinity* (Cambridge, Mass.: Harvard University Press, 2001) (hereafter *RI*): "Rule-Following, Meaning, and Constructivism," 53–80; "Wittgenstein's Argument as It Struck Kripke," 81–90; "On Making Up One's Mind: Wittgenstein on Intention," 116–42; "Wittgenstein's Rule-Following Considerations and the Central Project of Theoretical Linguistics," 170–214; "Wittgenstein's Later Philosophy of Mind: Sensation, Privacy, and Intention," 291–318.

4. The papers by McDowell in which he develops these ideas and which are referred to in this paper are collected in *Mind, Value, and Reality* (Cambridge, Mass.: Harvard University Press, 1998) (hereafter *MVR*): "Wittgenstein on Following a Rule," 221–62; "Intentionality and Interiority in Wittgenstein," 297–324. See also John McDowell, *Mind and World* (Cambridge, Mass.: Harvard University Press, 1994).

5. McDowell notes that Wright equivocates here between the idea that first-person avowals are groundless—there is nothing on which such claims are based—and the idea that they idea that they are noninferential, i.e. that they are based on observation of a private realm. If one thinks that the question "How can you tell?" is simply inappropriate in connection with knowledge of one's own intentions, then Wright's account of the benefits of the Cartesian picture collapses. See John McDowell, "Response to Crispin Wright," in *Knowing Our Own Minds*, ed. Crispin Wright, B. C. Smith, and Cynthia MacDonald (Oxford: Oxford University Press, 1998), 48–49.

6. Wright makes the affinity between the view he attributes to Wittgenstein and the views of Davidson explicit: "Against . . . Platonism I want to set what I take to be an idea of Wittgensteinian authorship, although it is also familiar from the writings of Donald Davidson: that the content of a subject's intentional states is not something which may merely be *accessed*, as it were indirectly, by interpretative methods . . . but as something which is intrinsically sensitive to the deliverances of best interpretative methodology" (*RI*, 239–40). As Wright sees it, both Wittgenstein and Davidson explain the authoritative status of first-person avowals of intentional states as deriving from the fact that first-person authority is partially constitutive of the interpretative principles by means of which meaning is determined.

7. McDowell's view that Wright's "straight solution" to the skeptical paradox is "only an insignificant divergence from Kripke" (McDowell, "Response to Crispin Wright," 268) is an expression of his more general commitment to the idea that what Sellars calls "the logical space of reasons" is "*sui generis*, by comparison with the realm of law" (McDowell, *Mind and World*, 72). That is to say, we cannot un-

derstand, i.e., reconstruct, normative notions naturalistically if "naturalistically" is construed in a way that excludes normative notions.

8. Quoted in *MVR*, 314.

9. That is, it depends upon our making sense of the idea of a rule which "forced a particular use on me" (*PI* 140).

10. McDowell, *Mind and World*, 95.

11. McDowell insists that in crediting Wittgenstein with "naturalized platonism," he is not flouting Wittgenstein's "insistence that he is not in the business of offering philosophical doctrine." "Naturalized platonism" is not, McDowell argues, "a label for a bit of constructive philosophy." He sees this form of naturalism as purely descriptive, as opposed to the constructive naturalism put forward by Wright; its aim is to assemble "reminders" as a means to relieving philosophical perplexity. If the reminders work, then, although we will not have been provided with a substantial account of what meaning consists in, nor will we be left with a sense of genuine philosophical questions that have not been answered. There is, therefore, no equivalent to Wright's idea that Wittgenstein "teach[es] us to know when philosophy can tell us nothing further."

12. McDowell, *Mind and World*, 95.

13. Ibid.

14. Quoted in *MVR*, 318.

15. Ludwig Wittgenstein, *Remarks on the Foundations of Mathematics* (hereafter *RFM*), ed. G. H. von Wright, Rush Rhees, and G. E. M.Anscombe (Oxford: Blackwell, 1978), 1:6.

16. Cf. *PI*, 520: "So does it depend wholly on our grammar what will be called (logically) possible and what not,—i.e. what that grammar permits?—But surely that is arbitrary!—Is it arbitrary?—It is not every sentence-like formation that we know how to do something, not every technique has an application in our life."

17. Wittgenstein clearly intends to prioritize the *activities* that human beings naturally engage in: "It is sometimes said that animals do not talk because they lack the mental capacity. And this means: 'they do not think, and that is why they do not talk.'—But—they simply do no talk. Or to put it better: they do not use language— if we except the most primitive forms of language.—Commanding, questioning, recounting, chatting, are as much a part of our natural history as walking, eating, drinking, playing" (*PI*, 25).

18. Thus, in the case of the rule "(x)fx→fa," Wittgenstein writes: "Cut down all these trees!—But don't you understand what 'all' means? (He had left one standing.) How did he learn what 'all' means? Presumably by practice. . . . One learns the meaning of 'all' by learning that 'fa' follows from '(x)fx'.—The exercises which drill us in the use of this word, which teach us its meaning, always make it natural to rule out any exception" (*RFM*, 1:10).

19. Cf. Ludwig Wittgenstein, *Lectures on the Foundations of Mathematics, Cambridge 1939* (hereafter *LFM*), ed. Cora Diamond (Chicago: Chicago University Press, 1989), 183: "Is one's criterion for meaning a certain thing by the rule [for the series

of cardinal numbers] the using of the rule in a certain way, or is it a picture of another rule or something of the sort? In that case, it is still a symbol—which can be interpreted in any way whatsoever. This has often been said before. And it has often been put in the form of an assertion that the truths of logic are determined by a consensus of opinions. Is this what I am saying? No. There is no *opinion* at all; it is not a question of *opinion*. They are determined by consensus of *action*: a consensus of doing the same thing, reacting in the same way. There is a consensus but it is not a consensus of opinion. We all act the same way, walk the same way, count the same way. . . . We express opinions by means of counting."

20. Ludwig Wittgenstein, *Last Writings on the Philosophy of Psychology: The Inner and the Outer*, vol. 2, ed. G. H. von Wright and H. Nyman (Oxford: Blackwell, 1992), 72.

21. It is difficult to see how a response that is in itself subrational can be rendered rational by being regarded as such by another, especially where that other is himself merely responding, subrationally, as it strikes him. This is, of course, one of the principal themes of McDowell's objections to Wright's reading.

22. McDowell is well aware that Wittgenstein draws our attention to the fact that our "concepts would not be the same if the facts of (first) nature were different, and the facts help to make it intelligible that the concepts are as they are" ("Two Sorts of Naturalism," in *MVR*, 193). McDowell's emphasis has been on arguing that this "does not mean that correctness and incorrectness in the application of the concepts can be captured by requirements spelled out at the level of the underlying facts" (ibid.). This is, as we've seen, a fundamental theme of Wittgenstein's remarks. However, the role of first nature—of how we naturally take or see or employ or respond to something—is also an important theme in Wittgenstein's discussion of rule following, and it might be regarded as distinctive of the form of naturalism that is present in his work.

23. See Lars Hertzberg, "Very General Facts of Nature," in *The Wittgenstein Handbook*, ed. M. E. McGinn (Oxford: Oxford University Press, forthcoming). Hertzberg argues very convincingly that this aspect of Wittgenstein's naturalism has been neglected at the expense of an emphasis on our all going on naturally in the same way, e.g., in developing a mathematical series.

24. Ludwig Wittgenstein, *Philosophical Occasions, 1912–1951*, ed. J. C. Klagge and Alfred Nordmann (Indianapolis: Hackett, 1993), 395 and 397.

25. Cf. ibid., 377: "What we are doing here above all is to *imagine* a basic form [of our language-game]: a possibility, indeed a *very important* possibility. (We very often confuse what is an important possibility with historical reality.)". The importance of the possibility lies in its serving as an object of comparison, one that helps us see our ordinary language-game in a new light.

26. Wittgenstein also emphasizes that doubt plays no role in this primitive form: "*The basic form* of our game must be one in which there is no such thing as doubt" (*Philosophical Occasions*, 377). This is a central theme of his remarks on both the

foundations of language generally and on our acquisition of psychological concepts in particular. He uses an aphorism from Goethe's *Faust* to make the point: "In the beginning was the deed" (quoted in Ludwig Wittgenstein, *On Certainty*, ed. G. E. M. Anscombe and G. H. von Wright [Oxford: Blackwell, 1969], 402). This is a further distinctive element in his naturalism.

CONTRIBUTORS

AKEEL BILGRAMI is the Johnsonian Professor of Philosophy at Columbia University and the director of the Heyman Center for the Humanities at Columbia University. He is the author of *Belief and Meaning* (Blackwell, 1992), *Self-Knowledge and Resentment* (Harvard University Press, 2006), and *Politics and the Moral Psychology of Identity* (Harvard University Press, forthcoming). He writes on ethics, philosophy of mind, philosophy of action, and philosophy of politics as well as on some broader cultural and political issues.

MARIO DE CARO is associate professor of moral philosophy at University Rome 3. He has been a Fulbright Fellow at Harvard University and, since 2000, he has also taught part-time at Tufts University. Besides authoring four books and editing three anthologies in Italian, he has edited *Interpretations and Causes* (Kluwer 1999), *Naturalism in Question* (with David Macarthur, Harvard University Press, 2004), and *Cartographies of the Mind* (with M. Marraffa and F. Ferretti, Springer, 2007). With David Macarthur he is now editing *Philosophy in an Age of Science: Physics, Mathematics, and Skepticism*, Hilary Putnam's next two volumes of essays (Harvard University Press, forthcoming).

JOHN DUPRÉ is a philosopher of science whose work has focused especially on issues in biology. He is currently professor of philosophy of science at the University of Exeter and since 2002 he has been director of the ESRC Centre for Genomics in Society (Egenis). He has formerly held posts at Oxford, Stanford, and Birk-beck College, London. In 2006 he held the Spinoza Visiting Professorship at the University of Amsterdam. His publications include *The Disorder of Things: Meta-physical Foundations of the Disunity of Science* (Harvard, 1993); *Human Nature and the Limits of Science* (Oxford, 2001); *Humans and Other Animals* (Oxford, 2002); and *Darwin's Legacy: What Evolution Means Today* (Oxford, 2003).

PETER GODFREY-SMITH grew up in Sydney, Australia, and studied at the University of Sydney and the University of California at San Diego. He has taught at Stan-ford University and the Research Schools at the Australian National University and since 2006 has been professor of Philosophy at Harvard. His main interests are in the philosophy of biology and the philosophy of mind. He has written three books, *Complexity and the Function of Mind in Nature* (Cambridge Univer-sity Press, 1996), *Theory and Reality: An Introduction to the Philosophy of Science* (Chicago University Press, 2003), and *Darwinian Populations and Natural Selec-tion* (Oxford University Press, 2009).

ERIN I. KELLY is associate professor of philosophy at Tufts University. Her research interests are in moral and political philosophy and the philosophy of law. She focuses on questions about justice, the nature of moral reasons, moral responsi-bility and desert, and theories of punishment. Her publications include "Doing Without Desert," *Pacific Philosophical Quarterly* (2002); and "Equal Opportunity, Unequal Capability," in *Measuring Justice: Capabilities and Primary Goods* (Cam-bridge University Press, forthcoming). She is editor of John Rawls, *Justice as Fair-ness: A Restatement* (Harvard University Press, 2001).

DAVID MACARTHUR is senior lecturer in philosophy at the University of Sydney. He has published articles on skepticism, naturalism, neo-pragmatism, metaphysical quietism, and Wittgenstein. His research interests also include the philosophy of art. He is coeditor (together with Mario De Caro) of *Naturalism in Question* (Harvard University Press, 2004). With Mario De Caro he is now editing *Phi-losophy in an Age of Science: Physics, Mathematics, and Skepticism*, the next two volumes of Hilary Putnam's philosophical papers (Harvard University Press, forthcoming).

MARIE MCGINN is professor emeritus at York University and part-time professor of philosophy at the University of East Anglia. She has written articles on Witt-genstein's early and later philosophy, skepticism, and the philosophy of mind. Her books include the Routledge Guidebook *Wittgenstein and the Philosophi-cal Investigations* (Routledge, 1997) and *Elucidating the Tractatus: Wittgenstein's Early Philosophy of Language and Logic* (Oxford University Press, 2006).

LIONEL K. MCPHERSON is associate professor of philosophy at Tufts University. His areas of research are ethics, political philosophy, and social philosophy. More spe-cifically, he writes on issues that, at varying levels of abstraction, revolve around the theme of special relationships and projects in relation to impersonal duties. His publications include "Normativity and the Rejection of Rationalism," *Journal of Philosophy* (2007) and "Is Terrorism Distinctively Wrong?" *Ethics* (2007).

PETER MENZIES is professor of philosophy at Macquarie University, Sydney. He has held positions at the University of Sydney and the Research School of Social Sciences, Australian National University. He works in metaphysics, philosophy of science, and philosophy of mind. He is a coeditor of the *Oxford Handbook of Causation* (Oxford University Press, 2009). He is a fellow of the Australian Academy of Humanities.

HUW PRICE is Challis Professor in Philosophy at the University of Sydney, having previously been professor of logic and metaphysics at the University of Edinburgh. He is a two-time ARC Federation Fellow, and a past president of the Australasian Philosophical Association. He has published widely on the philosophy of physics, time, causation, naturalism and neo-pragmatist approaches to language and metaphysics. His books include *Facts and the Function of Truth* (Blackwell, 1988) and *Time's Arrow and Archimedes Point* (Oxford University Press, 1996). A collection of his essays, *Naturalism Without Mirrors*, is forthcoming from Oxford University Press.

HILARY PUTNAM is Cogan University Professor Emeritus at Harvard University. He is the author of many books and articles, including *The Collapse of the Fact/Value Dichotomy and Other Essays* (Harvard University Press, 2002); *Ethics Without Ontology* (Harvard University Press, 2004); *Jewish Philosophy as a Guide to Life: Rosenzweig, Buber, Levinas, Wittgenstein* (Indiana University Press, 2008); and *Philosophy in an Age of Science: Physics, Mathematics and Skepticism* (Harvard University Press, forthcoming). His wide-ranging interests include philosophy of mathematics, philosophy of logic, metaphysics, philosophy of science, ethics, philosophy of language, epistemology, and history of philosophy. He is a past president of the American Philosophical Association, a fellow of the American Academy of Arts and Sciences, and a corresponding fellow of the British Academy and of the French Académie des Sciences Politiques et Morales, and he holds a number of honorary degrees.

PAUL REDDING is professor of philosophy at the University of Sydney. He is the author of *Hegel's Hermeneutics* (Cornell University Press, 1996), *The Logic of Affect* (Cornell University Press, 1999), *Analytic Philosophy and the Return of Hegelian Thought* (Cambridge University Press, 2007), and *Continental Idealism: Leibniz to Nietzsche* (Routledge, 2009). His interests include the relation of nineteenth-century German idealism to analytic philosophy and the viability of contemporary idealist alternatives to mainstream philosophical naturalism.

RICHARD RORTY (1931-2007) taught philosophy at Wellesley College, Princeton University, University of Virginia, and Stanford University. He received a Guggenheim Fellowship, a MacArthur Fellowship (1981–1986) and was president of the American Philosophical Association. He was the author of many books, including the world famous *Philosophy and the Mirror of Nature* (Princeton University Press, 1979), *Consequences of Pragmatism* (University of Minnesota Press, 1982), and the four volumes of *Philosophical Papers* (all published with Cambridge University Press) .

CAROL ROVANE is professor and chair in the Department of Philosophy at Columbia University. Her publications include *The Bounds of Agency: An Essay in Revisionary Metaphysics* (Princeton University Press, 1998), *For and Against Relativism* (Columbia University Press, forthcoming), and articles on various subjects in metaphysics, philosophy of language, philosophy of mind, and ethics.

T. M. SCANLON is Alford Professor of Natural Religion, Moral Philosophy, and Civil Polity at Harvard University. His writings include *What We Owe to Each Other* (Harvard University Press, 1998), *The Difficulty of Tolerance* (Cambridge University Press, 2003), and *Moral Dimensions: Permissibility, Meaning, Blame* (Harvard University Press, 2008).

ALBERTO VOLTOLINI is associate professor in philosophy of mind at University of Turin. He is a philosopher of language and mind whose works have focused mainly on fiction, intentionality, and Wittgenstein. From 2002 to 2008 was a member of the Steering Committee of the European Society for Analytic Philosophy. He was educated at Scuola Normale Superiore (Pisa) and has had scholarships at the Universities of Geneva and Sussex. He has been visiting professor at the Universities of California at Riverside, Auckland, and Australian National University, Canberra. His publications include *How Ficta Follow Fiction* (Springer, 2006).

STEPHEN L. WHITE teaches philosophy at Tufts University. He is the author of *The Unity of the Self* (MIT Press, 1991) and *The Necessity of Phenomenology* (forthcoming) and of a number of articles in philosophy of language, philosophy of mind, moral psychology, moral theory, and aesthetics published in *The Journal of Philosophy*, *The Philosophical Review*, *The Monist*, *Philosophical Topics*, and other journals. His research interests include narrow content, the subjective perspective, agency, value, the self, and expressive properties.

INDEX

beliefs, 32–33, 181; aboutness and, 319; within communities, 237–38; debates on, 35; direction of fit of, 191n20–192n21; dispositions and, 31; expressions of, 238; Humean moral psychology and, 205–6; logical connection argument and, 146–47; meaning vs., 65–66; moral judgments vs., 179; perceptual, 175; web of, 265

Bell, John Stewart, 94

Bell's Theorem, 94

Ben-Menahem, Yemima, 99n13

Bentley, Richard, 39–40

Bergman, Ingmar, 220

Berkeley, George, 108, 274

Berlin, Isaiah, 56

Bildung, 313–15, 317

Bilgrami, Akeel, 6–7, 77, 116–17n3

biology, 128–29, 273; cell, 294–300; Dewey and, 307; irreducibility in, 128–29; methodological pluralism and, 132

birfurcation thesis, 262n1

Blackburn, Simon, 184

Blumenberg, Hans, 56

Boghossian, Paul, 118

Bohr, Niels, 94

Boyle, Robert, 38

Boyle Lectures, 39–40, 52n14

brains, 294–97

Brandom, Robert, 5, 64–66, 144, 265, 274, 279; pragmatic phenomenalism and, 285n73

broad scientific naturalism, 7–8, 123, 126, 133–34; normative phenomena and, 134–35

Broome, John, 190n7

Buber, Martin, 90

Burke, Kenneth, 40–41

Camp, Liz, 320

Carnap, Rudolf, 90–91, 100–101, 265

Cartesianism, 297

causal efficacy, 153

causal explanations, 145–46; deductive-nomological model of, 154–56; model-based approach and, 162–67; predictions vs., 161

causal fundamentalism, 130–31, 133

causality: of intentional states, 143; mental properties and, 153

causal powers, 319–20

causation, 83n14–84n14; compatibilism and, 149; deviant, 151–52

Cavell, Stanley, 92–93

charity, principle of, 113–14

chemistry: reducibility of, 302n2; reductionism and, 290

Chomsky, Noam, 73, 115

Christianity, 62

Church, Alonzo, 66

Church's Translation Test, 66

circularity, 250–51n25

civility, 53n23

Clarke, Samuel, 38–40

Cobb, Richard, 211

cognitive abilities, 115

cognitivism, 186; moral, 193, 196–97; moral problem and, 218–19

cognitivist understanding, 176–77

coherence, 45

Collins, Anthony, 42–43

Colyvan, Mark, 141n45

commitments: behavioral, 242; factual, 315

commodification, 49

compatibilism, 145–46; one-tier, 159–67; two-tiered, 149–53

Compte, Auguste, 139n21

conceptual analysis, 60

conceptual-schemes argument, 110–12, 119n8, 120n11

conflicts, irresoluble, 104

consciousness, 56; reflective model of, 284n61; subject naturalism and, 62

consensus, 338–39

contractualism, 278–79

convergence, 37

ethics: logical positivism and, 90–91; ontology and, 84n15

ethnography, 276

evaluation, motivation vs., 212

evaluative statements, 218, 224n28

Evans, Gareth, 6, 28

evolution, 294–95, 302n8; of culture, 316

examples, 343–44

experience, 56

experiences: McDowell on, 313; objects and, 312

experiential grounding, 215

explanation, justification through, 193

expressions: of belief, 238; of desire, 234, 237; rules and, 331; of self-conception, 334

expressivism, 185–87, 203

extreme scientific naturalism, 7, 126; metaphysics and, 129–31; reductionism and, 131

facts: authority and, 186; institutional, 275; law vs., 282n32; reasons and, 208–9

fact/value dichotomy, 135

fallibility, 138n9

falsificationism, 140n33

falsity, 233–36, 242

Feyerabend, Paul, 140n33

Fichte, Jacobi, 275, 279

fictionalism, 246–48; mathematical, 18n9

Fine, Kit, 5, 63

first nature, 346

first-person perspective, 25, 267; avowals of intentions from, 327

Fodor, Jerry, 5, 64–66; intentional psychology and, 143; on reductionism, 131

Foot, Phillippa, 96

form of life, 322

frameworks, 314–15

Frankfurt, Harry, 184, 199, 209–13

Frazer, Alexander Campbell, 274

freedom, 320

freethinkers, 40, 42

free will, 79–80

Frege, Gottlob, 56

Frege-Geach problem of embeddings, 184

Friedman, Michael, 263, 280n3

Geist, 276

Geisteswissenschaften, 134, 273–74, 278

Gell-Man, Murray, 94

genealogy, 37–38

generalizations: folk psychology and, 158; in intentional psychology, 147–49; normativity of, 167; rational agency and, 160–61; two-tiered compatibilism and, 150–53

genetics, 128–29, 290, 294–300

Gibbard, Allan, 185

Gibson, J. J., 215

Giere, Ronald, 154–59

God: deracination of, 36; idea of, 89–90; interpretation of, 98n11; representation of, 271

Godfrey-Smith, Peter, 15–16, 157–59

Goethe, Johann Wolfgang von, 347

Goetz, Stewart, 82n2

Goldman, Alvin, 305

Habermas, Jürgen, 56

Hampshire, Stuart, 51n6

Hegel, Georg Wilhelm Friedrich, 56, 98n2, 265, 275, 278

Hegelianization, 274

Heidegger, Martin, 56, 97

Hertzberg, Lars, 346

Herz, Marcus, 270

heterodox pantheism, 52n15

Hill, Christopher, 44

historiography, 37

Hitchcock, Christopher, 162

Hobbes, Thomas, 58–59

Hobsbawm, Eric, 211

holism, 56–57; Brandom and, 64; concept-belief, 110–11; criticisms of, 119n10;

meaning-belief, 110; normative insularity and, 111, 120n11

Hornsby, Jennifer, 124

Horwich, Paul, 232

human behavior: context of language and, 293–94; intentional states and, 8, 142–43, 144; reductionism and, 294; subject naturalism and, 62

humanistic disciplines, 277, 283n56–284n56

human nature, 11–12; abstract nature of, 301–2; normative authority of reasons and, 199

human sciences, 125, 283n56–284n56; irreducibility of, 18n12; legitimacy of, 132–33; limitations of, 133–34; subject naturalism and, 273. *See also* social sciences

Hume, David, 24, 55; moral naturalism and, 194; moral sentiments and, 26–27

Humean moral psychology, 14, 194, 205–6, 209; actions and, 215; moral problem and, 218–19

Husserl, Edmund, 56, 90

idealism: analytic nonnaturalism and, 274–80; German, 275; McDowell and, 63; object naturalism and, 274; pragmatism vs., 63; self-consciousness and, 284n61; transcendental, 100

ideals, 318–20

identity, practical, 183, 199

ignorance, error vs., 114

immaterialism, 274–75

imprudence, 206–7

indeterminacy, 159

indispensability argument, 135

individualism, promiscuous, 301

inductive logic, myth of, 95–96

inference, rules of, 336–37, 342

inferentialism, 64–66

inflation, 217, 224n26

inquiry: boundaries of, 108; relativism and, 107

intelligence: Dewey on, 307; forms of, 114

intelligent design, 71–72

intelligibility, 111–12, 115

intention: judgment-sensitive nature of, 201; prediction vs., 24–26

intentionality, 215, 283n54; irreducibility of, 32; self-knowledge vs., 6; subject naturalism and, 62

intentional psychology, 142–46; generalizations of, 147–49; as model, 154–59; one-tier compatibilism and, 159–67; two-tiered compatibilism of, 149–53

intentional states, 50n3, 142–45; actions and, 149–50; causality of, 143; desire and, 216; human behavior and, 8; meaning and, 324; social sciences and, 135; Wittgenstein on, 332–33

intentions: avowals of, 327; complex, 347

internalism, 218–19

internal properties, 341

interpersonal relations: disagreements in, 185–86; relativism and, 107

interpretation: determinacy of meaning and, 324–25; obedience without, 344

irrationality, 181–82; of desires, 212; extreme imprudence and, 206; structural, 183

Jablonka, Eva, 302n8

Jackson, Frank, 5, 18n8, 60–61, 64, 129, 147–48, 190n1

Jacobi, Friedrich, 266

James, William, 233

Jamesian pragmatism, 232

judgments, 189; about reasons for action, 173, 187, 189; correctness of, 338–39; intrapersonal rational significance of, 185; normative significance of, 187, 189; of norms, 342; practical, 185; value, 221–22; weakness of will and, 213–14 justification, 254; through explanation, 193; intersubjective, 255; of moral reasons, 200; truth vs., 243; truth with, 229, 233; Wittgenstein on, 332

Kant, Immanuel, 55, 98n2; Herz and, 270; idealism and, 275; *Methodenstreit* and, 127; on morals, 24; sensibility and, 116n1; transcendental idealism and, 100
Kantianism, 11; naturalized, 264; positivism and, 263; subject naturalism and, 271–74
Kantian rationalism, 143–46
Kelly, Erin I., 13–14
Kelly, Thomas, 190
Kennett, Jeanette, 208–9
Kim, Jaegwon, 78
Kitcher, Philip, 305
knowledge: Dewey and, 308; epistemological reflection and, 268; moral, 175; relations and, 311; revisable a priori, 18n8; tacit, 158
Kobel, Max, 117–18n5
Korsgaard, Christine, 181–83, 190n6, 199; moral obligations and, 204n13; normativity objection and, 208–9, 213
Kripke, Saul, 5, 63, 166, 323–25
Kuhn, Thomas, 134, 140n33

Lacey, Alan, 72
Lamb, M. J., 302n8
language, 58, 256–57; context of, 293–94; cooperation and, 259, 261; evaluative, 258; formal ordering of, 346; object naturalism and, 281n28; reality vs., 61; understanding, 344
language-game, 322, 337–38, 343
Leiter, Brian, 5, 57–58, 63
Lenman, James, 82n2
Levellers, 52n15
Lewis, David, 5, 63, 129, 209–13, 268; intentional psychology and, 143

Lewontin, Richard, 131
liberal naturalism, 3, 8, 127–28; characterization of, 12; dilemma of, 70–72; doctrine of, 9–16; feasibility of, 79–82; legitimacy of, 71; possibility of, 75–79
Lilburne, John, 43
linguistic frameworks, 100–101, 117n4, 264
linguistic practices: assertoric, 242; coexistence of, 58; culture and, 61; subject naturalism and, 269; truth in, 230–31
linguistics, 65–66, 273
Locke, John, 270
logical connection argument, 145–49
logical fictions, 80
logical positivism, 90–91, 93, 100
Luce, A. A., 283n49
Lycan, William, 80

Macarthur, David, 7, 59, 63
MacFarlane, John, 117–18n5
Mach, Ernst, 90
Maibom, Heidi, 157–58
manifest images, 76, 97
Marx, Karl, 49
materialism: mechanistic, 56; nonreductive, 50n3
mathematics, 337–34; ontology and, 84n15
McDowell, John, 1–2, 5, 16, 24, 50nn2–3, 77, 96, 124, 133, 218, 220; antinaturalism and, 29; Dewey and, 305–6; disenchantment and, 35; free will and, 80; Kantian rationalism and, 144; Lenman and, 82n2; *Mind, Value, and Reality*, 51n5; *Mind and World*, 312–18; naturalism of second nature and, 127–28; secular enchantment and, 48; Williamson and, 62–64; on Wittgenstein, 323, 329–34
McGinn, Colin, 73
McGinn, Marie, 15–16
McLaughlin, Brian, 152
McPherson, Lionel, 13–14, 203n5

meaning, 139n22; belief vs., 65–66; determinacy of, 324–25; Dewey on, 307; intentional states and, 324

medicine, democratization of, 43

Menger, Carl, 138n12

mentality, 24

mental properties, 153

Menzies, Peter, 8–9

merely-opinionated assertion (MOA), 238–43, 255

Merleau-Ponty, Marcel, 220

metaphysical glue, 271

metaphysics: Dewey and, 310; disenchantment and, 42; dogmatic, 264; empiricism and, 139n22; empiricism vs., 56; extreme scientific naturalism and, 129–31; location and, 60; logical positivism and, 90–91; moral judgments and, 175–76; Royal Society and, 38–39; serious, 60; truth without, 261–62

Methodenstreit, 2, 127, 138n12

methodical discontinuity, 78

methodological doctrines, 17n4

methodological pluralism, 132

methodology, scientific naturalism and, 125

Michotte, Albert, 215

microbes, 299, 303n11

mind, 58; Dewey and, 312–18; nature and, 306–12

minimalism, 232

mistakes, 260

Mitchell, Braddon, 147–48

MOA. *See* merely-opinionated assertion

Mo'ans, 238–43, 255–56

modal properties, 79–82

model-based approach, 155–59; causal explanations and, 162–67; rational agency and, 160–61

molecules, 291–92

monism, 152

Moore, G. E., 51n8, 274

morality: motivation and, 201–2; noumenal status of, 24; psychology and, 196; rational bindingness of, 196; requirements for, 196; secularism and, 66–67n1; structural accounts and, 198–99

moral judgments, 14; acceptance of, 179; beliefs vs., 179; causal claims of, 174–75; cognitivist understanding of, 176–77; metaphysics and, 175–76; normative significance of, 201; normativity of, 193; ordinary criteria for, 174; perspective and, 195; reasons and, 193; reason-sensitive nature of, 197–98; truth vs., 173

moral naturalism, 193

moral obligations, 204n13

moral problem, 218–19

moral psychology, 14

moral reasons, 198–203

moral sentiments, 26–27

moral standards, 194

moral theory, purpose of, 202

motivation, 34, 179; evaluation vs., 212; morality and, 201–2; problems of, 182; rationality vs., 188; respect and, 210

motives, monetary, 41

Müller Lyer illusion, 218

multimundialism, 107–9; scientific realism vs., 112; unimundialism vs., 109–10

Murdoch, Iris, 96

Nagel, Thomas, 206–7, 267

narrow scientific naturalism, 7, 126, 132

naturalism: art and, 71; basic, 124–26; broad scientific, 7–8, 123, 126, 133–34; catholic, 124; defining, 3–4, 71; of Dewey, 305–6; ethical, 71; evolution of, 23–24; expansive, 124; extreme scientific, 7, 126, 129–31; methodological, 4; naïve, 124; narrow scientific, 7, 126, 132;

naturalism (*continued*)

object, 269, 274, 281n28; object vs. subject, 59; ontological aspect of, 289–90;
pluralistic, 272; pragmatic, 59–60;
religion and, 83n8; without representationalism, 59; scientific, 3–5, 123–41;
of second nature, 124, 127–28, 314, 331;
subject, 59–62, 265, 268–74. *See also*
liberal naturalism; moral naturalism;
scientific naturalism

naturalization, 125–26

natural philosophers, 38

nature: appealing to, 289; discontinuities
of, 274; disenchantment and, 42; disenchantment of, 39; mind and, 306–12;
reciprocation with, 42

Naturwissenschaften, 134

Neta, Ram, 12, 18n17, 70, 72, 82n1, 83n7

neurophysiology, 213

neuroscience, 57

Newton, Isaac, 38–39

Newtonianism, 43

Nietzsche, Friedrich, 268

nihilism: dialogical, 246; relativistic, 278

nihilism problem, 266, 268

nominalization, 255–56

noncontradiction, law of, 104–5

nonfactualist theories, 2–3

nonnaturalism, analytic, 274–80

normative insularity, 106; holism and, 111,
120n11

normative phenomena, 126–27; broad scientific naturalism and, 134–35; place of,
2; science and, 123

normative requirements, 190n7

normative significance, 178–79, 195, 197;
expressivism and, 185; of judgments
about reasons, 189

normativity: of generalizations, 167; irreducible, 31; of moral judgments, 193;
nonscientific understanding and, 137;
of practical reasoning, 208, 218; science
and, 95–96; scientific naturalism and,

127, 136; substantive vs. structural, 183;
of truth, 244; ubiquity of, 9

normativity objection, 213

norms: adoption of, 256; of assertion,
234–43; distinguishing between,
254–55; explicit vs. implicit, 258–59;
judgment of, 342; McDowell on, 315,
330; obedience to, 259; social, 65

obedience: without interpretation, 344; to
norms, 259; to rules, 324

Oberheim, Eric, 137

objects: causal powers of, 319–20; elusive,
63

observations: terms, 91; theory-ladenness
of, 33–34; values and, 26

omniscient interpreter argument, 112–13

ontogeny, 295

ontological doctrine, 17n4

ontological naturalism, 4

ontology: defining, 138n10; ethics and,
84n15; formation of, 72; mathematics
and, 84n15; scientific naturalism and,
125; tolerance of, 77–78

ontotheology, 89–90, 97–98

Oppy, Graham, 190n1

otherness, 108

overdetermination, 139n25

Overton, Richard, 43

Oyama, Susan, 302n6

pain, 208, 223n12

pantheism, 39; atheism vs., 39–40; heterodox, 52n15; of Spinoza, 42–43

Papineau, David, 71, 124, 305; causal fundamentalism and, 130

Parfit, Derek, 190, 206–7

partial translatability, 111

patriotism, 199–200

Pearl, Judea, 162

Peirce, Charles Sanders, 56, 233, 243–46

perception: priority of, 224n24; rich, 215;
systems, 219–20; value, 217

perspective, 25–26; agential, 76; distinguishing, 27; moral judgments and, 195; Williams and, 267–68
Pettit, Philip, 58
phenomenalism, pragmatic, 285n73
phenomenology: of agency, 214–17; inflationary/deflationary, 217–19; virtue theory and, 220–21
philosophical method, 323
philosophy: Anglophone, 50n1; assimilation into science of, 271–72; culture and, 93; influence on science of, 94–95; naturalization of, 258; need for, 89–90; normal, 313; Pettit on, 58; progress in, 97; as science, 99n10; science vs., 5; semantics and, 277; theocentric, 266; value of, 92–94
phylogeny, 297
physical description, 311
physicalism, 130, 139n19, 289–90
physics, 126, 128–29, 273; causal fundamentalism and, 130–31; metaphysical claims in, 130; reductionism and, 290, 302n2
Pippen, Robert, 56
placement problem, 2, 123, 269
Platonism, 330, 335–36; naturalized, 345–47, 349n11; rejection of, 340
point of view. See perspective
positivism: Kantianism and, 263; logical, 90–91, 93, 100
postmodernism, 92; value of, 93–94
practical reasoning, 178, 180; Humean conception of, 205–6; normativity of, 208, 218
pragmatism, 67n19–68n19; Dewey and, 305; idealism vs., 63; identification of truth, 243–45; Jamesian, 232; nonreductive, 233; realism vs., 230; truth with justification and, 229
prediction: causal explanations vs., 161; intention vs., 24–26; vocabulary of, 58

Price, Huw, 5, 10–11, 59–61, 268–71
proto-subjectivity, 317
psychology: comparative, 316; evolutionary, 294–95, 302n8; folk, 142, 157–59; morality and, 196; social, 307; structural accounts vs., 199. See also intentional psychology
psychophysical laws, 150
punishment, 210–11
Putnam, Hilary, 5–6, 9, 18n19, 19n22, 77, 84n15, 89–90, 271; convergence and, 37; fact/value dichotomy and, 135; intentional psychology and, 143; Kantian rationalism and, 144; scientism and, 277

quantum mechanics, 94
quasi-dispositional states, 325
quietism, 57
Quine, Willard, 5, 73, 110, 269; on Carnap, 265; indispensability argument, 135; minimalism and, 232; ontology and, 138n10; principle of charity of, 113–14

radical enlightenment, 46–47
Ramberg, Bjorn, 59–60
rational agency, 160–61; structural accounts and, 200
rational benevolence, 222n6
rationality: claims about, 180–81; desires and, 212; disenchantment and, 42; instrumental, 182; moral reasons and, 203; motivation vs., 188; reason vs., 202; scientific, 36, 45–47, 49; structural, 187, 190n7; structural claims about, 201
rationalizing, 145–46; two-tiered compatibilism and, 151–53
realism: practical, 182; pragmatism vs., 230; promiscuous, 301; relation to relativism of, 109–16; relativism vs., 101–2; representationalism vs., 92; spiritual, 274. See also scientific realism

reality: language vs., 61; mind-independent, 101, 108, 114; oneness of, 106–7; portrayal of, 101

reasonableness, scientific inquiry and, 95

reasons: acknowledging vs. accepting, 195, 202; for action, 173, 187, 189; binding-ness of, 196; conclusive, 201; disagreements about, 185–86; facts and, 208–9; McDowell on, 312; McPherson and, 203n5; moral, 198–203; moral judgments and, 193; moral obligations and, 204n13; normative authority of, 199; normativity of, 13–14; rationality vs., 202; space of, 76

receptivity, 312

reciprocal recognition, 279

reciprocation, with nature, 42

Redding, Paul, 11

reductionism, 131; criticism of, 290–91; hierarchy of, 299; human behavior and, 294; object naturalism and, 269; physics and, 290, 302n2

reductionist theories, 2

Rée, Jonathan, 249n8

reflection, 266–67; epistemological, 268

regulative ideals, 119n7

relativism: doctrine of, 102–9; objection to, 245; realism vs., 101–2; scientific realism vs., 8; true, 118n6; values and, 34. See also cultural relativism

relativity, 99n13

religion: disenchantment and, 42; naturalism and, 83n8

representation, 282n32; capacity for, 270–71; subject naturalism and, 62

representationalism: naturalism without, 59; realism vs., 92; subject naturalism and, 269

representations, stabilized, 320

responsibility, 279, 320; assignment of, 58; free will and, 79–80

revelation, 266

Rice, Grantland, 252n32

rights, 53n23

Rorty, Richard, 5, 10–11, 92–94, 229–30, 233, 248, 310, 320

Rovane, Carol, 8, 50

Royal Society, 38–39

rules: establishing, 341; expressions and, 331; grasping, 326; of inference, 336–37; obeying, 324, 344–45; of translation, 336; usage, 240

Russell, Bertrand, 80, 90, 274

Sacks, Mark, 264–65

Scanlon, Thomas, 13, 140n43–141n43, 196–98, 278–79

Schelling, Friedrich von, 266–67, 275, 280n10

scheme-content distinction, 101

Schmoller, Gustav von, 138n12

scholasticism, medieval, 56

science: abstractions in, 298; assimilating philosophy into, 271–72; behavioral, 30; boundaries of, 140n33; causal explanations by, 4–5; deductive-nomological conception of, 131; disenchantment and, 42; disunity of, 132; fallibility of, 138n9; human, 125; influence on philosophy of, 94–95; moral judgments and, 174–75; natural, 125; normativity and, 95–96; object naturalism and, 272; philosophy as, 99n10; philosophy vs., 5; scope of, 273; success of, 90; unity of, 131; values and, 134–35. See also human sciences; social sciences

scientific inquiry, 95

scientific method, 140n34

scientific naturalism, 3–5; characterizing, 73; claims of, 125; doctrines of, 11; intentional psychology and, 142–43; liberal interpretation of, 124; liberalization of, 128–33; normativity and, 127, 136; supernaturalism vs., 69, 72–75;

viability of, 9. *See also* broad scientific naturalism; extreme scientific naturalism; narrow scientific naturalism

scientific realism: convergence and, 37; multimundialism vs., 112; relativism vs., 8. *See also* realism

scientific theories: convergence and, 37; model-based, 155–56; one-tier compatibilism and, 160–61

scientism, 30, 50n2, 277

second nature, 127–28, 313–15; naturalism of, 331

secularism: enchantment and, 48; morality and, 66–67n1

Sein und Zeit (Heidegger), 97

self-conceptions, 334

self-consciousness, 284n61

self-constitution, 199–200

self-knowledge, intentionality vs., 6

self-respect, 212

self-sufficiency, 212

Sellars, Wilfrid, 4, 5, 58, 93, 144, 279

semantics, 277; inferentialist, 64–66

sense-data theories, 14

sense-datum theories, 217

Sheldon, Jane, 320

Sidgwick, Henry, 206–7, 222–23n6

Sidwickean objection, 206–7

skepticism, 56–57, 221; omniscient interpreter argument and, 112–14. *See also* quietism

Skyrm, Brian, 318–19

Smith, Michael, 190n1, 190n6, 191n20–192n21, 218–19, 224n28

social sciences: intentional states and, 135; irreducibility of, 18n12; opportunities and, 30. *See also* human sciences

sociolinguistics, 277

sociology, 132–33, 273

soul, immateriality of, 89, 98n2

Spinoza, Baruch, 98n2, 275; pantheism of, 42–43

spontaneity, 312

Sterelny, Kim, 305

Strauss, Leo, 56, 67–68n19

Strawson, Peter, 50n2, 124

Stroud, Barry, 124

structural accounts, 198–99

structural claims, 180–81, 201

structural rationality, 187, 190n7

Sturgeon, Scott, 190

substantive claims, 180–81

supernaturalism, 3, 9; defining, 83n12; scientific naturalism vs., 69, 72–75; theistic, 73. *See also* antisupernaturalism; crypto-supernaturalism

supervenience, 50n3, 77, 291–93

Svavarsdottir, Sigrun, 203–4n5

symbiosis, 300

sympathy, 26–27

systematic dependency relations, 24

Taliaferro, Charles, 82n2

Taylor, Charles, 133

Taylor, Kenneth, 64, 68n22

technique, 339

teleology, 37–38

teleosemanticism, 129

Teller, Paul, 156

telos, 295

theism, 231–32; rejecting, 247

theoretical hypotheses, 156

third-person perspective, 25–26

time, 281n29

Toland, John, 39, 42–43, 52n12

transcendental idealism, 100

transcendental philosophy, 265–66

transcendental reasoning, 264

translation, rules of, 336

triangulation argument, 113–14

truth, 254; alternativeness and, 103; cognitivist understanding and, 177; as fictionalism, 246–48; identifying, 235–36, 243–45; with justification, 229, 233;

Printed in the USA
CPSIA information can be obtained
at www.ICGtesting.com
JSHW021435221024
72172JS00002B/9

9 780231 134675